Structure and Function of Collagen Types

Biology of Extracellular Matrix: A Series

Editor

ROBERT P. MECHAM

A list of books in this series is available from the publisher on request.

STRUCTURE AND FUNCTION OF COLLAGEN TYPES

Edited by **RICHARD MAYNE**
Department of Cell Biology and Anatomy
The University of Alabama at Birmingham
Birmingham, Alabama

ROBERT E. BURGESON
Shriners Hospital for Crippled Children and
Departments of Biochemistry, Cell Biology, and Anatomy
Oregon Health Sciences University
Portland, Oregon

1987

ACADEMIC PRESS, INC.
Harcourt Brace Jovanovich, Publishers
Orlando San Diego New York Austin
Boston London Sydney Tokyo Toronto

ACADEMIC PRESS, INC.
Orlando, Florida 32887

United Kingdom Edition published by
ACADEMIC PRESS INC. (LONDON) LTD.
24–28 Oval Road, London NW1 7DX

Library of Congress Cataloging in Publication Data

Structure and function of collagen types.

(Biology of extracellular matrix)
Includes bibliographies and index.
1. Collagen. I. Mayne, Richard. II. Burgeson, Robert E. III. Series.
QP552.C6S77 1987 599'.01852 86-32065
ISBN 0–12–481280–5 (alk. paper)

PRINTED IN THE UNITED STATES OF AMERICA

87 88 89 90 9 8 7 6 5 4 3 2 1

Contents

TYPE V COLLAGEN

JOHN H. FESSLER and LISELOTTE I. FESSLER

TYPE VI COLLAGEN

RUPERT TIMPL and JÜRGEN ENGEL

TYPE VII COLLAGEN

ROBERT E. BURGESON

TYPE VIII COLLAGEN

HELENE SAGE and PAUL BORNSTEIN

TYPE IX COLLAGEN

MICHEL VAN DER REST and RICHARD MAYNE

TYPE X COLLAGEN

THOMAS M. SCHMID and THOMAS F. LINSENMAYER

Preface

During the past few years, several different collagenous proteins have been isolated from connective tissues, and the collagen family is now expanded from type I to type XI. In this volume, each article reviews one of the "new" collagens in turn and provides a valuable source of information for not only newcomers to the collagen field but also all scientists investigating the structure and function of the extracellular matrix. All of the articles are written by leading investigators directly involved with the appropriate collagen and therefore represent the current "state of the art."

In the introductory article, Kühn provides a review of the fibril-forming collagens (types I, II, and III) and traces the early work on the structure of these collagens to our present knowledge of the structure of the collagen genes. This article is followed by a detailed description of type IV (basement membrane) collagen by Glanville, who presents the evidence that this collagen forms a network of molecules and publishes for the first time the complete amino acid sequence of the α1(IV) chain. For type V collagen, Fessler and Fessler primarily address the complex issues of the biosynthesis and chain assembly of the three or perhaps four genetically different chains which form this collagen type. Timpl and Engel present the evidence that type VI collagen is assembled to form tetramers, which subsequently undergo end-to-end associations to form the microfibrils present in many connective tissues. The subsequent article by Burgeson shows that molecules of type VII collagen are assembled to form partially overlapping dimers that subsequently undergo lateral associations giving rise to the anchoring fibers of the dermis. Endothelial cells, and perhaps some other cell types, synthesize a unique collagen called type VIII, and Sage and Bornstein present the available information for the structure of this collagen.

Cartilagenous tissues are now known to synthesize at least three collagens in addition to type II collagen, and these have been designated type IX, type X, and type XI. The structure of type IX collagen is described in the article by van der Rest and Mayne, and for type X collagen, Schmid and Linsenmayer show that this collagen is unique to hypertrophic cartilage. Type XI collagen was previously called 1α2α3α collagen, and Eyre and Wu now present the evidence that this collagen can be regarded as a new collagen type. Finally, Vasios *et al.* review the recombinant DNA techniques presently being used to investigate collagen structure and show that it is possible to recognize a new collagen type from a cDNA library.

This volume of *Biology of Extracellular Matrix: A Series* is the first time that such a comprehensive review of the different collagen types has been presented, and it is hoped that further research will be stimulated by having all of this information easily available within a single volume.

RICHARD MAYNE
ROBERT E. BURGESON

The Classical Collagens: Types I, II, and III

Klaus Kühn

Max-Planck-Institut für Biochemie, Abteilung Bindegewebsforschung, D-8033 Martinsried bei München, Federal Republic of Germany

I. Introduction

Modern research on collagen at the molecular level began in the 1950s. At this time, Highberger *et al.* (1950) and Schmitt *et al.* (1953) characterized the collagen molecule in the electron microscope, Ramachandran and Kartha (1954) and Rich and Crick (1955) developed models for the triple-helical structure of collagen, and finally Boedtker and Doty (1956) determined the physicochemical properties of the collagen molecule in solution. By comparing the cross-striation period of 70 nm along the collagen fibril with the length of a collagen molecule of 280 nm, Schmitt (1956) first developed the concept of a quarter staggered array of molecules in the native fibril. Further protein chemistry and biochemical investigations led to a detailed knowledge of collagen structure and eventually resulted in the discovery, by Miller and colleagues (Miller and Matukas, 1969; Miller *et al.*, 1971), that the 70-nm-banded collagen fibrils observed in the electron microscope are formed by three different but homologous collagen types. These are type I collagen, the most abundant and, for a long time, the only one known; type II collagen, the main collagenous protein of cartilage; and type III collagen, which accompanies type I collagen in different ratios in almost all tissues. Kefalides (1971) observed that basement membranes also contain a specific collagen (type IV), and later, Kühn *et al.* (1981) and Timpl *et al.* (1981) showed that type IV collagen differs markedly in structure from the fiber forming collagens. Presently, we know that in addition to type IV collagen, which forms a nonfibrillar network, the structures of type VI, VII, IX, and X collagens are all different from each other and from the fibril-forming collagens (Table I) (for a review, see Martin *et al.*, 1985). In recent years, abundant knowledge of the structure of these new collagens was

STRUCTURE AND FUNCTION OF
COLLAGEN TYPES

TABLE I
THE COLLAGENS

Type	Chains	Molecular structure	Supermolecular structure	Localization
I	α1(I), α2(I)	300 nm	67-nm banded fibrils	Skin, tendon, bone, etc.
II	α1(II)	300 nm	Small 67-nm banded fibrils	Cartilage, vitreous humor
III	α1(III)	300 nm	Small 67-nm banded fibrils	Skin, muscle, etc.
IV	α1(IV), α2(IV)	390 nm C globular domain	Nonfibrillar network	All basement membranes
V	α1(V), α2(V), α3(V)	300 nm N globular domain	Small fibers	Most interstitial tissues
VI	α1(VI), α2(VI), α3(VI)	150 nm N+C globular domains	Microfibrils, 100 nm banded fibrils	Most interstitial tissues
VII	?	450 nm	Dimer	Anchoring fibrils
VIII	α1(VIII)	?	?	Some endothelial cells
IX	α1(IX), α2(IX), α3(IX)	200 nm N globular domain	?	Cartilage
X	α1(X)	150 nm C globular domain	?	Hypertrophic and mineralizing cartilage
XI	α1(XI), α2(XI), α3(XI)	300 nm	Small fibers	Cartilage

provided by biosynthetic approaches, protein chemical, electron microscopic, immunological, and molecular biological methods. All collagens have in common a triple-helical domain, which is combined differently with globular and nonhelical structural elements. They all are synthesized intracellularly as individual molecules or precursor molecules. After secretion into the extracellular space, they aggregate with one another into supramolecular structures, which subsequently may become stabilized by intermolecular covalent bonds. The different collagen types seem to have developed during evolution to provide the organism with a variety of supramolecular structures necessary to adapt connective tissue to its manyfold mechanical and physiological functions.

This article will deal with the classical fiber-forming collagens: types I, II, and III. A chronological description will be made of the important discoveries during the course of collagen research in order to give an overview of how our knowledge of collagen has evolved in the last 30 years.

II. Early Investigations

Collagen is the main constituent of skin, the raw material of leather and gelatin, and was first investigated by chemists involved in leather and gelatin production. They described the characteristic amino acid composition, the high glycine content, and the presence of hydroxyproline and hydroxylysine. In addition, the swelling capacity, the extraordinary tensile strength of collagen fibrils, and the unusual resistance to proteolytic attack were all investigated (for a review, see Grassmann *et al.*, 1951). The first models for the primary structure as well as for the arrangement of the polypeptides in the fibril were developed by Astbury in 1940. Based on the amino acid composition of collagen, he assumed that every third residue along the peptide chain is glycine and every ninth residue is hydroxyproline. The X-ray reflection pattern of collagen fibrils, which showed a strong meridian reflection of 2.89 Å (Herzog and Gonnel, 1924) and an equatorial reflection which varied between 10.5 and 16 Å depending on the degree of swelling of the fibers (Küntzel and Prakke, 1933), led to models with parallel oriented, fully extended peptide chains.

Another characteristic property of collagen fibers was independently discovered by Hall *et al.* (1942) and Wolpers (1943). In the electron microscope, the fibers showed a cross-striation period of 60–70 nm, similar to the identity period of 67 nm observed in low-angle X-ray reflection patterns (Bear, 1944). Electron micrographs of collagen fibrils at higher resolution after staining with phosphotungstic acid (PTA), uranyl salts, and chromium complexes later showed a subdivision of the identity period with a characteristic cross-striation pattern containing 13 dark bands (Nemetschek *et al.*, 1955). Several theories were developed to explain this pattern, but only Bear's (1952) came near the truth. He assumed that the dark bands consist of accumulated polar amino acids with long side chains, able to bind heavy metal ions, and that the light interbands contain mainly apolar residues with short side chains.

III. Electron Microscopic Investigations

A. The Collagen Molecule

Three parallel developing branches of collagen research led finally to an understanding of collagen structure at the molecular level. Exact high-angle X-ray diffraction patterns of stretched tendons obtained by Cowan *et al.* (1953) and the exact determination of the coordinates of the peptide chain by Pauling *et al.* (1951), such as the atomic distances and the angles between the individual bonds, enabled Ramachandran and Kartha (1954) and Rich and Crick (1955) to develop a model of the triple helix. It was proposed that each polypeptide chain is arranged in a stretched polyproline helix whose mobility is restricted by the proline rings. The fact that every third position in the collagen chains is occupied by a glycine residue allows the arrangement of three chains around a common axis to the triple helix, which is stabilized by hydrogen bonds between the chains perpendicular to the helix axis (reviewed by Ramachandran and Ramakrishnan, 1976).

Nageotte (1927) observed that acidic extracts of embryonic connective tissue contained collagenous proteins which could be reaggregated to fibrillar structures. Such collagen extracts were reinvestigated by Schmitt and colleagues with the help of the electron microscope. They found that acid-soluble collagen will aggregate at neutral pH to fibrils with the typical cross striation of native fibrils (Schmitt *et al.*, 1942). Depending on the experimental conditions used, two additional aggregates absent in tissues were observed. In the presence of highly negatively charged compounds, such as hyaluronic acid or chondroitin sulfate, the collagen molecules do not form fibrils of the native type. Because of their interactions with the negatively charged compound, they are forced to aggregate to the so-called long spacing fibrils (FLS) with an axial period of 180–250 nm and a centrosymmetrical cross-striation pattern (Schmitt *et al.*, 1953). By dialyzing neutral phosphate extracts of connective tissue against acetic acid, another artificial aggregate, the segment long spacing (SLS) crystallites (Schmitt *et al.*, 1953) were observed. They were 250–280 nm long and, after staining with phosphotungstic acid, they depicted an asymmetric cross-striation pattern (see Fig. 1b). Highly negatively charged substances, such as nucleic acid or nucleotides present in the phosphate extract, were responsible for the precipitation of collagen at acidic pH. ATP was found to be the best reagent for the formation of SLS crystallites from collagen in acetic acid solution. An important finding was that all three macromolecular forms, the native fibrils, FLS, and SLS aggregates, were interconvertible by dissolution and subsequent reaggregation under certain conditions (Schmitt *et al.*, 1953). Thus, it is possible for the three different forms to be assembled by the same molecule with a length of about 250–280 nm.

These assumptions were confirmed by the physicochemical characterization of collagen in solution by Boedtker and Doty (1956). They also described the collagen molecule as a thin rod with a length of 280 nm. Putting together the X-ray, electron microscopic, and physicochemical investigations, it followed that the collagen molecule consists of three polypeptide chains each with a length of about 1000 amino acid residues. They are arranged in a triple-helical structure that gives the molecule both its rod-like form and its rigid properties. The diameter of the collagen molecule, 1.4 nm calculated from the triple-helical structure, was in agreement with the physicochemical investigations and the X-ray data. Finally, Hall and Slayter (1959) observed individual collagen molecules in the electron microscope using the shadowing technique. The molecules were visible as thin and flexible rods with a length of 280–300 nm (see Fig. 1a).

Particularly helpful for further investigation of the molecular and macromolecular structure of collagen were the SLS crystallites, which turned out to be the electron microscopic image of the collagen molecule. Positive staining with PTA and uranyl acetate produced an asymmetric cross-striation pattern with 58 bands (Fig. 1b). Since PTA reacts with the positively charged side chains and uranyl cations react with negatively charged side chains of the polar amino acid residues (Kühn *et al.*, 1958; Gebhardt and Kühn, 1963), it was suggested that the cross striation pattern of SLS crystallites reflects the distribution of these residues along the polypeptide chains. Therefore, SLS crystallites were used for an initial comparison of the primary structure of the different collagens before the actual amino acid sequence was

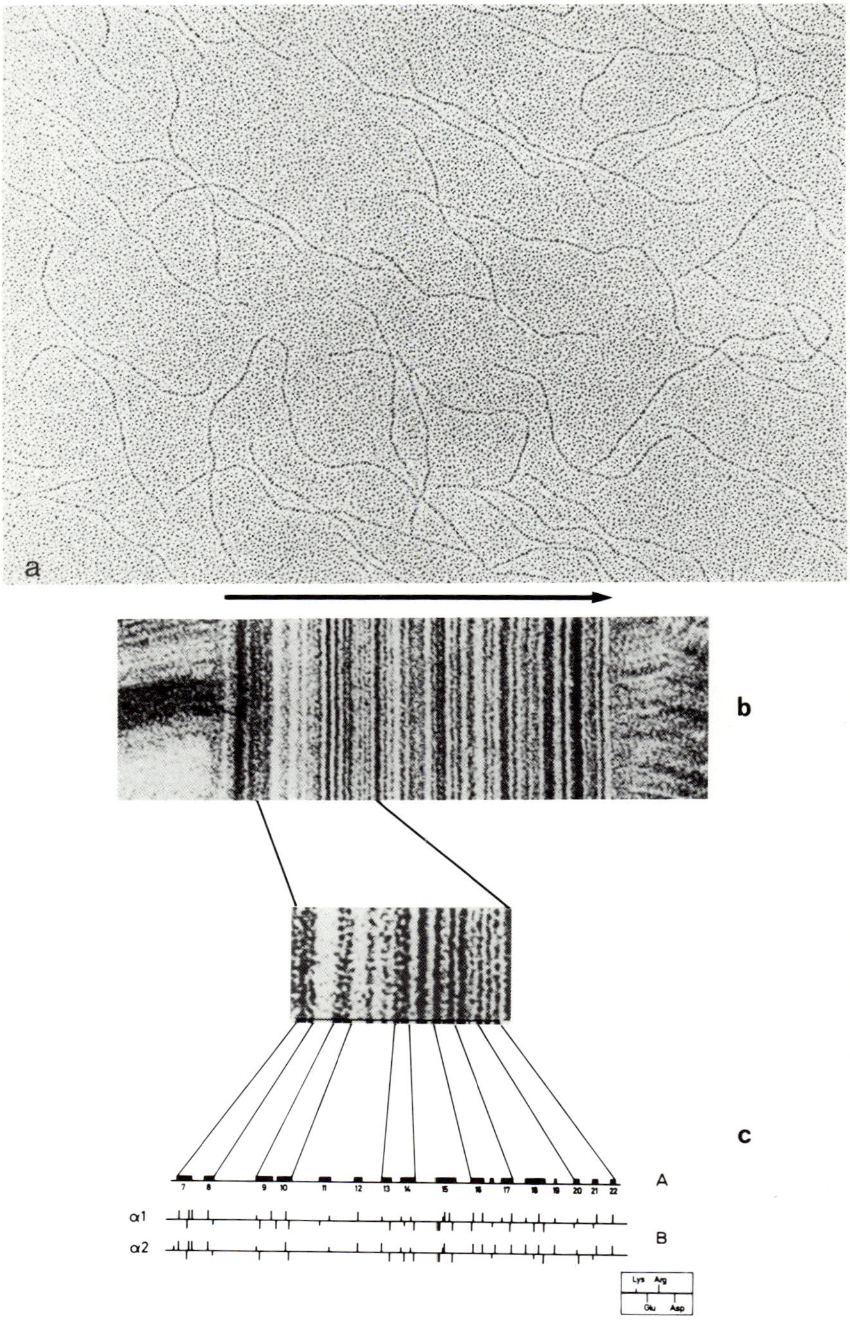
a
b
c
7 8 9 10 11 12 13 14 15 16 17 18 19 20 21 22
A
α1
α2
B
Lys Arg
Glu Asp

known. Initially, collagens from animals of various phylogenetic stages were compared. Crystallites were prepared from pepsin-dissolved collagen of sea anemone, liver fluke, and carp. Surprisingly, their cross-striation patterns were identical with those prepared from bovine collagen. This revealed a remarkable stability of the amino acid sequence during evolution, at least in as far as the distribution of the polar charged residues (Nordwig and Hayduk, 1969) was concerned. This notion was verified when the first continuous stretch of amino acid sequence of some length of the α1(I) chain became known (von der Mark *et al.*, 1970). An excellent correlation between the amino acid sequence and the cross-striation pattern was observed. Each band could be attributed to a distinct cluster of polar amino acids. The correlation was so exact that, in some cases, individual arginine residues were reflected by a single dark band in the cross-striation pattern (Fig. 1c).

B. The Structure of the Native Fibril

In 1956, Schmitt suggested that the 70-nm-long cross-striation period of the native collagen fibril is caused by a parallel but quarter-staggered array of the 280-nm-long molecule, so that the band pattern of one period is caused by the superimposed position of four different quarters of the collagen molecule. This hypothesis was experimentally confirmed by Hodge and Schmitt (1960) and Kühn *et al.* (1960). By superimposing the cross-striation pattern of four quarter staggered SLS crystallites in a way in which the dark bands matched best, a composite cross-striation pattern corresponding to that obtained with native fibrils was achieved. Since the dark bands represent areas with polar charged amino acid residues, it was assumed that the aggregation of the molecules is determined by the sequence of the charged residues with the help of electrostatic forces (summarized by Kühn, 1982).

In 1964, Petruska and Hodge found that this kind of construction

---

FIG. 1. Electron microscopic image of the collagen type I molecule. (a) Individual molecules visualized by the rotary-shadowing method. The length of the molecule is 300 nm. (b) Segment long spacing crystallites stained with phosphotungstic acid and uranyl acetate. The molecules are parallel with their ends in register. (c) Comparison of the amino acid sequence of the α1(I) and α2(I) chain (position 120–405) with the corresponding cross-striation pattern of SLS crystallites. There is an exact correlation between the distribution of the polar charged amino acids and the dark bands in the cross-striation pattern (from Fietzek and Kühn, 1976).

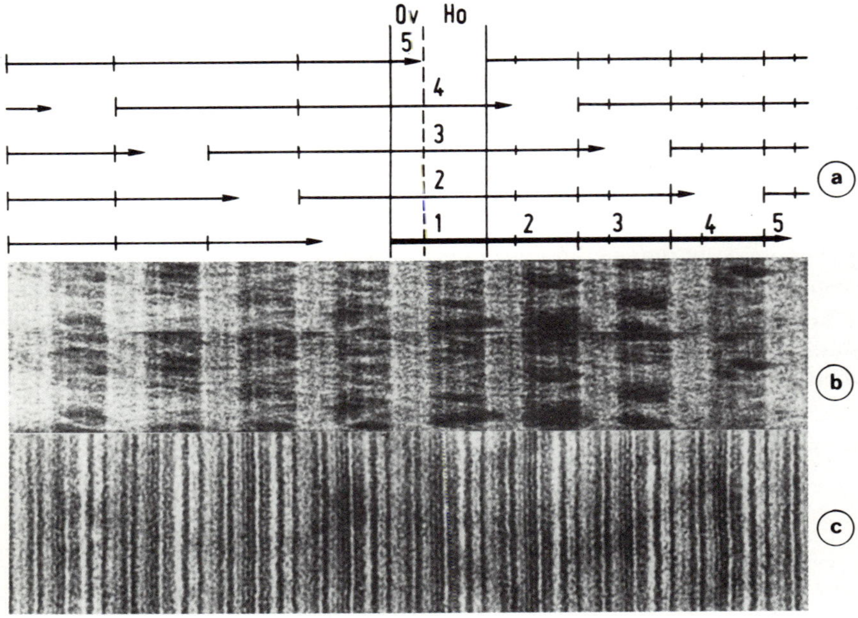

FIG. 2. The array of the collagen molecules in a native fibril. (a) Schematic representation of the parallel, D staggered arrangement of the collagen molecules in the fibril. (b) Collagen fibril, negatively stained with phosphotungstic acid. The dark and light areas reflect loosely and densely packed hole (Ho) and overlap zones (Ov), respectively. (c) Collagen fibril, positively stained with phosphotungstic acid and uranyl acetate, depicting the characteristic cross-striation pattern with the identity period of 67 nm.

did not reflect the exact arrangement of the end regions of the collagen molecules. In negatively stained fibrils, the 67-nm-long period is divided into a dark and a light area reflecting loosely and densely packed regions, respectively. With more exact determination of the length of SLS crystallites after negative staining, they found that the collagen molecule is not 4 but 4.4 D units (D = the length of one cross-striation period of 67 nm) and that, in the dense area of the period, the molecules overlap with their ends by about 30 nm (Fig. 2). It is striking that these electron microscopic investigations led to an exact determination of the arrangement of the molecules in a native fibril without a detailed knowledge of the primary structure. A decade later these results were confirmed by calculation of the interaction patterns of the molecules using a three-dimensional analysis of the amino acid sequence (Trus and Piez, 1976; Hofmann *et al.*, 1978) and new interpretation of X-ray diffraction patterns (Hulmes and Miller, 1979).

IV. Protein Chemical Investigation

In the 1950s, investigation on the primary structure of collagen also began. However, the elucidation of the amino acid sequence of a protein as large as collagen, with an M_r of 360,000, appeared to be an almost unsolvable problem with the available protein chemical methods. First attempts were made by Schroeder *et al.* (1953, 1954). Collagen was fragmented by partial alkaline and acidic hydrolysis, and tri- and tetrapeptides were isolated and sequenced. The results gave the first indications of both a regular distribution of glycine along the peptide chain and the presence of the sequence Gly-Pro-Hyp. These experiments were extended by Grassmann and co-workers who investigated peptides obtained after tryptic digestion of denatured bovine collagen, and who observed a prevalence of the common sequence Gly-X-Y (for a review, see Grassmann *et al.,* 1965).

The basis for a systematic investigation of the amino acid sequence of collagen was provided by Piez and co-workers in 1963. It was shown that during denaturation the collagen molecules disintegrate into a complex mixture of monomeric α chains, di- and trimeric components, and higher fractions, which reflect the presence of polymeric collagen molecules in solution. In a now classical article, Piez *et al.* (1963) described the separation of these components by chromatography on carboxymethylcellulose and showed that the collagen molecule consists of two α1 and one α2 chain. Further progress was achieved by Piez and colleagues after applying the cyanogen bromide (CNBr) cleavage method to the purified α chains, which splits polypeptides specifically at the methionine residues (Bornstein and Piez, 1966). This resulted in a clear mixture of relatively few fragments which, after separation, was used as starting material for a systematic sequence analysis of the α1 and the α2 chains (summarized by Fietzek and Kühn, 1976). Initial sequence studies were carried out mainly on the small cyanogen bromide peptides of rat collagen with the manual Edman degradation procedure (Kang *et al.,* 1967; Bornstein and Piez, 1966). These studies were accelerated by the introduction of the automated degradation procedure developed by Edman and Begg (1967). In 1972 the primary structure of an entire α1(I) chain was known (see Fietzek and Kühn, 1976). The sequence data, partly from rat and partly from bovine collagen, revealed a peptide chain with a continuous triple-helical domain of 1014 residues and nonhelical telopeptides at the N- and C-terminus of 16 and 25 residues in length, respectively (see below).

The exact knowledge of the CNBr pattern of the α1 and the α2 chain provided a basis of information for the discovery of type II and III collagens by Miller and co-workers. During an investigation of extracts of collagen from lathyritic chick cartilage, Miller and Matukas (1969) found a much higher proportion of α1 to α2 than the expected ratio of 2:1. Cyanogen bromide cleavage of the isolated α1 chain from cartilage depicted a peptide map different from that of the α1 and α2 chains investigated previously. This new, genetically distinct collagen type is now known to be the major collagenous component of hyaline cartilage. It was first called type II collagen by Miller, whereas the "old" collagen, the main constituent of skin, bone, and tendon, became type I collagen. The type II collagen molecule consists of three identical chains which were designated as α1(II) (Miller, 1971). The SLS crystallites of type II collagen showed a band pattern similar to that of type I collagen, indicating a strong sequence homology of both collagen types (Stark *et al.,* 1972). A characteristic difference was found in the higher content of hydroxylysine residues (see Miller, 1976) and the higher rate of glycosylation of α1(II).

Type III collagen was found by Miller *et al.* (1971) in pepsin-dissolved skin collagen. They observed an α chain which differed in its chromatographic behavior on carboxymethylcellulose as well as in its CNBr peptide map from α1(I), α2(I) and α1(II). It turned out that three of these α1(III) chains form the molecule of type III collagen which accompanies type I collagen in many connective tissues except tendon and bone (see Miller, 1976). Electron-microscopic investigation of the band pattern of type III collagen SLS crystallites indicated a clear homology to type I and type II collagen (Wiedemann *et al.,* 1975).

TABLE II
STEPS OF POSTTRANSLATIONAL PROCESSING OF COLLAGEN

Modification	Biological importance
Removal of the signal sequence	Unknown
4-Hydroxylation of proline	Regulation of the stability of the triple helix
3-Hydroxylation of proline	Unknown
Hydroxylation of lysine	Stabilization of intermolecular cross-links and hydroxylysyl glycosylation
Glycosylation of hydroxylysine	Regulation of fiber diameter?
Chain association, disulfide bonding, and triple-helix formation	Formation of the triple-helical procollagen molecule
Conversion of procollagen to collagen	Prerequisite of fibril formation and regulation of fibril formation
Self assembly of collagen molecules	Fibril formation
Cross-link formation	Stabilization of fibrils

Amino acid sequence analysis of the newly discovered types II and III collagens was finally accelerated by applying recombinant DNA technology. The first amino acid sequences of collagen deduced from $\alpha 1$(I)- and $\alpha 2$(I)-specific cDNA were published by Fuller and Boedtker (1981). Today, the covalent structure of all three fiber-forming collagens is elucidated. However, before the relationship of the amino acid sequence, higher structures, and function of collagen can be discussed, it will be necessary to describe the posttranslational processing events which collagen undergoes on its way from the biosynthesis of the polypeptide chain to the formation of macromolecular structure in the extracellular space (Table II).

V. Biochemical Investigations

A. *Hydroxylation of Proline and Hydroxylysine and Glycosylation of Hydroxylysine*

The first indications for a posttranslational modification of the amino acids in collagen was the occurrence of the specific amino acid residues hydroxyproline and hydroxylysine, and the observation of Stetten (1949) that radioactively labeled hydroxyproline cannot be incorporated into collagen. These results were confirmed by Peterkofsky and Udenfriend (1963, 1965), who could not find specific amino acyladenylate synthetases for hydroxyproline and hydroxylysine. They were able to show that hydroxylation did not take place at the stage of the amino acyl complexes of the tRNA but after incorporation of the residues in the peptide chain. Meanwhile, it became clear that the hydroxylation of not only proline but also of lysine is catalyzed by two separate enzymes, the proline 4-hydroxylase and the lysyl hydroxylase (see Kivirikko and Myllylä, 1980, 1985; Majamaa *et al.*, 1985). Both are located on the inner side of the membrane of the rough endoplasmic reticulum. The prolyl as well as the lysyl hydroxylases require Fe^{2+}, 2-oxoglutarate, O_2, and ascorbate. They all decarboxylate 2-oxoglutarate into succinate, one atom of the O_2 molecule being incorporated into succinate while the other is transferred as a hydroxyl group to the amino acid residues. Both enzymes catalyze the oxidation of residues in position Y of the tripeptide unit Gly-X-Y. An exception has been observed in the non-triple-helical telopeptides of the α chains, where lysine residues are hydroxylated without being present in a tripeptide region. These residues are particularly important for the formation of stabilized intermolecular cross-links.

There is also a prolyl-3-hydroxylase which oxidizes proline residues mainly in position X (see Cardinale and Udenfriend, 1974). In the fiber-forming collagens, this has been observed only once in the α1(I) chain in the sequence Gly-3Hyp-4Hyp (Fietzek and Kühn, 1976). In invertebrate collagens, 3Hyp has been found more often (Adams, 1978).

The presence of D-galactose and D-glucose covalently linked to collagen was shown by Grassmann *et al.* (1957), Gross *et al.* (1958), and Kühn *et al.* (1959). In 1966, Butler and Cunningham demonstrated that the carbohydrate component of collagen is bound to the hydroxyl group of hydroxylysine in the form of a disaccharide. Its structure, a 2-*O*-α-D-glucopyranosyl-*O*-β-D-galactopyranosylhydroxylysine was elucidated by Spiro (1967). The formation of this collagen-specific carbohydrate group is catalyzed by two specific enzymes. First, galactose is transferred to a hydroxylysine residue by the enzyme galactosyltransferase. Then, a glucose is transferred to the galactosyl monosaccharide by the enzyme galactosylhydroxylysyl glucosyltransferase (Spiro and Spiro, 1971a,b; Kivirikko and Myllylä, 1979). Both enzymes use as a donor the corresponding UDP-glycosides and require the presence of a bivalent cation, such as manganese.

All five posttranslational enzymes, the three hydroxylases as well as the two glycosyltransferases can only modify the collagen polypeptide chain before formation of the triple helix occurs. In this connection, it is striking that the 4-hydroxyl group of the hydroxyproline residues in position Y of the tripeptide Gly-X-Y has an important function in the stabilization of the triple-helical structure, and that the degree of hydroxylation of proline appears to regulate triple-helix formation. A triple helix formed by nonhydroxylated α chains exhibits a transition temperature of about 15°C lower than the t_m of a triple helix of normal, hydroxylated collagen (Berg and Prockop, 1973). This indicates that new or underhydroxylated collagen is not able to form a triple-helical molecule under physiological conditions at a temperature of 37°C. When hydroxylation has reached a certain degree, then triple-helix formation takes place and stops the enzymes. It has been speculated that, in this way, the t_m value of the collagen molecules are adjusted to the body temperature.

Taking into account the important role of hydroxyproline, attempts have been made to interfere in fibrotic diseases with an uncontrolled overproduction of collagen, such as liver cirrhosis or lung fibrosis. One approach was to inhibit proline 4-hydroxylase with inhibitors specifically designed for the active site of this enzyme (Hanauske-Abel and Gunzler, 1982). Another approach was to give experimental animals an excess of proline analogs, such as *cis*-4-fluoro-L-proline or 3,4-dehydroproline, which are incorporated in place of proline into the peptide chains during protein synthesis and which cannot be hydroxylated at position 4 (Uitto and Prockop, 1975).

The role of the collagen-specific disaccharide units attached to the hydroxylysine residues of the triple-helical domain remains unclear. Their number varies from collagen to collagen. Type II collagen, for example, often forms fibrils of small diameter and contains more carbohydrates than type I collagen, which preferentially builds thick fibrils (see Miller, 1976). A high number of the bulky disaccharide groups appears to hinder the formation of highly ordered fibrils with a large diameter. Type IV collagen, which forms a nonfibrillar network (Timpl *et al.,* 1981), is characterized by a high content of hydroxylysine residues, which are all occupied by the galactosylglucosyl disaccharide (see Glanville, this volume).

B. The Precursor Form

For a long time, it remained unnoticed that the α chains of the fiber-forming fibrils are synthesized in a precursor form. First indications came from biosynthetic investigations of collagen in cultures of fibroblast and rat calvaria (Layman *et al.,* 1971; Bellamy and Bornstein, 1971), when collagenous polypeptide chains with an M_r of 120,000 instead of 95,000 were observed. A similar precursor form was isolated from the skin of calves with the inherited disease dermatosparaxis by Lenaers *et al.* (1971). SLS crystallites from procollagen isolated from dermatosparactic calves and from the medium of fibroblast cultures revealed an additional extension peptide at the amino-terminus of the collagen molecule. The carboxy end of the molecule, however, did not appear to be changed (Stark *et al.,* 1971). It turned out later that these dermatosparactic animals had a defect in the amino-terminal procollagen peptidase, a proteolytic enzyme involved in the removal of the amino-terminal propeptide during conversion of procollagen. Further biosynthetic studies using tendon fibroblasts from chick embryos revealed that this precursor form did not reflect the genuine procollagen molecule. Procedures which prevented the action of endogenous proteolytic enzymes by adding protease inhibitors and avoiding acidic pH during the isolation procedure resulted finally in the isolation of an intact precursor, a precursor which contained propeptides not only at the amino-terminal but also at the carboxy-terminal end (Hoffmann *et al.,* 1976).

The discovery of the two propeptides completed our knowledge of the structure of collagen. It was thus shown that for the construction of the genuine procollagen molecule not only triple-helical areas but also globular domains and nonhelical sections are used. However, in the case of the fiber-forming collagens, the globular end regions are not deposited in the final fibrillar structure. They may be used as control elements important during the posttranslational development of collagen. However, in other collagen types, for instance in type IV and type VI collagens, such globular domains are used as integral parts of the final macromolecular structure of the collagen (Timpl *et al.*, 1981; von der Mark *et al.*, 1984).

The bulky globular domains of the propeptides of the intact collagen molecule may function as a transport form and prevent fiber formation inside the cell. An important posttranslational event is therefore the conversion of the precursor form to the active molecule. Only when the propeptides are split off extracellularly by two specific enzymes—the procollagen amino and carboxy proteinases—can the collagen molecules aggregate in an ordered D staggered array to form native fibrils. In the skin of dermatosparactic calves or sheep, where the amino propeptide is not removed because of a defective enzyme, the pN molecules do not form intact fibrils but aggregate into distorted ribbons which are not able to provide the skin with the required biomechanical stability. As will be discussed below, the timing of the cleavage of the two propeptides seems to be important for the regulation of fiber diameters. Also, other functions of the propeptides have been proposed. The amino propeptide appears to play a role in the feedback control of collagen biosynthesis (Wiestner *et al.*, 1979), whereas the carboxy propeptide is involved in the alignment of three α chains to facilitate the formation of the triple helix (Fessler *et al.*, 1981) (see below).

C. *Formation of Cross-Links*

An important posttranslational modification, which occurs outside the cell after aggregation of the molecules, is the formation of intermolecular cross-links responsible for the mechanical stability of fibrils and the fiber network. Investigation of this complex event at the molecular level was initiated by the observation of Gross *et al.* (1960) that a marked increase in the amount of soluble collagen occurs when β-aminopropionitrile (β-APN) is administered to growing animals. β-APN was found to be the active compound of the sweet pea, *Lathyrus odoratus,* which causes the connective tissue disorder osteolathyrism. Amino acid sequence analysis of rat collagen later revealed that, in lathyritic collagen, the single lysine or hydroxylysine residue in the amino-terminal telopeptide, in contrast to the collagen from healthy animals, is not oxidized to the lysine-derived allysine. Therefore, the intramolecular cross-link which leads to the β_{12} component linking two collagen chains cannot be formed (Bornstein *et al.*, 1966; Bornstein and Piez, 1966; see also Piez, 1968). The same is true for the carboxy-terminal telopeptide which also contains an oxidizable lysine or hydroxylysine residue (Rauterberg *et al.*, 1972). This suggested that β-APN inhibits the formation of aldehydes from lysyl residues and, thus, the formation of cross-links.

Investigation of some of the cross-link compounds formed was facilitated by reductive labeling with tritiated sodium borohydride (see Bailey *et al.*, 1974; Tanzer, 1976). The major cross-links found in borohydride-reduced collagen were hydroxylysinonorleucine and dihydroxylysinonorleucine, produced by the condensation of allysine with hydroxylysine or hydroxyallysine with hydroxylysine to form a Schiff's base. It is an important fact that the degree of hydroxylation of the one lysine residue in each telopeptide determines the stability of the cross-link. If an aldehyde derived from hydroxylysine is involved in cross-linking, the Schiff's base initially produced undergoes an Amadori rearrangement to a stable ketoimine form (Fig. 3). If a lysine-derived allysine is the partner, a rearrangement of the formed aldimine bond is not possible; the cross-link remains labile and can, for example, be opened again by acidic pH (Robins and Bailey, 1975). This fact explains the differences in solubility of collagen from different tissues in dilute organic acids. Skin collagen, with a relatively low degree of hydroxylation of lysine in the telopeptides, is more readily soluble than collagen from bone and cartilage, which contains more hydroxylysines in the telopeptides. In addition to the described divalent cross-links, a number of tri-and tetravalent condensation products have been isolated which also can be reduced by sodium borohydrate (see Tanzer, 1976; Robins and Bailey, 1977). During maturation and aging of collagen, the reducible cross-links are gradually converted into nonreducible compounds in a manner that is as yet not clear. The possibility of *in vivo* reduction has been excluded (see Eyre, 1980). Other stabilization mechanisms have been suggested more recently involving oxidation to yield a peptide bond (Bailey *et al.*, 1977) or to form a multivalent cross-link of unknown nature (Light and Bailey, 1979).

From the D staggered array of the collagen molecules in the fibril, the allysine or hydroxyallysine residues in the amino- and carboxy-terminal telopeptides are in the neighborhood of four different areas of the triple helix, where they are able to find a lysine or hydrox-

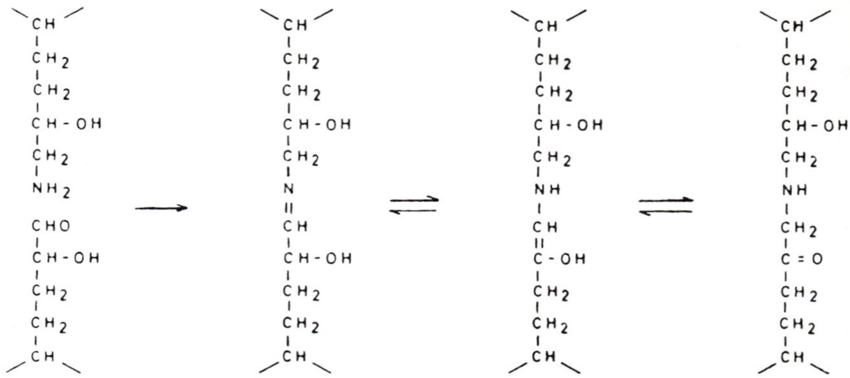

FIG. 3. Cross-link formation between a hydroxyallysine and a hydroxylysine. The resulting dehydrodihydroxylysinonorleucine is subsequently stabilized by an Amadori rearrangement to a ketoimine compound. If an allysine is the reaction partner so that the hydroxyl group is missing, then the rearrangement cannot take place and the aldimine bond can be opened again, e.g., by acidic pH (Robins and Bailey, 1973).

ylysine residue as reaction partner to form an aldimine cross-link. Inspection of the amino acid sequence of type I, II, and III collagens revealed, for the amino-terminal allysine, that a reaction partner existed only in the 4D staggered array, namely, hydroxylysine-930. The situation is similar with the carboxy terminal allysine. The 4D staggered arrangement provides hydroxylysine-87 as partner in the 1D position, and lysine-327 may be a putative partner too (see Fietzek and Kühn, 1976). In order to localize the cross-links, insoluble collagen was fragmented by cyanogen bromide or by proteolytic attack, and cross-linked peptides were isolated and characterized (summarized by Fietzek and Kühn, 1976; Light and Bailey, 1980). In accordance with the amino acid sequence of types I, II, and III collagen, only those peptides which can be formed between 4D staggered molecules were found. As will be discussed below, the two helical cross-link regions around the hydroxylysine residues 87 and 930 represent a special, evolutionarily stable amino acid sequence which may also act as an attachment site for the enzyme lysyl oxidase.

The initial step for the formation of cross-links, the oxidation of lysine or hydroxylysine, is catalyzed by the copper-dependent lysyl oxidase (reviewed by Siegel, 1979). Siegel (1974) found that this enzyme has no effect on dissolved collagen molecules. The prerequisite for its catalytic function is aggregated molecules. This observation led to the suggestion that lysyl oxidase does not bind to the molecule which is oxidized but to distinct sites of the neighboring molecule (see below) and, therefore, that a fibrillar collagen is required for enzyme action.

VI. The Covalent Structure of the Procollagens

In this paragraph those areas of the primary structure of procollagen chains, which determine and control the formation of the molecular and macromolecular structure and which provide a better understanding of the complex sites of posttranslational events and metabolism of collagen, will now be discussed. Figure 4 depicts schematically the covalent structure of the α1(III) chain. Because of the strong homology between the fiber-forming collagen types I, II, and III, it can be taken as representative for all α chains [α1(I), α2(I), α1(II), and α1(III)] (Hofmann *et al.*, 1980). The triple-helical domains of the α chains were mainly sequenced at the protein level (summarized by Fietzek and Kühn, 1976, and Galloway, 1982). This is also true for the amino propeptides of α1(I) and α1(III), which could be extracted from the skin of dermatosparactic or embryonic calves, respectively (Hörlein *et al.*, 1979; Brandt *et al.*, 1984). The sequences of the carboxy propeptides of all α chains (Fuller and Boedtker, 1981; Yamada *et al.*, 1983a; Sandell *et al.*, 1984), as well as of the amino propeptides of α2(I) and α2(II), which were available only in small amounts, have been deduced from the DNA sequences (Tate *et al.*, 1983; Kohno *et al.*, 1984). The triple-helical domain, inclusive of the non-triple-helical short appendices at both ends, contains all the struc-

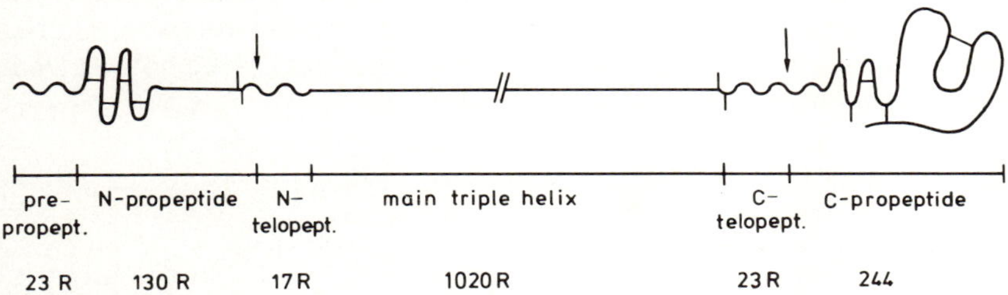

FIG. 4. Schematic representation of the prepro α1(III) chain of collagen. The amino propeptide, the amino telopeptide, and the triple-helical domain of bovine (Brandt *et al.*, 1984; Fietzek *et al.*, 1979; Dewes *et al.*, 1979a,b; Bentz *et al.*, 1979; Lang *et al.*, 1979; Allmann *et al.*, 1979) as well as the triple helix of human (Seyer and Kang, 1977, 1978, 1981; Seyer *et al.*, 1980) collagens were sequenced at the protein level. The sequence of the signal peptide, carboxyl telopeptide, and the carboxyl propeptide were deduced from chick genomic DNA (Yamada *et al.*, 1983a,b). The arrows indicate the cleavage sites of the amino and carboxyl procollagen peptidase. R, Amino acid residues.

tural information for the construction of a cross-linked fibril in the extracellular space. The amino- and the carboxy-terminal propeptides, which are cleaved off before or during fibrillogenesis, appear to be important as control elements for the posttranslational development of collagen, and possibly also for a feedback regulation of collagen biosynthesis (Fessler *et al.*, 1981; Wiestner *et al.*, 1979).

A. *The Triple-Helical Domain*

The sequence of the main and central part of the α chain is essential for the formation of the molecular and macromolecular structure of collagen. It has the ability to form a triple helix with two other α chains, which gives the molecule a rodlike shape, and subsequently each molecule has the capability to aggregate in a highly ordered fashion to form fibrils. The formation of a continuous triple helix is dependent on an uninterrupted tripeptide sequence Gly-X-Y, whose length varies between 1014 residues in α1(I) and 1023 residues in α1(III) (see Galloway, 1982). The presence of the imino acids proline and hydroxyproline is important for the stability of the triple helix. There are three types of tripeptides containing imino acids, which contribute to the stability of the helix in the following order: Gly-Pro-Hyp > Gly-Pro-Y and > Gly-X-Hyp (Segal, 1969). The rigidity and the flexibility of distinct areas along the molecule can thus be adjusted by varying amounts of proline and hydroxyproline. A typical example is the carboxy-terminal end of the triple-helical domain of the α1(I) and α2(I) chain, where the helix is sealed by a 5-fold repeat of the triplet Gly-Pro-Hyp. Two other regions of the molecule, the collagenase cleavage site and the carboxy-terminal helical cross-link area, both of which require a more relaxed and flexible structure (see below), do not contain tripeptide units with Pro or Hyp.

1. Formation of Fibrils

Early electron microscopic investigations showed that the molecules in the native fibril are staggered from one another by a distance of 60–70 nm (Petruska and Hodge, 1964). This specific arrangement is mainly brought about by electrostatic forces determined by the sequence of the basic and acidic amino acids. Knowledge of the amino acid sequence allowed a more accurate analysis of the intermolecular interactions. Statistical evaluation of the sequence of α1(I), α2(I), and α1(III) revealed that the 234-amino-acid-residue-long D unit was the most prominent repeat. It is formed mainly by polar charged and hydrophobic residues and divides the α chains into four homologous parts, D1, D2, D3, and D4 (Hulmes *et al.*, 1973; Hofmann *et al.*, 1980). Three-dimensional calculations of the interaction between collagen I and collagen III molecules, respectively, indicated not only the highest polar but also the highest hydrophobic contact when the molecules were shifted against each other by the distance of 234 amino acid residues (Hofmann *et al.*, 1978). These results confirmed early electron microscopic investigations and revealed, in addition, an important role of the amino acid residues with hydrophobic character. The fact that the polar as well as the hydrophobic amino acid residues are accumulated in polar and hydrophobic areas, respectively, seems to be important for stabilization of the collagen fibril structure. The interaction with neighboring molecules allows the formation of ordered clusters of ions (Piez and Torchia, 1975) and of hydrophobic side chains, which results in a higher stability than is possible if only single residues interacted with one another.

Beside the axial arrangement of molecules in the fibril, their lateral array in a fibril is also important. Here, no clear relationship between the amino acid sequence and lateral arrangement of the molecules can be found, which indicates that the lateral array is not as strongly determined by the amino acid sequence as is the axial stagger of the molecules. This is in agreement with experimental observations. According to observations of the X-ray reflection pattern of tendon, the molecules seem to have a pseudohexagonal array (Hulmes and Miller, 1979) which can be easily disturbed, for example, by the swelling and unswelling of tendon (Nemetschek, 1969) or by staining with PTA (Nemetschek and Hosemann, 1973; Nemetschek *et al.*, 1979). The possibility of varying the azimuthal orientation of the molecules in a fibril appears to be an important structural feature of collagen, preventing the formation of crystal-like, rigid, nonfunctional fibrillar structures, and perhaps facilitating fiber formation by interaction with other extracellular matrix proteins.

2. The Cross-Link Sites

In addition to the amino acid sequence, which regulates the formation of higher structures of collagen, there are also specific regions along the triple helix, which are important for the stabilization of the collagen fibril by cross-linking, for the degradation of the fibrils by collagenase and for the interaction of collagen with cells and other

a

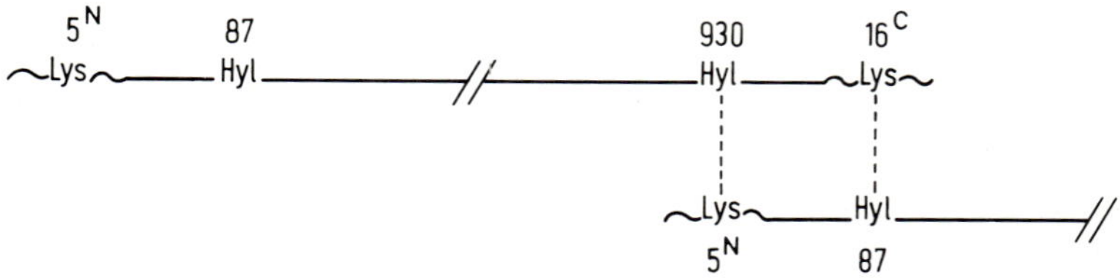

b

	82	87	93
α1(I)	-Gly-Leu-Hyp-Gly-Met-Hyl-Gly-His-Arg-Gly-Phe-Ser-		
α2(I)	-Gly-Leu-Hyp-Gly-Phe-Hyl-Gly-Ile-Arg-Gly-His-AsN-		
α1(III)	-Gly-Phe-Hyp-Gly-Met-Hyl-Gly-His-Arg-Gly-Phe-Asp-		

	925	930	936
α1(I)	-Gly-Asp-Arg-Gly-Ile-Hyl-Gly-His-Arg-Gly-Phe-Ser-		
α2(I)	-Gly-His-Glu-Gly-Leu-Pro-Gly-Leu-Hyl-Gly-His-AsN-		
α1(III)	-Gly-Ala-Met-Gly-Ile-Hyl-Gly-His-Arg-Gly-Phe-Ser-		

FIG. 5. (a) Schematic representation of the cross-link sites in the D/3-long end-to-end overlap region of 4D-staggered collagen molecules (see Kühn, 1982). Lysine (Lys) or hydroxylysine (Hyl) 5^N and 16^C are the active sites of the N- and C-terminal nonhelical regions. They have to be oxidized before cross-link formation can occur. Hyl-87 and -930 are the active sites of the triple-helical cross-link regions which serve as reaction partners for the lysine- or hydroxylysine-derived aldehydes. (b) The amino acid sequences in the vicinity of the hydroxylysine residues 87 and 930 are evolutionarily very stable, as is shown for the α-chains of the bovine collagens I and III.

matrix constituents. Cross-linking sites are characterized by low proline and hydroxyproline content and the presence of the sequence Hyl-Gly-His-Arg, which appears to be important as an attachment site for the enzyme lysyl oxidase (Fig. 5). It is thought that, after ordered assembly of the collagen molecules, this enzyme attaches first to the helical cross-link region before it can oxidize the hydroxylysine in the telopeptide of a D staggered adjacent molecule. Another characteristic feature of this area is the relaxed triple helix, caused by the absence of proline and hydroxyproline residues. This is important for the packing of the molecules into fibrils. Because of the end-to-end overlap of the 4D staggered molecules, a tight packing requires bending of the molecules right in the cross-link area, which comes to lie at the borders between the hole and the overlap zone (Hofmann *et al.*, 1980).

3. The Collagenase Cleavage Site

The intact triple helix is resistant to proteolytic attack. Incorporation of the molecules into fibrils and the subsequent cross-link formation result in a further stabilization of the molecules, so that the denaturation temperature of an individual molecule rises from about 40 to more than 60°C. In the fibril the triple-helical molecule can be cleaved only by a specific enzyme, the mammalian collagenase, which cleaves the molecule at a distinct position between glycine 775 and isoleucine 776 into two pieces, TC^A and TC^B (three-fourths and one-fourth of the length of the molecule, respectively) (Gross *et al.*, 1974; Miller *et al.*, 1976; Gross and Nagai, 1965). The cleavage site is located in an area with a relatively low content of imino acids, and the relaxed triple helix is thought to facilitate the collagenase action (for a review, see Gross, 1976). In highly cross-linked collagen, the molecules are so extensively stabilized that even collagenase can act only slowly. The breakdown of collagen fibrils seems to be accelerated by the presence of unspecific proteolytic enzymes which split the α chains at the non-helical telopeptide between the end of the triple helix and the hydroxylysine residue involved in cross-linking, resulting in a loosening of the fiber structure. Cooperation with other proteases is also necessary for further degradation of the triple-helical fragments PC^A and PC^B after denaturation (reviewed by Weiss, 1976). The rate of collagenase attack on types I, II, and III collagen was found to be different. Thus, rheumatoid synovial collagenase splits types I and III collagen five times faster than type II collagen, although comparison of the amino acid sequence of the three collagen types around the cleavage site does not provide an explanation for this fact (Woolley *et al.*, 1975).

B. The Amino Propeptides

At the amino-terminus of all α chains of types I, II, and III collagen, there is a 23-amino-acid-residue-long signal peptide which enables the nascent α chains to penetrate the membrane of the endoplasmic reticulum and is subsequently cleaved off (Fig. 6). The structure of the remaining amino propeptide differs from α chain to α chain. In α1(I) and α1(III) they consist of two main domains: a globular one 66 and 68 residues in length stabilized by five disulfide bridges, and a triple-helical one of 48 and 39 residues. In the amino propeptide of α2(I) the globular domain is missing, and the signal peptide is directly connected to the triple-helical domain. The same is true for the α1(II) chain. Its triple-helical domain is, however, 12 tripeptide units longer than that of the α1(III) chain (reviewed by Kühn, 1986).

Two important functions are proposed for the globular amino-terminal part; (1) the control of fiber formation in the extracellular space, which may also involve the carboxy propeptide and will be described later, and (2) a function in the control of collagen biosynthesis. It has

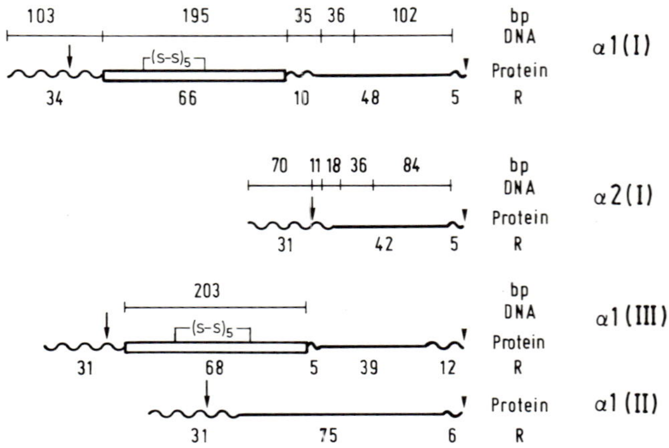

FIG. 6. Schematic representation of the amino-terminal prepro region of the α1(I), α2(I), α1(III), and α1(II) chains. From α1(I) and α2(I) not only the protein structure but also the gene structure is known. The subdivision of prepro area in exons is indicated. R, Amino acid residues; bp, base pairs present in each exon. Arrows indicate the cleavage site for the signal peptide, and arrowheads the cleavage site for the amino propeptide. Data were taken from the following references: α1(I), Chu *et al.* (1984), Hörlein *et al.* (1979); α2(I), Tate *et al.* (1983); α1(III), Yamada *et al.* (1983b), Brandt *et al.* (1984); α1(II), Kohno *et al.* (1984).

been observed that the globular part of the amino propeptide specifically inhibits collagen biosynthesis of fibroblasts in culture and also in cell-free translation systems (Wiestner *et al.*, 1979; Paglia *et al.*, 1979). These experimental results led to the proposal that the amino propeptide may be involved in a feedback control of types I and III collagen biosynthesis. However, it remains unclear as to both how the extracellular amino propeptide is taken up again by the fibroblast and how it is transported to the polysomes where protein synthesis occurs. Since the α1(II) chain does not contain a globular domain in the amino propeptide region, this putative feedback control discussed for types I and III collagen cannot occur for type II collagen.

C. *The Carboxy Propeptides*

The carboxy propeptides of all the four α chains [α1(I), α2(I), α1(II), and α1(III)] are 244 to 245 residues in length. They contain eight cysteine residues, of which six are involved in intrachain disulfide bridges, which stabilize the globular structure. Two cysteine residues form interchain bridges which are important for the stabilization of the complex of three α chains. These interchain bridges are specifically formed at the carboxy-terminus and are a prerequisite for the formation of a triple-helical molecule (Uitto and Prockop, 1973). In agreement with their similar function, the carboxy propeptide regions of the α chains are strongly homologous to one another (Yamada *et al.*, 1983a). However, there must be characteristic sequence regions in the propeptides which determine the type-specific interactions of the carboxy propeptides and ensure, for instance, that two α1(I) and one α2(I) chains regularly come together to form a type I collagen molecule and prevent formation between α1(III) and α1(I) or α2(I) chains.

D. The Junction Regions

The flexible junction regions which connect the amino and carboxy propeptides with the main triple helix are also important structural elements of collagen. They contain the cleavage site for the specific procollagen amino and carboxy proteinases as well as the nonhelical crosslink site (reviewed by Kühn, 1984). After cleavage of the junction regions, the telopeptides remain with the triple helix, providing the molecule with available oxidizable lysine or hydroxylysine residues essential for the formation of the intermolecular bonds. The telopeptides of all α chains contain a hydroxylysine with the exception of α2(I). In the carboxy-terminal telopeptide of α2(I) a stretch of 11 amino acid residues is missing which includes the hydroxylysine residue, so that the carboxy end of the α2(I) cannot be involved in cross-linking (see Yamada *et al.*, 1983a).

E. Control of Fiber Formation by Propeptides

There is experimental evidence that the amino and carboxy propeptides may play an important role in the control of fiber formation and fiber growth. With immunoelectron microscopy, Fleischmajer *et al.* (1981) have observed pN type I and III collagen molecules, which contain the amino propeptide on the surface of thin fibrils. Collagen type I fibrils with increasing diameter lack the amino propeptides, whereas collagen type III fibrils, which do not grow, keep the amino propeptide on their surface. These results, which were confirmed by biochemical investigations, led to the proposal that the amino propeptide specifically prevents the formation of thick collagen type III fibrils.

Fibrillogenesis experiments *in vitro* using pN and pC type I collagen, from which during fibrillogenesis the amino or the carboxy propeptides were cleaved off by the specific procollagen amino and carboxy

pN α1(I)	-Gly-Gly-Asn-Phe-Ala-Pro↓Gln-Leu-Ser-Tyr-Gly-Tyr-
pN α1(III)	-Gln-Asn-Asn-Tyr-Ala-Pro-Gln-Tyr-Glu-Ala-Tyr-Asp-

FIG. 7. The cleavage site of the procollagen amino proteinase in the amino-terminal junction region of the α1(I) and the α1(III) chain. The enzyme splits the proline–glutamine bond (arrow). The phenylalanine residue three positions amino-terminal to the cleavage site seems to be important for the α1(I) chain-cleaving amino proteinase. Substitution by tyrosine in α1(III) prevents cleavage (Morikawa *et al.*, 1980).

proteinases, supported the proposal that fiber formation is controlled by the propeptides. pN collagen itself formed thin cross-striated fibrils, and cleavage of the amino propeptide during fiber formation resulted in better structured and tighter packed fibrils of the same diameter. pC collagen is, because of its bulky carboxy propeptide, unable to aggregate. After removing the carboxy propeptide by carboxy proteinase, highly ordered thick fibrils and fiber bundles were formed (Miyahara *et al.*, 1984). In light of these experiments, the sequence of removal of the propeptides during posttranslational modifications seems to be important. The conclusions of biosynthetic studies are that type III collagen first loses the carboxy propeptide and later the amino peptide (Fessler *et al.*, 1981), whereas in type I collagen, the amino propeptide is cleaved off before the carboxy propeptide (Fessler and Fessler, 1978). This strongly suggests that a controlling influence of the propeptides during fibrillogenesis also occurs *in vivo*.

The different rates of cleavage of the amino propeptide in types I and III collagen can be explained by a difference of the amino acid sequences around the procollagen amino proteinase cleavage site Pro-Gln (Fig. 7). Using a variety of synthetic peptides as inhibitors for the type I collagen-specific amino proteinase, it was shown that a phenylalanine residue three positions amino-terminal to the cleavage site appears to be particularly important for the enzyme. In type III collagen, however, which contains the same cleavage site Pro-Gln but which cannot be cleaved by the type I collagen-specific enzyme, this phenylalanine residue is replaced by a tyrosine (Morikawa *et al.*, 1980).

F. Interaction of Collagen with Cells

For a long time, only the biomechanical properties of collagen and its function as a stabilizing scaffold of connective tissues for the organism were of research interest. In the 1960s, it was realized that differentiation and behavior of cells during embryonic development are not only controlled by humoral factors but also by interactions with constituents of the surrounding extracellular matrix (for a review, see Hay, 1984). Thus Hauschka and Konigsberg (1966) showed that a collagen substrate promotes the differentiation of chick embryo myoblasts to myotubes. Another example is the observation of Meier and Hay (1974) that gels of type I or type II collagen enhance the differentiation of corneal epithelium to the same extent as found for frozen and killed lenses.

Collagen can interact directly with cells via specific receptors at the cell surface (Mollenhauer and von der Mark, 1983), or the interaction can be mediated by other extracellular matrix components. The best investigated is the glycoprotein fibronectin which can serve as a mediator between collagen and the cell surface. Fibronectin has a particularly high affinity for denatured collagens (Jilek and Hörmann, 1978). Using cyanogen bromide peptides from α chains of types I, II, and III collagen, it was found that in all collagens the same region between residues 751 and 791 is responsible for collagen binding (Kleinman *et al.,* 1976; Dessau *et al.,* 1978). This region overlaps with the cleavage site of vertebrate collagenase, and it has been shown that the integrity of the collagenase-sensitive bond between residue 775 and 776 is required for fibronectin binding (Kleinman *et al.,* 1979). Fibronectin itself contains binding sites for collagen and for the cell surface (for reviews, see Yamada, 1983; Yamada *et al.,* 1985). Important for the discussion of the physiological role of fibronectin as a mediator between collagen and the cell is the fact that it has a higher affinity for denatured than for native collagen (Jilek and Hörmann, 1978). Heparin, however, appears to stabilize the binding of native collagen to fibronectin (Jilek and Hörmann, 1979).

Rubin *et al.* (1977) suggested that hepatocytes do not require fibronectin for binding to collagen. Hepatocytes attach equally well to all fiber-forming collagens, whereas native collagen was more efficient in promoting cell attachment than denatured collagen. Since all cyanogen bromide peptides of type I collagen showed similar binding activities, it was suggested that cell binding sites in collagen have a fairly simple structure and occur in multiple copies along the α chains. This is in agreement with observations that collagen-like synthetic peptides also show a low but significant cell attachment effect (Rubin *et al.,* 1984, 1986).

Mollenhauer and von der Mark (1983) and Mollenhauer *et al.* (1984) have isolated a collagen-binding glycoprotein from chondrocyte membranes, anchorin CII, which, *in vitro,* preferentially reacts with native but not with denatured type II collagen. Treatment of type II collagen with pronase, which removes the telopeptides on both ends of the molecules, abolished the reaction with anchorin CII, indicating that the binding sites for anchorin CII are located at the end regions of the collagen molecule (von der Mark *et al.*, 1985). Since denatured type II collagen does not react with anchorin CII, the binding site is conformation dependent. A similar collagen receptor has, in the meantime, been isolated from bone and was designated as anchorin CI. It is similar to anchorin CII. Both proteins have the same M_r of 34,000 and cross-react immunologically.

Important for hemostatic events is the interaction of collagen with platelets. When subendothelial components of the vascular wall are exposed after loss of the endothelium, as may occur during atherosclerosis, platelets will adhere rapidly, undergo a release reaction, and subsequently aggregate to form a primary hemostatic plug. It has been shown that collagen is the most thrombogenic subendothelial component (Beachey *et al.*, 1979). The collagen–platelet interaction has been investigated in an aggregometer. Whereas *in vivo* native cross-linked fibrils become exposed to platelets, in the *in vitro* experiments, purified collagens were reacted with platelets as reformed fibrils of the native type, unspecific aggregates of native molecules, dissolved triple-helical molecules, denatured α chains, and CNBr-derived peptides. Platelets can only be aggregated by aggregates of native molecules (Jaffe and Deykin, 1974; Balleisen *et al.*, 1976). The activity of the fibrils seems to be independent of the array of molecules, either in a D staggered order or in an irregular manner.

In contrast to aggregation, adhesion of platelets does not need collagen in an aggregated form but requires an intact triple-helical structure (Balleisen *et al.*, 1975a). If denatured collagen or CNBr peptides are used, the adhesion of platelets is drastically reduced, but there are some CNBr peptides (for example, the carboxy-terminal α1(I)CB6 and the α1(III)CB4 (position 403–551), which exhibit a small but significant adhesion activity (Fauvel *et al.*, 1978a,b). If these *in vitro* experiments are relevant also for the *in vivo* situation, then in types I and III collagens different areas of the molecule may be responsible for platelet adhesion. In type I collagen it appears to be the carboxy-terminal telopeptide (Fauvel *et al.*, 1978a) and in type III collagen a sequence region (position 478–486) in the middle of the molecule (Fauvel *et al.*, 1979). The carboxy-terminal telopeptide of type I collagen does not seem, however, to be important for platelet aggregation. Pepsin-treated type I collagen, where the carboxy-terminal telopeptide has been removed, as well as type I collagen, with intact telopeptides, exhibits in fibrillar form the same activity in aggregating platelets

(Balleisen *et al.*, 1976). It has also been suggested that the first step of collagen-platelet interaction involves the carbohydrate moiety of collagen (Jamieson *et al.*, 1971). The fact, however, that the collagen types with a high content of carbohydrates, such as types II and IV collagen, are less active than types I and III collagen, with a lower carbohydrate content, argues against an important role of these structures (Balleisen *et al.*, 1975b).

G. *The Antigenic Determinants of Collagen*

Antibodies against fiber-forming collagens and their precursors play an important role as a tool for the investigation of structure, biosynthesis, and metabolism of the collagens, their distribution in tissues, and any alterations which may occur in genetic and acquired connective tissue diseases. A short discussion of the relationship between structure and antigenicity of the fiber-forming collagens follows. At least five distinct groups of antigenic determinants have been recognized (for a review, see Furthmayr and Timpl, 1976) (Fig. 8). Two of these groups of determinants are specific for the precursor form and are localized in the globular domains at the ends of the molecule. Three further groups of determinants have been identified in the collagen molecule itself. The helical determinants are located in the triple-helical body of the molecule and are dependent on an intact triple-helical conformation. The terminal determinants, found at the telopeptides, are independent of an intact triple-helical conformation. Finally, unfolding of the triple helix exposes new determinants in the α chains which are called central determinants. Recognition of these groups of determinants during the antibody response depends on various parameters. Procollagens are better immunogens than collagens, and immunization with procollagen usually leads to production of precursor-specific antibodies. It has also been observed that the choice of the animal species used for immunization may determine the recognition sites. Rabbits, for example, produce high antibody levels to the nonconformational terminal determinants of collagen, while rats, guinea pigs, and mice show a preferential response to triple-helical structures (Furthmayr and Timpl, 1976).

A precise characterization of the antigenic determinants was carried out for the terminal telopeptides. The method involved degradation of the respective CNBr peptides by proteases and comparison of the fragments in inhibition assays. For the amino-terminal region of $\alpha 2(I)$ from rat and $\alpha 1(I)$ from calf collagen as well as the carboxy-terminal region of calf collagen, a possible maximal size was found to involve

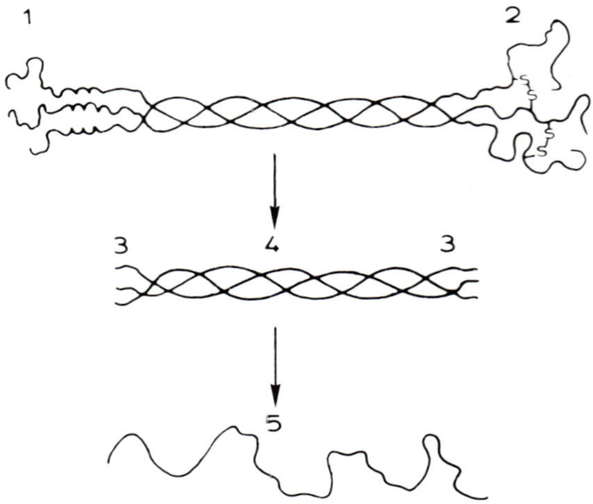

FIG. 8. Schematic representation of the antigenic determinants of procollagen and collagen. Determinants are located at the amino propeptide (1), the carboxy propeptide (2), the amino- and carboxy-terminal telopeptides (3), the triple-helical region (4), and the central part of the denatured α chains (5).

five or six amino acids. Additional information could be obtained by comparison of the antigenic regions with the sequence of the non-cross-reacting collagen from rabbits, the species in which the antibodies had been raised. For each antigenic region, at least one substitution was found on rabbit collagen. The difference either may be as small as a single substitution of leucine by methionine or may involve the whole region (summarized by Timpl, 1976; Timpl *et al.*, 1977).

VII. THE GENE STRUCTURE OF COLLAGENS I, II, AND III

Eukaryotic genes are characterized by a frequent interruption of the coding sequence by noncoding intervening sections, which results in a characteristic exon structure for each gene. Comparison of gene structures with the respective protein structures revealed that exons often (but not always) code for structural or functional protein domains (Gilbert, 1978). Earlier, McLachlan (1976) proposed that there is a primordial gene unit, with the length of D, from which the collagen genes would have evolved by duplication. Closer investigation of the amino acid sequence of the triple-helical domain revealed a number of additional repeats (Table III) (Hofmann *et al.*, 1980). The D/3 repeat,

TABLE III
REPEATING UNITS OF THE AMINO ACID SEQUENCE OBSERVED IN THE TRIPLE-HELICAL DOMAIN OF THE α1(I), α2(I), AND α1(III) CHAINS[a]

D	=	234 Residues Formed by polar-charged and hydrophobic (interactive) amino acids; control of self-assembly, D-staggered array
D/3	=	78 Residues Formed by noninteractive amino acids; length of triple-helical domain 4D + D/3; hole zone 2/3D, overlap zone 1/3D
D/6	=	39 Residues Formed by polar-charged amino acids (pitch?)
D/11	=	21.3 Residues (2D/11 42.6 residues) Formed by hydrophobic amino acids (pitch?)
D/13	=	18 Residues (2D/13 36 residues) Formed by interactive and noninteractive amino acids; primordial gene unit

[a]From Hofmann *et al.* (1980).

for example, has also been considered as a functional unit since it correlates with the length of the overlap regions of two 4D staggered molecules. In the fibril, therefore, each period of 67 nm is divided into a hole and an overlap zone 2 D/3 and 1 D/3 long, respectively. In contrast to the D repeat, which is formed mainly by the polar charged and hydrophobic amino acids (the interactive residues), the D/3 repeat is mainly formed by the group of noninteractive residues, particularly by prolines and alanines. Because the noninteractive residues have not been exposed to a specific selection pressure directed toward an improved aggregation of the molecules into the fibril, it was suggested that the original genetic unit from which the molecules developed by gene duplication may be better reflected by the D/3 rather than by the D repeat (Hofmann *et al.*, 1980). No obvious function has been found for the shorter repeats D/6, D/11, and D/13 present in the triple-helical area. They were thought to determine the pitch of the triple helix (Hulmes *et al.*, 1973). The fact that the pro α chains consist of three structural and functional domains, the triple helix and the two terminal procollagen peptides, suggests that the gene may also be composed of three sections of different exon structure, and the questions arise of whether or not they have evolved differently and when during the course of evolution they were combined to one gene.

The application of recombinant DNA technology resulted initially in the completion of the covalent structure of all four α chains. This is especially true for the carboxy-terminal propeptide regions, where

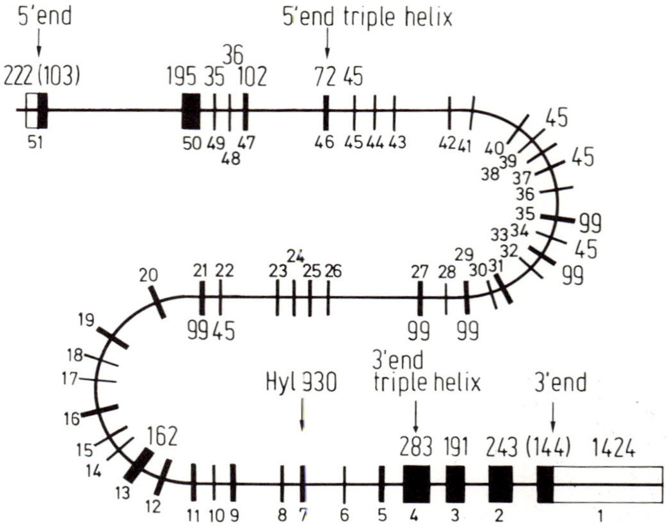

FIG. 9. Schematic representation of the approximately 18-kb-long human α1(I) gene. Data are taken from Chu *et al.* (1984). The bars reflect the positions of the exons along the α1(I) gene determined by R-loop experiments. The length of the individual exons of the carboxy and amino propeptide regions are indicated. In the triple-helical domain, thin bars reflect exons with 54 nucleotides and thick bars reflect exons with 108 nucleotides. Exons with the length of 45, 99, and 162 bp are indicated. The length of the exons, particularly of the triple-helical domain and the carboxy propeptide region, are almost identical in all investigated genes of the fiber-forming collagens I, II, and III. The longest gene is α2(I) with approximately 40 kb (Boedtker *et al.*, 1983).

only few protein sequences were available (Fuller and Boedtker, 1981; Dickson *et al.*, 1981; Yamada *et al.*, 1983a; Sandell *et al.*, 1984). The extreme amino-terminal part of the α chains, the signal peptides, and to some extent the amino propeptide would also not be known today without the DNA structure (Tate *et al.*, 1983; Vogeli *et al.*, 1981; Yamada *et al.*, 1983b; Kohno *et al.*, 1984).

Finally, these investigations led to the gene structure of α2(I) (Boedtker *et al.*, 1983) and α1(III) (Yamada *et al.*, 1984) of chicken, and α1(I) from human (Chu *et al.*, 1984) (Fig. 9). The size of collagen genes varies from 39,000 bp in the case of α2(I) chicken (Boedtker *et al.*, 1983) to 18,000 bp for α1(I) human (Chu *et al.*, 1984). The coding sequences, about 10–30% of the gene, are frequently interrupted. The number and size of exons are very similar in all genes of the fiber-forming collagens. The size of the introns, however, can vary to a much higher degree (for a review, see Boedtker *et al.*, 1983). The comparison of the subdivision of the coding sequence in exons with the multifunctional domain structure of collagen proteins led, on the one hand, to a better definition of the functional domains important for the formation of the molecular and macromolecular structures. On the other hand there are periodic structures, mainly in the triple-helical domain, which do not easily relate to the exon composition of this area.

A. The Triple-Helical Domain

The coding sequence of the triple-helical domain of α1(I) gene from human is divided into 41 exons, plus two exons at the end region that include in addition the flanking regions of the triple helix (Fig. 9). The basic exon unit of the triple-helical domain is 54 bp in length (Table IV). Exons with a length of 108 or 162 bp seem to have originated by the fusion of a 54-bp unit to form dimers and trimers. The occurrence of exons 45 and 99 bp long can be explained by the deletion of nine nucleotides corresponding to a tripeptide unit Gly-X-Y (Yamada *et al.*, 1980).

Knowing the exon structure of the gene as well as the protein structure, attempts can now be made to analyze how these two periodic structures correlate with each other. It is clear that the 18-amino-acid-residue-long D/13 repeat found in the protein structure reflects the 54-bp exon unit of the triple helix. There also appears to be a relationship between the D repeat and the D/13-long exon unit. Thus, the carboxy-terminal half of the triple-helical domain consists of a tandem repeat of 26 exon units whereby the borders of the exon units correlate exactly with the end of the D repeats (H. Hofmann and K. Kühn, unpublished results). Molecules staggered against one another by integer multiples of an exon unit should show a preferred interaction between homologous regions. The optimal interaction between 26-exon unit-

TABLE IV
SIZES OF THE 41 EXONS (5–45) OF THE TRIPLE-HELICAL DOMAIN[a]

Exon length (bp)	Number of exons
54	21 (1×, primordial gene unit)
108	9 (2×)
162	1 (3×)
45	5 (1×, deletion of 9 bp)
99	5 (2×, deletion of 9 bp)

[a]See Fig. 9.

long hypothetical molecules to fibrillar structures would occur by a stagger of 13 exon units, which in fact correlates with D. In this scheme, the fact that exon units with a distance of D exhibit the highest homology between one another also fits. The situation in the amino-terminal half of the triple-helical domain is, however, different. Because of a series of deletions of tripeptides, the exon units are out of phase with the D units in this area. But here it can be observed also that exon units, almost in phase with the D scheme, show the highest interaction with the neighboring molecules.

It is likely that the triple-helical domain has evolved by repeated duplication of the exon unit of 54 bp. Once, by chance, a unit of 13 exons may have duplicated as such, and thus the origin of the repeating D unit was created. As this pattern proved favorable for an ordered assembly into fibrils, it was further strengthened by mutation and selective pressure as it can be observed in the amino terminal half of the triple-helical domain. Although the evolution of collagen cannot be reconstructed in detail by the comparison of the gene and the protein structure, it seems to be obvious that the D repeat of the triple helix has not evolved independently of the preexisting exon unit of 54 bp (H. Hofmann and K. Kühn, unpublished results).

B. The Amino-Terminal Region

The gene structure of this area is known for α2(I) from chick (Tate *et al.*, 1983), α1(I) from human (Chu *et al.*, 1984), and, in part, α1(III) from chick (Yamada *et al.*, 1983b) (Fig. 6). The subdivision of the DNA into coding exons agrees well with the structural elements at the protein level. In the α1(I) chain, the amino-terminal signal peptide, the globular domain, and the short, flexible connection between the globular and the triple-helical portion are each encoded by one exon. The triple-helical portion consists of two exons of 36 and 102 bp whose length does not match the exon unit of 54 bp found in the main triple helix. The amino-terminal end of the α2(I) chain does not contain a globular domain. Here, the signal peptide is directly attached to the triple helix and is encoded by a 70-bp-long exon. The two following short exons, 11 and 18 bp in length, appear to be remnants of the globular part and its connection to the triple helix. This exon composition suggests that the α2(I) chain originally possessed a globular domain similar to α1(I) and that it was lost during the course of evolution.

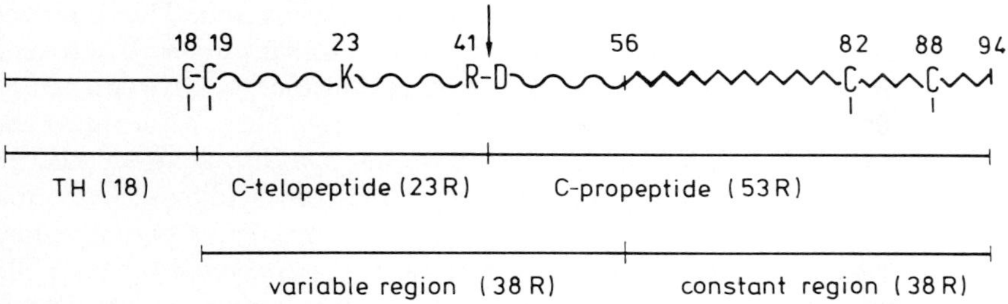

FIG. 10. Scheme of the 283-nucleotide-long exon 4 of the α1(III) gene encoding the carboxy junction area. Numbers give the position of amino acid residues in the area of exon 4. Arrow indicates the carboxy propeptidase cleavage site between arginine and aspartic acid. Cysteines in positions 18, 19, 82, and 88 are involved in interchain disulfide bridges (Yamada *et al.*, 1983a). R, Amino acid residues.

C. *The Junction Regions*

It seems likely that the periodic structure of the triple-helical domain of the α chains evolved independently and according to different rules than the two globular propeptides. The triple helix, containing almost all information for the self-assembly of the molecules into fibrils, had to be completed by control elements, the two propeptides. The two regions used to connect the propeptides with the triple helix have gained important functions at the protein level, which is preserved at the gene level using the two exons 46 and 4. Figure 10 depicts, as an example, the carboxy-terminal function encoded by exon 4, the longest continuous coding sequence of the collagen gene (Yamada *et al.*, 1983a; Boedtker *et al.*, 1983). In addition to the functions which can also be attributed to the amino-terminal junctions, such as to serve as a flexible connection, to provide the cleavage sites for the procollagen proteinases, and to contain a hydroxylysine residue involved in cross-linking, the carboxy-terminus is important for the formation of the triple helix initiated by a specific alignment of the carboxy propeptides of three α chains. Exon 4 includes the end part of the triple helix, the carboxy telopeptide, and the amino end region of the carboxy propeptide. The end region of the triple helix is characterized by a particularly high stability. In α1(I) and α2(I) it consists of five consecutive tripeptide units Gly-Pro-Hyp, and in type III collagen the very end of the helix is sealed by interchain disulfide bridges. The structure is thus adapted to a special function of this region, namely to facilitate the initiation of triple-helix formation at the carboxy-terminus by stabilizing the short helical section first formed. The adjacent flexible connection to the carboxy propeptide of 38 amino acid residues exhibits a large number of amino acid substitutions between the α chains (Yamada *et al.,* 1983a). It has been shown that, after aggregation of three carboxy propeptides, triple-helix formation can only occur after the stabilization of the trimer by disulfide bridges (Uitto and Prockop, 1973; Fessler *et al.,* 1981). Exon 4, which has been formed during the fusion process of two gene sections coding for the triple helical domain and the carboxy propeptide, respectively, is an excellent example of how a functionally important area is preserved by one exon although it is composed of three structurally different elements.

VIII. Synopsis

The three fiber-forming collagens, types I, II, and III, described in this article have, in principle, the same molecular structure. In their final processed form, more than 97% of the molecules consist of a continuous triple helical domain along which the main information for an ordered parallel and D staggered self-assembly is located. Thus, fibrils of the three collagens when reconstituted *in vitro* do not show striking differences. *In vivo,* however, the macromolecular structure of the three collagen types is variable. Comparative electron microscopy of tendon and cornea, both of which contain type I collagen, and hyaline cartilage, which consists mainly of type II collagen, revealed that the latter generally forms fibrils of smaller diameter than type I collagen. In addition, the often observed tendency of type I fibrils to assemble in parallel bundles is not found in type II collagen. Immunoelectron microscopy of human skin with ferritin-labeled antibodies against types I and III collagen revealed thick fiber bundles of type I and thin fibrils of type III collagen. Although these electron microscopic investigations indicate that type I and type III collagens may form separate fibrils (Fleischmajer *et al.,* 1981), there is chemical evidence that both collagens can be present in the same fibrils (Henkel and Glanville, 1982).

Variations in the macromolecular structure of the three collagen types probably modify the biomechanical properties of the extracellular matrix to the various physiological functions of different connective tissues. However, these variations cannot be explained by the differences in the amino acid sequence of the triple-helical molecules. It is, however, likely that some of the posttranslational modification steps are responsible, at least to some extent, in controlling fiber diameter. Thus, it is striking that the degree of glycosylation of type II collagen, which generally forms small fibrils, is higher than in type I collagen. Another possible influence on fibril assembly is by controlling the time course in the removal of the amino and carboxy propeptides (Miyahara *et al.,* 1984).

In this context, type V collagen, which also belongs to the family of the fiber forming collagens, and is described in detail by Fessler and Fessler (this volume), should be mentioned. The primary structure of the carboxy propeptide, as well as of the triple-helical domain of type V collagen, is homologous to that of the types I, II, and III collagen. Major differences exist in the amino propeptide region, which is processed in several steps and remains part of the molecule when incorporated into a fibril (Broek *et al.,* 1985). The macromolecular structure of type V collagen has not been fully established. However, it appears to form small fibrils and is often found in a pericellular arrangement, for instance in the vicinity of smooth muscle cells (Broek *et al.,* 1985).

During the last five years it has become clear that the ability of the fiber-forming collagens to vary their macromolecular organization is not enough to build stabilizing scaffolds for all of the different extracellular matrices. To introduce a higher variability, triple-helical elements are often combined with globular domains on the one hand; on the other hand, the length and some structural features of the triple helical domain itself were often altered (see Table I). A typical example is type IV collagen, the collagenous component of all basement membranes (see Glanville, this volume). In order to adapt the macromolecular organization of this collagen to the sheetlike flexible structure of basement membranes, the 390-nm-long molecules of type IV collagen do not form fibrils but a network in which they are connected to one another via their like ends (Timpl *et al.,* 1981). To give the network the necessary flexibility, the triple helix was frequently interrupted by nonhelical areas. This reduces the stability of the triple helix in these areas and is compensated by an altered distribution of the proline residues between the positions X and Y in favor of position Y. Proline residues are only hydroxylated in position Y, and thus the triple helix is stabilized. Another important difference from the fiber-forming collagens was observed for the first time in type IV collagen. The C-terminal globular domain of the molecule is not removed but is used as an integral part of the final macromolecular structure of the molecule. Like most of the presently known collagens, type IV collagen contains the triple-helical domain as a major structural element. There is one exception, type VI collagen, where the triple-helical domain represents only one-third of the molecule. It contains two large terminal, globular domains which are also incorporated into the macromolecular structure of type VI collagen (see Timpl and Engel, this volume).

Some of the newly discovered collagens (types IV, VI, and VII) fall into the classical definition of collagenous proteins. After biosynthesis of the triple-helical or partially triple-helical molecules, they are secreted into the extracellular space where they aggregate to macromolecular structures which are subsequently stabilized by covalent, intermolecular bonds. Some of the new collagens, however, do not fall into this category. Type IX collagen, for example, does not appear to form its own macromolecular structure and is thought to aggregate with type II collagen and to influence the formation of the fibers of type II collagen in hyaline cartilage (Müller-Glauser *et al.,* 1986). Another new aspect of the collagen family is represented by type X collagen. It is synthesized only by chondrocytes from hypertrophic cartilage (Gibson and Flint, 1985) and is thought to act as a factor involved in endochondrial calcification (see Schmid and Linsenmayer, this volume). Triple-helical segments are also found in other, less-characterized collagenous components (Sage, 1985; Yamauchi *et al.,* 1986) and in other proteins, such as the complement subcomponent C1q (Reid, 1979) acetylcholinesterase (Mays and Rosenberry, 1981) and conglutinin (Strang *et al.,* 1986). For these latter proteins, the triple-helical elements are combined with globular domains which are functionally active. In addition, short collagenous sequences are present in other proteins such as the mannose-binding proteins of liver (Drickamer *et al.,* 1986) and the pulmonary surfactant apoproteins (Floros *et al.,* 1986). These results all indicate that the triple helix, the main structural element of the classical fiber-forming collagen, is more widely used than previously thought.

The question arises whether all the triple helices derived from the different collagenous proteins have evolved from a common ancestor. The continuous triple-helical domain of the fiber-forming collagens is formed by duplication of a primordial gene unit of 54 bp (Yamada *et al.,* 1980). Present information on the gene structure of other collagen genes, such as type IV collagen (Kurkinen *et al.,* 1985; Soininen *et al.,* 1986; Sakurai *et al.,* 1986), type IX collagen (Lozano *et al.,* 1985), and also collagens from *Drosophila* (Monson *et al.,* 1982) and nematodes (Kramer *et al.,* 1982) revealed exon units which obviously do not fit into the 54-bp scheme. This would mean that there is no simple relationship between the triple-helical domains of different collagens and that they may have evolved in different ways. The present knowledge of the gene structure of the individual collagens does not yet allow a more detailed discussion of their evolutionary development.

References

Adams, E. (1978). *Science* **202,** 591–598.

Allmann, H., Fietzek, P. P., Glanville, R. W., and Kühn, K. (1979). *Hoppe-Seyler's Z. Physiol. Chem.* **360,** 861–868.

Astbury, W. T. (1940). *J. Soc. Leather Trades' Chem.* **24,** 69–92.

Bailey, A. J., Robins, S. P., and Balian, G. (1974). *Nature (London)* **251,** 105–109.

Bailey, A. J., Ranta, M. H., Nicholls, A. C., Partridge, S. M., and Elsden, D. F. (1977). *Biochem. Biophys. Res. Commun.* **78,** 1403–1410.

Balleisen, L., Marx, R., and Kühn, K. (1975a). *Blut* **31,** 95–106.

Balleisen, L., Gay, S., Marx, R., and Kühn, K. (1975b). *Klin. Wochenschr.* **52,** 903–912.

Balleisen, L., Marx, R., and Kühn, K. (1976). *Haemostasis* **5,** 155–164.

Beachey, E. H., Chiang, T. M., and Chan, E. H. (1979). *Int. Rev. Connect. Tissue Res.* **8,** 1–21.

Bear, R. S. (1944). *J. Am. Chem. Soc.* **66,** 1297–1305.

Bear, R. S. (1952). *Adv. Protein Chem.* **7,** 69–160.

Bellamy, G., and Bornstein, P. (1971). *Proc. Natl. Acad. Sci. U.S.A.* **68,** 1138–1142.

Bentz, H., Fietzek, P. P., and Kühn, K. (1979). *Hoppe-Seyler's Z. Physiol. Chem.* **360,** 833–840.

Berg, R. A., and Prockop, D. J. (1973). *Biochem. Biophys. Res. Commun.* **52,** 115–120.

Boedtker, H., and Doty, P. (1956). *J. Am. Chem. Soc.* **78,** 4267–4280.

Boedtker, H., Fuller, F., and Tate, V. (1983). *Int. Rev. Connect. Tissue Res.* **10,** 1–63.

Bornstein, P., and Piez, K. A. (1966). *Biochemistry* **5,** 3460–3473.

Bornstein, P., Kang, A. H., and Piez, K. A. (1966). *Proc. Natl. Acad. Sci. U.S.A.* **55,** 417–424.

Brandt, A., Glanville, R. W., Hörlein, D., Bruckner, P., Timpl, R., Fietzek, P. P., and Kühn, K. (1984). *Biochem. J.* **219,** 625–634.

Broek, D. L., Madri, J., Eikenberry, E. F., and Brodsky, B. (1985). *J. Biol. Chem.* **260,** 555–562.

Butler, W. T., and Cunningham, L. W. (1966). *J. Biol. Chem.* **241,** 3882–3888.

Cardinale, G. J., and Udenfriend, S. (1974). *Adv. Enzymol. Relat. Areas Mol. Biol.* **41,** 245–300.

Chu, M.-L., de Wet, W., Bernard, M., Ding, J.-F., Morabito, M., Myers, J., Williams, C., and Ramirez, F. (1984). *Nature (London)* **310,** 337–340.

Cowan, P. M., North, A. C. T., and Randall, J. T. (1953). *In* "Nature and Structure of Collagen" (J. T. Randall, ed.), pp. 241–249. Butterworth, London.

Dessau, W., Adelmann, B. C., Timpl, R., and Martin, G. R. (1978). *Biochem. J.* **169,** 55–59.

Dewes, H., Fietzek, P. P., and Kühn, K. (1979a). *Hoppe-Seyler's Z. Physiol. Chem.* **360,** 821–832.

Dewes, H., Fietzek, P. P., and Kühn, K. (1979b). *Hoppe-Seyler's Z. Physiol. Chem.* **360,** 851–860.

Dickson, L. A., Ninomiya, Y., Bernard, M. P., Pesciotta, D. M., Parsons, J., Green, G., Eikenberry, E. F., de Crombrugghe, B., Vogeli, G., Pastan, I., Fietzek, P. P., and Olsen, B. R. (1981). *J. Biol. Chem.* **256,** 8407–8415.

Drickamer, K., Dordal, M. S., and Reynolds, L. (1986). *J. Biol. Chem.* **261,** 6878–6887.

Edman, P., and Begg, G. (1967). *Biochemistry* **6,** 80–91.

Eyre, D. R. (1980). *Science* **207,** 1315–1322.

Fauvel, F., Legrand, Y. J., and Caen, J. P. (1978a). *Thromb. Res.* **12,** 273–285.

Fauvel, F., Legrand, Y. J., Bentz, H., Fietzek, P. P., Kühn, K., and Caen, J. P. (1978b). *Thromb. Res.* **12,** 841–850.

Fauvel, F., Legrand, Y. J., Kühn, K., Bentz, H., Fietzek, P. P., and Caen, J. P. (1979). *Thromb. Res.* **16,** 269–273.

Fessler, J. H., and Fessler, L. I. (1978). *Annu. Rev. Biochem.* **47,** 129–162.

Fessler, L. I., Timpl, R., and Fessler, J. H. (1981). *J. Biol. Chem.* **256,** 2531–2537.

Fietzek, P. P., and Kühn, K. (1976). *Int. Rev. Connect. Tissue Res.* **7,** 1–60.

Fietzek, P. P., Allmann, H., Rauterberg, J., Henkel, W., Wachter, E., and Kühn, K. (1979). *Hoppe-Seyler's Z. Physiol. Chem.* **360,** 809–820.

Fleischmajer, R., Timpl, R., Tuderman, L., Raisher, L., Wiestner, M., Perlish, J. S., and Graves, P. N. (1981). *Proc. Natl. Acad. Sci. U.S.A.* **78,** 7360–7364.

Floros, J., Steinbrink, R., Jacobs, K., Phelps, D., Kriz, R., Recny, M., Sultzman L., Jones, S., Taeusch, H. W., Frank, H. A., and Fritsch, E. F. (1986). *J. Biol. Chem.* **261,** 9029–9033.

Fuller, F., and Boedtker, H. (1981). *Biochemistry* **20,** 996–1006.

Furthmayr, H., and Timpl, R. (1976). *Int. Rev. Connect. Tissue Res.* **7,** 61–99.

Galloway, D. (1982). *In* "Collagen in Health and Disease" (J. B. Weiss and M. I. V. Jayson, eds.), pp. 528–557. Churchill-Livingstone, Edinburgh.

Gebhardt, E., and Kühn, K. (1963). *Z. Anorg. Allg. Chem.* **320,** 71–77.

Gibson, G. J., and Flint, M. H. (1985). *J. Cell Biol.* **101,** 277–284.

Gilbert, W. (1978). *Nature (London)* **271,** 501.

Grassmann, W., Schneider, F., and Truppke, J. (1951). *In* "Physiologische Chemie" (B. Flaschenträger and E. Lehnartz, eds.), pp. 484–499. Springer-Verlag, Heidelberg.

Grassmann, W., Hörmann, H., and Hafter, R. (1957). *Z. Physiol. Chem.* **307,** 87–96.

Grassmann, W., Engel, J., Hanning, K., Hörmann, H., Kühn, K., and Nordwig, A. (1965). *Fortschr. Chem. Org. Naturst.* **13,** 195–314.

Gross, J. (1976). *In* "Biochemistry of Collagen" (G. N. Ramachandran and A. H. Reddi, eds.), pp. 275–317. Plenum, New York.

Gross, J., and Nagai, Y. (1965). *Proc. Natl. Acad. Sci. U.S.A.* **54,** 1197–1204.

Gross, J., Dumsha, B., and Glazer, N. (1958). *Biochim. Biophys. Acta* **30,** 293–297.

Gross, J., Levene, C. I., and Orloff, S. (1960). *Proc. Soc. Exp. Biol. Med.* **105,** 148–150.

Gross, J., Harper, E., Harris, E. D., Jr., McCroskery, P., Highberger, J. H., Corbett, C., and Kang, A. H. (1974). *Biochem. Biophys. Res. Commun.* **61,** 555–602.

Hall, C. E., and Slayter, H. S. (1959). *J. Biophys. Biochem. Cytol.* **5,** 11–17.

Hall, C. E., Jakus, M. A., and Schmitt, F. O. (1942). *J. Am. Chem. Soc.* **64,** 1234.

Hanauske-Abel, H. M., and Gunzler, V. (1982). *J. Theor. Biol.* **94,** 421–455.

Hauschka, S. D., and Konigsberg, I. R. (1966). *Proc. Natl. Acad. Sci. U.S.A.* **55,** 119–126.

Hay, E. D. (1984). *In* "The Role of Extracellular Matrix in Development" (R. L. Trelstad, ed.), pp. 1–31. Liss, New York.

Henkel, W., and Glanville, R. W. (1982). *Eur. J. Biochem.* **122,** 205–213.

Herzog, R. O., and Gonnel, H. W. (1924). *Naturwissenschaften* **12,** 1153–1155.

Highberger, J. H., Gross, J., and Schmitt, F. O. (1950). *J. Am. Chem. Soc.* **72,** 3321–3322.

Hodge, A. J., and Schmitt, F. O. (1960). *Proc. Natl. Acad. Sci. U.S.A.* **46,** 186–197.

Hoffmann, H. P., Olsen, B. R., Chen, H. T., and Prockop, D. (1976). *Proc. Natl. Acad. Sci. U.S.A.* **73,** 4304–4308.

Hofmann, H., Fietzek, P. P., and Kühn, K. (1978). *J. Mol. Biol.* **125,** 137–165.

Hofmann, H., Fietzek, P. P., and Kühn, K. (1980). *J. Mol. Biol.* **141,** 293–314.

Hörlein, D., Fietzek, P. P., Wachter, E., Lapiere, C. M., and Kühn, K. (1979). *Eur. J. Biochem.* **99,** 31–38.

Hulmes, D. J., and Miller, A. (1979). *Nature (London)* **282,** 878–880.

Hulmes, D. J. S., Miller, A., Parry, D. A., Piez, K. A., and Woodhead-Galloway, J. (1973). *J. Mol. Biol.* **79,** 137–148.

Jaffe, R., and Deykin, D. (1974). *J. Clin. Invest.* **53,** 875–883.

Jamieson, G. A., Urban, C. L., and Barber, A. J. (1971). *Nature (London), New Biol.* **234,** 5–7.

Jilek, F., and Hörmann, H. (1978). *Hoppe-Seyler's Z. Physiol. Chem.* **359,** 247–250.

Jilek, F., and Hörmann, H. (1979). *Hoppe-Seyler's Z. Physiol. Chem.* **360,** 597–603.

Kang, A. H., Bornstein, P., and Piez, K. A. (1967). *Biochemistry* **6,** 788–795.

Kefalides, N. A. (1971). *Biochem. Biophys. Res. Commun.* **45,** 226–234.

Kivirikko, K. I., and Myllylä, R. (1979). *Int. Rev. Connect. Tissue Res.* **8,** 23–72.

Kivirikko, K. I., and Myllylä, R. (1980). *In* "The Enzymology of Post-Translational Modification of Proteins" (R. B. Freedman and H. A. Hawkins, eds.), pp. 53–104. Academic Press, London.

Kivirikko, K. I., and Myllylä, R. (1985). *In* "Biology, Chemistry, and Pathology of Collagen" (R. Fleischmajer, B. R. Olsen, and K. Kühn, eds.), pp. 187–201. New York Academy of Sciences, New York.

Kleinman, H. K., McGoodwin, E. B., and Klebe, R. J. (1976). *Biochem. Biophys. Res. Commun.* **72,** 426–432.

Kleinman, H. K., McGoodwin, E. B., Rennard, S. I., and Martin, G. R. (1979). *Anal. Biochem.* **94,** 308–312.

Kohno, K., Martin, G. R., and Yamada, Y. (1984). *J. Biol. Chem.* **259,** 3668–3673.

Kramer, J. M., Cox, G. N., and Hirsh, D. (1982). *Cell* **30,** 599–606.

Kühn, K. (1982). *Collagen Relat. Res.* **2,** 61–80.

Kühn, K. (1984). *Collagen Relat. Res.* **4,** 309–322.

Kühn, K. (1986). *Rheumatology* **10,** 29–69.

Kühn, K., Grassmann, W., and Hofmann, U. (1958). *Z. Naturforsch., B: Anorg. Chem., Org. Chem.* **13,** 154–160.

Kühn, K., Grassmann, W., and Hofmann, U. (1959). *Z. Naturforsch., B: Anorg. Chem., Org. Chem.* **14,** 436–443.

Kühn, K., Grassmann, W., and Hofmann, U. (1960). *Naturwissenschaften* **47,** 258–259.

Kühn, K., Wiedemann, H., Timpl, R., Risteli, J., Dieringer, H., Voss, T., and Glanville, R. W. (1981). *FEBS Lett.* **125,** 123–128.

Küntzel, A., and Prakke, F. (1933). *Biochem. Z.* **267,** 243–295.

Kurkinen, M., Bernhard, M. P., Barlow, D. P., and Chow, L. T. (1985). *Nature (London)* **317,** 177–179.

Lang, H., Glanville, R. W., Fietzek, P. P., and Kühn, K. (1979). *Hoppe-Seyler's Z. Physiol. Chem.* **360,** 841–850.

Layman, D. L., McGoodwin, E. B., and Martin, G. R. (1971). *Proc. Natl. Acad. Sci. U.S.A.* **68,** 454–458.

Lenaers, A., Ansay, M., Nusgens, B. V., and Lapiere, C. M. (1971). *Eur. J. Biochem.* **23,** 533–543.

Light, N. D., and Bailey, A. J. (1979). *FEBS Lett.* **97,** 183–188.

Light, N. D., and Bailey, A. J. (1980). *In* "Biology of Collagen" (A. Viidik and J. Vuust, eds.), pp. 15–38. Academic Press, London.

Lozano, G., Ninomiya, Y., Thompson, H., and Olson, B. R. (1985). *Proc. Natl. Acad. Sci. U.S.A.* **82,** 4050–4054.

McLachlan, A. D. (1976). *J. Mol. Biol.* **107,** 159–174.

Majamaa, K., Turpeenniemi-Hujanen, T. M., Latipää, P., Günzler, V., Hanauske-Abel, H. M., Hassinen, I. E., and Kivirikko, K. I. (1985). *Biochem. J.* **229,** 127–133.

Martin, G. R., Timpl, R., Müller, P. K., and Kühn, K. (1985). *Trends Biochem. Sci.* **10,** 285–287.

Mays, C., and Rosenberry, T. L. (1981). *Biochemistry* **20,** 1810–1817.

Meier, S., and Hay, E. D. (1974). *Dev. Biol.* **38,** 249–270.
Miller, E. J. (1971). *Biochemistry* **10,** 1652–1659.
Miller, E. J. (1976). *Mol. Cell. Biochem.* **13,** 165–192.
Miller, E. J., and Matukas, V. J. (1969). *Proc. Natl. Acad. Sci. U.S.A.* **64,** 1264–1268.
Miller, E. J., Epstein, E. H., Jr., and Piez, K. A. (1971). *Biochem. Biophys. Res. Commun.* **42,** 1024–1029.
Miller, E. J., Harris, E. D., Chung, E., Finch, J. E., McCroskery, P. A., and Butler, W. T. (1976). *Biochemistry* **15,** 787–792.
Miyahara, M., Hayashi, K., Berger, J., Tanzawa, K., Njieha, F. K., Trelstad, R. L., and Prockop, D. J. (1984). *J. Biol. Chem.* **259,** 9891–9898.
Mollenhauer, J., and von der Mark, K. (1983). *EMBO J.* **2,** 45–50.
Mollenhauer, J., Bee, J. A., Lizarbe, A., and von der Mark, K. (1984). *J. Cell Biol.* **98,** 1572–1578.
Monson, J. M., Natzle, J., Friedman, J., and McCarthy, B. J. (1982). *Proc. Natl. Acad. Sci. U.S.A.* **79,** 1761–1765.
Morikawa, T., Tuderman, L., and Prockop, D. J. (1980). *Biochemistry* **19,** 2646–2650.
Müller-Glauser, W., Humbel, B., Glatt, M., Sträuli, P., Winterhalter, K. H., and Bruckner, P. (1986). *J. Cell Biol.* **102,** 1931–1939.
Nageotte, J. (1927). *C. R. Hebd. Seances Acad. Sci.* **184,** 115–117.
Nemetschek, T. (1969). *Virchows Arch. B* **2,** 255–264.
Nemetschek, T., and Hosemann, R. (1973). *Kolloid Z. Z. Polym.* **251,** 1044–1056.
Nemetschek, T., Grassmann, W., and Hofmann, U. (1955). *Z. Naturforsch. B: Anorg. Chem., Org. Chem.* **10,** 61–68.
Nemetschek, T., Riedl, H., and Jonak, R. (1979). *J. Mol. Biol.* **133,** 67–83.
Nordwig, A., and Hayduk, U. (1969). *J. Mol. Biol.* **44,** 161–172.
Paglia, L., Wilczek, J., de Leon, L. D., Martin, G. R., Hörlein, D., and Müller, P. K. (1979). *Biochemistry* **18,** 5030–5034.
Pauling, L., Corey, R. B., and Branson, H. R. (1951). *Proc. Natl. Acad. Sci. U.S.A.* **37,** 205–211.
Peterkofsky, B., and Udenfriend, S. (1963). *J. Biol. Chem.* **238,** 3966–3977.
Peterkofsky, B., and Udenfriend, S. (1965). *Proc. Natl. Acad. Sci. U.S.A.* **53,** 335–342.
Petruska, J. A., and Hodge, A. J. (1964). *Proc. Natl. Acad. Sci. U.S.A.* **51,** 871–876.
Piez, K. A. (1968). *Annu. Rev. Biochem.* **37,** 547–570.
Piez, K. A., and Tochia, D. A. (1975). *Nature (London)* **258,** 87.
Piez, K. A., Eigner, E. A., and Lewis, M. S. (1963). *Biochemistry* **2,** 58–66.
Ramachandran, G. N., and Kartha, G. (1954). *Nature (London)* **174,** 269–270.
Ramachandran, G. N., and Ramakrishnan, C. (1976). *In* "Biochemistry of Collagen" (G. N. Ramachandran and A. H. Reddi, eds.), pp. 45–84. Plenum, New York.
Rauterberg, J., Fietzek, P., Rexrodt, F., Becker, U., Stark, M., and Kühn, K. (1972). *FEBS Lett.* **21,** 75–79.
Reid, K. B. M. (1979). *Biochem. J.* **179,** 367–371.
Rich, A., and Crick, F. H. C. (1955). *Nature (London)* **176,** 915–916.
Robins, S. P., and Bailey, A. J. (1973). *FEBS Lett.* **33,** 167–171.
Robins, S. P., and Bailey, A. J. (1975). *Biochem. J.* **149,** 381–385.
Robins, S. P., and Bailey, A. J. (1977). *Biochem. J.* **163,** 339–346.
Rubin, K., Kjellen, L., and Öbrink, B. (1977). *Exp. Cell Res.* **109,** 413–422.
Rubin, K., Borg, T. K., Holmdahl, R., Klareskog, L., and Öbrink, B. (1984). *Ciba Found. Symp.* **108,** 93–116.
Rubin, K., Gullberg, D., Borg, T. K., and Öbrink, B. (1986). *Exp. Cell Res.* **164,** 127–138.
Sage, H. (1985). *Biochemistry* **24,** 7430–7440.

Sakurai, Y., Sullivan, M., and Yamada, Y. (1986). *J. Biol. Chem.* **25,** 6654–6657.
Sandell, L. J., Prentice, H. L., Kravis, D., and Upholt, W. B. (1984). *J. Biol. Chem.* **259,** 7826–7834.
Schmitt, F. O. (1956). *Proc. Am. Philos. Soc.* **100,** 476–486.
Schmitt, F. O., Hall, C. E., and Jakus, M. A. (1942). *J. Cell. Comp. Physiol.* **20,** 11–33.
Schmitt, F. O., Gross, J., and Highberger, J. H. (1953). *Proc. Natl. Acad. Sci. U.S.A.* **39,** 459–470.
Schroeder, W. A., Honnen, L., and Green, F. C. (1953). *Proc. Natl. Acad. Sci. U.S.A.* **39,** 23–30.
Schroeder, W. A., Kay, L. M., LeGette, J., Honnen, L., and Green, F. C. (1954), *J. Am. Chem. Soc.* **76,** 3556–3564.
Segal, D. M. (1969). *J. Mol. Biol.* **43,** 497–517.
Seyer, J. M., and Kang, A. H. (1977). *Biochemistry* **16,** 1158–1164.
Seyer, J. M., and Kang, A. H. (1978). *Biochemistry* **17,** 3404–3411.
Seyer, J. M., and Kang, A. H. (1981). *Biochemistry* **20,** 2621–2627.
Seyer, J. M., Mainardi, C., and Kang, A. H. (1980). *Biochemistry* **19,** 1583–1589.
Siegel, R. C. (1974). *Proc. Natl. Acad. Sci. U.S.A.* **71,** 4826–4830.
Siegel, R. C. (1979). *Int. Rev. Connect. Tissue Res.* **8,** 73–118.
Soininen, R., Tikka, L., Chow, L., Pihlajaniemi, T., Kurkinen, M., Prockop, D. J., Boyd, C. D., and Tryggvason, K. (1986). *Proc. Natl. Acad. Sci. U.S.A.* **83,** 1568–1572.
Spiro, R. G. (1967). *J. Biol. Chem.* **242,** 4813–4823.
Spiro, R. G., and Spiro, M. J. (1971a). *J. Biol. Chem.* **246,** 4899–4909.
Spiro, R. G., and Spiro, M. J. (1971b). *J. Biol. Chem.* **246,** 4910–4918.
Stark, M., Lenaers, A., Lapiere, C., and Kühn, K. (1971). *FEBS Lett.* **18,** 225–227.
Stark, M., Miller, E. J., and Kühn, K. (1972). *Eur. J. Biochem.* **27,** 192–196.
Stetten, M. R. (1949). *J. Biol. Chem.* **181,** 31–37.
Strang, C. J., Slayter, H. S., Lachmann, P. J., and Davis, A. E. (1986). *Biochem. J.* **234,** 381–389.
Tanzer, M. L. (1976). *In* "Biochemistry of Collagen" (G. N. Ramachandran and A. H. Reddi, eds.), pp. 137–162. Plenum, New York.
Tate, V. E., Finer, M. H., Boedtker, H., and Doty, P. (1983). *Nucleic Acids Res.* **11,** 91–104.
Timpl, R. (1976). *In* "Biochemistry of Collagen" (G. N. Ramachandran and A. H. Reddi, eds.), pp. 319–375. Plenum, New York.
Timpl, R., Wick, G., and Gay, S. (1977). *J. Immunol. Methods* **18,** 165–182.
Timpl, R., Wiedemann, H., van Delden, V., Furthmayr, H., and Kühn. K. (1981). *Eur. J. Biochem.* **120,** 203–211.
Trus, B. L., and Piez, K. A. (1976). *J. Mol. Biol.* **108,** 705–732.
Uitto, J., and Prockop, D. J. (1973). *Biochem. Biophys. Res. Commun.* **55,** 904–911.
Uitto, J., and Prockop, D. J. (1975). *In* "Collagen Metabolism in the Liver" (H. Popper and K. Becker, eds.), pp. 139–148. Stratton Intercontinental, New York.
Vogeli, G., Ohkubo, H., Sobel, M. E., Yamada, Y., Pastan, I., and de Crombrugghe, B. (1981). *Proc. Natl. Acad. Sci. U.S.A.* **78,** 5334–5338.
von der Mark, K., Rexrodt, F., Wendt, P., and Kühn, K. (1970). *Eur. J. Biochem.* **16,** 143–151.
von der Mark, H., Aumailley, M., Wick, G., Fleischmajer, R., and Timpl, R. (1984). *Eur. J. Biochem.* **142,** 493–502.
von der Mark, K., Mollenhauer, J., Müller, P. K., and Pfäffle, M. (1985). *In* "Biology, Chemistry, and Pathology of Collagen" (R. Fleischmajer, B. R. Olsen, and K. Kühn, eds.), pp. 214–223. New York Academy of Sciences, New York.

Weiss, J. B. (1976). *Int. Rev. Connect. Tissue Res.* **7,** 101–157.

Wiedemann, H., Chung, E., Fuji, T., Miller, E. J., and Kühn, K. (1975). *Eur. J. Biochem.* **51,** 363–368.

Wiestner, M., Krieg, T., Hörlein, D., Glanville, R. W., Fietzek, P., and Müller, P. K. (1979). *J. Biol. Chem.* **254,** 7016–7023.

Wolpers, C. (1943). *Klin. Wochenschr.* **22,** 624.

Woolley, D. E., Lindberg, K. A., Glanville, R. W., and Evanson, J. M. (1975). *J. Biochem.* **50,** 437–444.

Yamada, K. M. (1983). *Ann. Rev. Biochem.* **52,** 761–799.

Yamada, Y., Avvidimento, V. E., Mudryj, M., Ohkubo, H., Vogeli, G., Irani, M., Pastan, I., and de Crombrugghe, B. (1980). *Cell* **22,** 887–892.

Yamada, Y., Kühn, K., and de Crombrugghe, B. (1983a). *Nucleic Acids Res.* **11,** 2733–2744.

Yamada, Y., Mudryj, M., and de Crombrugghe, B. (1983b). *J. Biol. Chem.* **258,** 14914–14919.

Yamada, Y., Liau, G., Mudryj, M., Obici, S., and de Crombrugghe, B. (1984). *Nature (London)* **310,** 333–337.

Yamada, K. M., Akiyama, S. K., Hasegawa, T., Hasegawa, E., Humphries, M. J., Kennedy, D. W., Nagata, K., Urushihara, H., Olden, K., and Chen, W.-T. (1985). *J. Cell. Biochem.* **28,** 79–97.

Yamauchi, M., Kuboki, Y., Sasaki, S., and Mechanic, G. L. (1986). *Biochemistry* **25,** 1997–2002.

Type IV Collagen

Robert W. Glanville

Shriners Hospital for Crippled Children and Department of Biochemistry, Oregon Health Sciences University, Portland, Oregon 97201

I. Introduction

Type IV collagen is a highly specialized form of collagen found only in basement membranes. It is one of the major components of all basement membranes together with the glycoproteins laminin, nidogen, entactin, and heparan sulfate proteoglycan.

Basement membranes are ubiquitous, thin, sheetlike structures found frequently under epithelial and endothelial cell linings but also surrounding many cell types such as muscle, nerve, and fat. They function as a selective filtration barrier for macromolecules, for example, in the kidney, blood–brain barrier, and placenta, but also separate extracellular matrix from epithelial or endothelial cell layers as in gut, skin, cornea, lung, and blood vessels. For more information about the structure and function of basement membranes and their noncollagenous components, the reader is referred to the following books and reviews (Kefalides, 1978; Robert *et al.*, 1979; Kühn *et al.*, 1982; Timpl and Martin, 1982; Timpl and Dziadek, 1986).

Indications that basement membranes contained a collagen came from X-ray studies of intact basement membranes as early as 1951 (Pirie, 1951). Later, hydroxyproline and then hydroxylysine were detected in amino acid compositions of whole basement membranes (Kefalides and Winzler, 1966). Because of the insolubility of basement membrane components, attempts were made to solubilize the collagen using Pronase, a method that had proved useful for type I collagen (Kefalides, 1968). The material that was isolated and characterized was clearly different from the other interstitial collagens known at that time, i.e., α1(I), α1(II), and α1(III). Basement membrane collagen was therefore designated type IV collagen (Kefalides, 1973).

STRUCTURE AND FUNCTION OF
COLLAGEN TYPES

II. Structure

A. *Macromolecular Structure Using the Electron Microscope*

Because of the size of the collagen triple-helical molecule, it has been possible to use the electron microscope to investigate aggregates formed *in vivo* and *in vitro* to gain some insight into the macromolecular structure of the molecules. In the case of the interstitial collagens, observations on the banding patterns of native fibers, segment long spacing (SLS) crystallites (Kühn, 1982), and more recently rotary shadowed molecules, were instrumental in defining the detailed structure of these molecules (Furthmayr and Madri, 1982). In the case of type IV collagen, *in vivo* observations of macromolecular structures formed *in vivo* have been, as yet, of limited value because of the amorphous structure of basement membranes.

1. Segment Long Spacing Crystallites

Type IV collagen does not form well-ordered SLS crystallites. Extensive investigations on acid-extracted lens capsule material indicated that a globular noncollagenous domain was present (Schwartz and Veis, 1978, 1979). It was, however, incorrectly deduced that there was a globular domain at each end of the molecule. Even after mild pepsin digestion, SLS crystallites showing well-defined cross-striation banding patterns were difficult to prepare. It was possible to say that the banding pattern, and thus the primary structure of type IV, differed from that of the interstitial collagens (Schwartz and Veis, 1978; Timpl *et al.*, 1978; Glanville *et al.*, 1979). Using trypsin-resistant fragments, crystallites were formed that were considerably longer (350 nm) than those of the interstitial collagens (290 nm), indicating for the first time that the helical domain of type IV was longer (Timpl *et al.*, 1978).

2. Images of Rotary Shadowed Molecules

Applying the rotary shadowing technique to type IV collagen preparations revealed how different the tertiary structure of this molecule was to all other well-characterized collagens. Initial studies were done on pepsin-solubilized material which indicated for the first time that the basic structural unit of basement membrane collagen is probably a tetramer (Kühn *et al.*, 1981). This also showed that 7 S collagen, a disulfide cross-linked collagenase-resistant fragment that had been originally isolated from the EHS mouse tumor (Timpl *et al.,* 1979b) and later from other tissues (Risteli *et al.,* 1980a; Dixit *et al.,* 1981; Madri *et al.,* 1983; Qian and Glanville, 1984) was, in fact, part of the type IV collagen molecule.

As shown in Fig. 1A, the most intact form of type IV collagen present in a pepsin digest appeared to be four triple-helical molecules overlapping at one end which was later shown to be the amino-terminal end. Many of the molecules have one or more arms missing. The missing fragments were further degraded as evidenced by the presence of a number of monomeric fragments of various lengths (Glanville *et al.,* 1982). Increasing the temperature of pepsin digestion caused the removal of all four arms leaving the central overlap domain, called the short form of the 7 S domain. Under slightly milder conditions an intermediate, carrying vestiges of all four arms, can be prepared and is referred to as the 7 S long form (Fig. 1B). Both forms of the 7 S domain can also be prepared, under appropriate conditions, using bacterial collagenase in place of pepsin (Fig. 1C).

Careful measurements gave values of 356 ± 10 nm for the lengths of the arms and 30 ± 1 nm for the central overlap region, giving a total molecular length of 386 nm (Kühn *et al.,* 1981). Subsequently, a value of 417 nm was reported which was considered to be a significant difference in length (Bächinger *et al.,* 1982b). It was suggested that the longer molecule may represent a type IV procollagen which carries an amino-terminal extension. This total length is considerably longer than the helix of the interstitial collagens, which is around 300 nm. The electron microscope pictures not only gave indications of the size and macromolecular organization of type IV molecules but also flexible sites in the threadlike arms of the molecules could also be localized (Glanville *et al.,* 1982). Most of the detected flexible sites allowed random bending about the mean 0°, but in one case, at the border of the 7 S domain of tetrameric type IV, a flexible site with a preferential angle of 40° was found (Hofmann *et al.,* 1984). This is probably an important feature for the formation of higher aggregates of tetramers as discussed below.

Having characterized the pepsin-solubilized type IV material, the next step was to investigate more intact material produced in cell culture or, in one case, by direct extraction of a mouse tumor with acetic acid (Timpl *et al.,* 1981; Oberbäumer *et al.,* 1982; Bächinger *et al.,* 1982b). Examination of the rotary shadowed pictures of such intact molecules revealed a globular domain at the end of the molecule opposite to the 7 S domain as shown in Figure 2D. Furthermore, dimeric

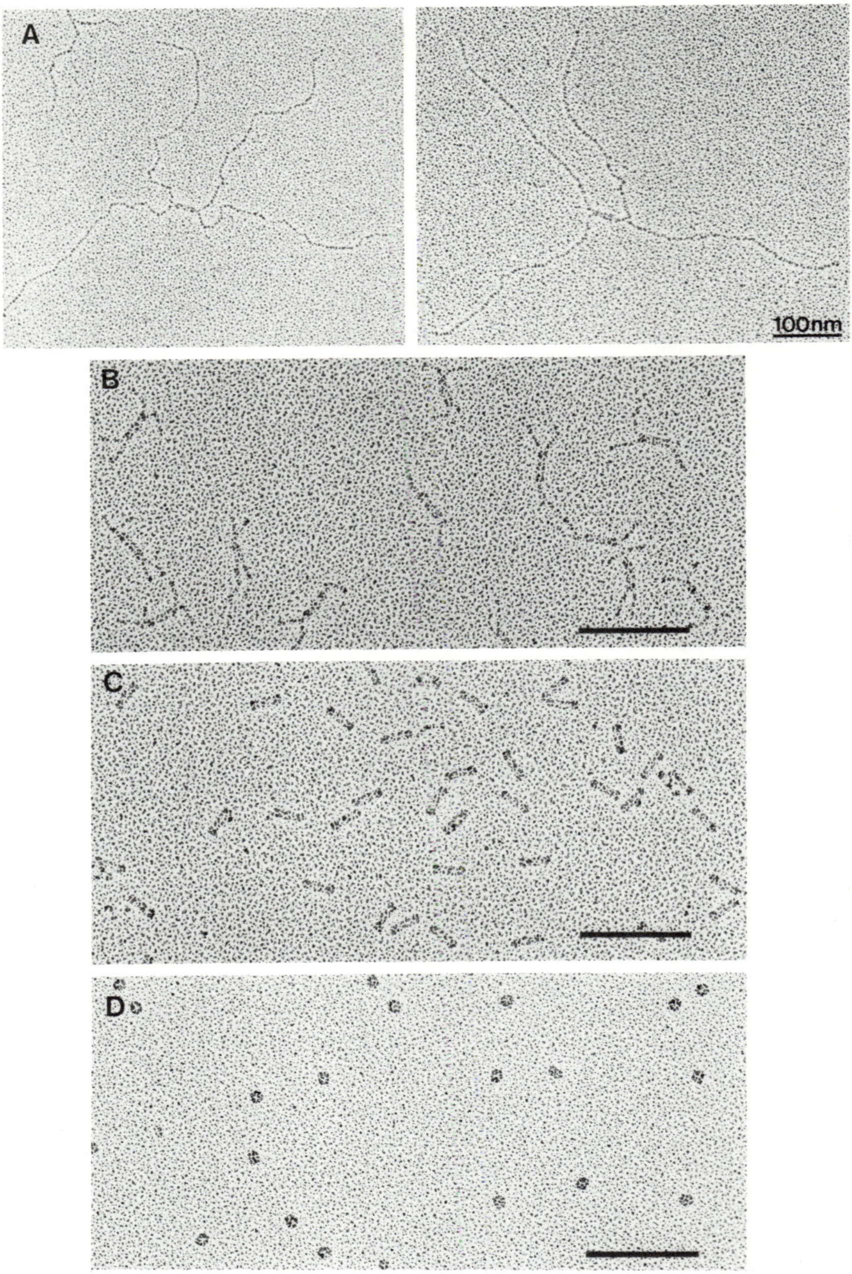
A
100nm
B
C
D

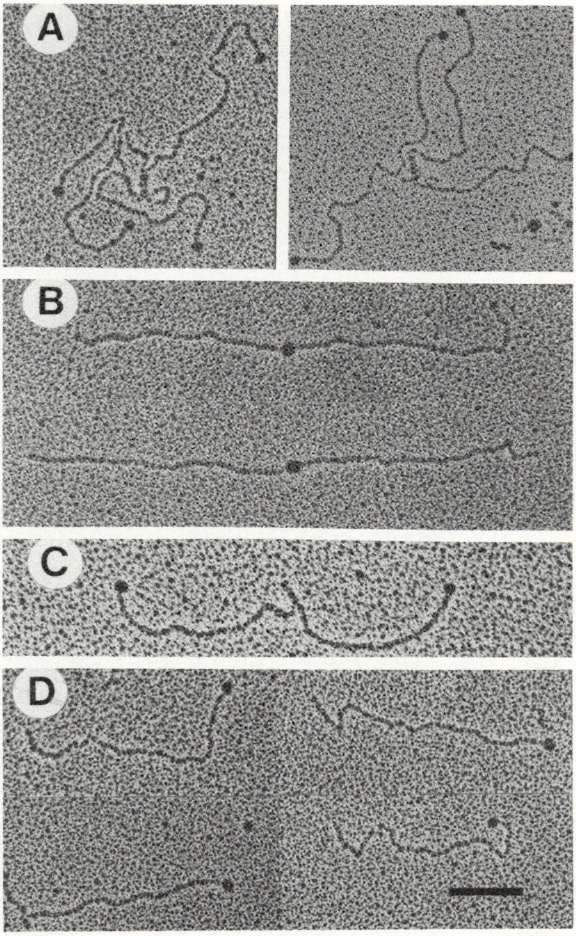

FIG. 2. Electron micrographs of intact type IV collagen molecules found in cell culture medium. (A) Tetramers from PF-HR9 cells (left), WISH cells (right). (B) Dimers from PF-HR9 cells formed by association of NC1 domains. (C) Dimers from PF-HR9 cells formed by association of amino-termini. (D) Monomers from PF-HR9 cells. Bar, 100 nm.

FIG. 1. Electron micrographs of proteolytic fragments of type IV collagen prepared by the rotary shadowing technique. (A) Tetrameric type IV collagen prepared from pepsin digests of the EHS mouse tumor (left) and human placenta (right). (B) Long form of the 7 S domain of human placenta type IV collagen from a pepsin digestion at 20°C. (C) Short form of the 7 S domain of EHS mouse tumor type IV collagen from a collagenase digest of pepsin-solubilized material at 37°C. (D) Hexameric globular domain NC1 of EHS mouse tumor type IV collagen from a collagenase digest of tissue at 37°C. Bars, 100 nm.

molecules were seen joined via their globular domains (Fig. 2B) as well as the expected dimers and tetramers by association of the helical regions that form the 7 S domain (Fig. 2A and 2C). The dimer globular domain (Fig. 1D) can be isolated from bacterial collagenase digests of basement-membrane-rich tissues (see NC1 domain below).

These electron microscope investigations, therefore, gave a good indication of the size and shape of the type IV collagen molecule as well as valuable information about the way in which molecules interact to form macromolecular structures. Both ends of the molecular appear to be cross-linking domains, and the arms function as flexible spacers.

Two macromolecular organizations of type IV collagen molecules have been proposed, both compatible with the observations described above. The difference between the two possibilities seems to depend upon whether the tetrameric aggregate is formed first by self assembly of the 7 S domain or whether dimers are first formed by interaction of two globular domains (Fig. 3). The former results in a network model which involves only interaction of the ends of the molecules (Timpl *et al.,* 1981) while the latter involves lateral interactions giving rise to a regular polygonal structure (Yurchenco and Furthmayr, 1984).

In a cell culture system, the monomers (Fig. 2D) were observed to associate mainly into tetramers (Fig. 2A) by formation of the 7 S domain (Duncan *et al.,* 1983). Only small amounts of material were in the form of dimers, formed by association of the carboxy-terminal (Fig. 2B) or amino-terminal (Fig. 2C). These data support the network model. The polygonal model was based upon rotary shadowing studies of the association products of preformed dimers, similar to those shown in Fig. 2B, that had been reduced under nondenaturing conditions. The physiological significance of these observations is, therefore, in question. Nevertheless, a regular polygonal structure is supported by direct observation of Descemet's membrane (Sawada, 1982) and renal tubular basement membranes in which a hexagonal array could be seen (Carlson *et al.,* 1981). It was, however, not demonstrated that the observed structure was in fact type IV collagen.

A third model, which is based upon the network model but incorporates globular domain/helix interactions and possibly helix/helix interactions, was developed based upon assumptions about cross-linking (Bailey *et al.,* 1984). The assumptions made were that the cross-linking site sequence Arg-His-Gly-Hyl, which has been conserved in interstitial collagens, is also present in type IV collagen and that these particular sites are at certain positions along the molecule, thus allowing formation of the proposed structure. However, these interstitial

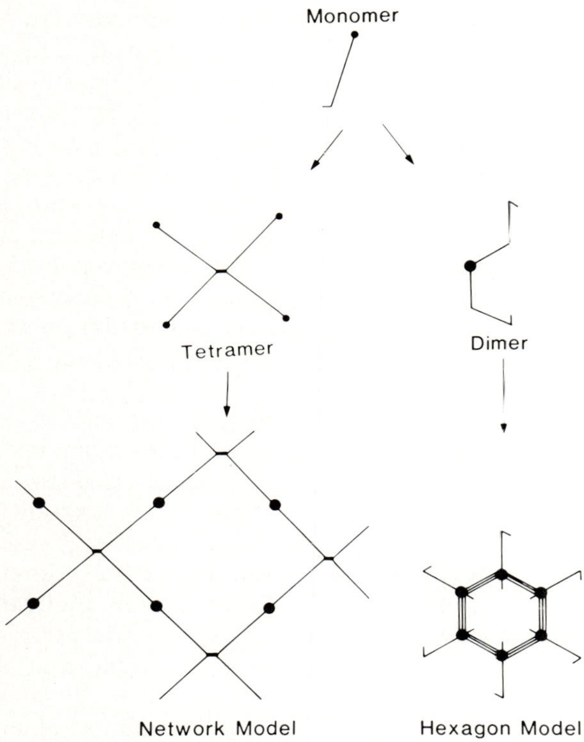

FIG. 3. Schematic diagram illustrating two possible ways for monomer type IV collagen molecules to associate to form macromolecular structures. Formation of the 7 S domain through association of four molecules at their amino-termini gives rise to tetramers that can aggregate further into a regular tetragonal or irregular polygonal meshwork. Formation of dimers through association of two NC1 domains could lead to the formation of hexagons containing 12 molecules. These could aggregate further into a regular hexagonal or irregular polygonal network. The distinguishing features of these models are that the first possibility requires no overlap or lateral interaction of the triple-helical portion of the molecules whereas the second does. The "pore size" of the regular networks would also clearly be different.

cross-linking sites are not present in type IV collagen and so the proposed structure remains a model without justification.

B. *Characterization of Proteolytic Fragments*

It has long been known that the collagenous component of basement membranes is insoluble under nondenaturing conditions. Original attempts to isolate basement membrane collagen were first made using glomeruli which were either extracted after reduction and alkylation under denaturing conditions or treated with proteases such as Pronase and collagenase (Marquardt *et al.*, 1973a,b; Hudson and Spiro, 1972; Sato and Spiro, 1976; Levine and Spiro, 1979). The former method produced small amounts of high-molecular-weight materials, while the latter produced complex mixtures of fragments. These investigations indicated that type IV collagen contained cysteine, lysine that was extensively hydroxylated and glycosylated, and that type IV collagen was a homotrimer (Kefalides, 1973). A glycopeptide, carrying both asparagine-linked and hydroxylysine-linked carbohydrates, isolated from a collagenase/pronase digest of glomerular basement membrane (Levine and Spiro, 1979) was later localized to the 7 S domain of type IV (Glanville *et al.*, 1985).

The first indications that type IV collagen did, in fact, contain two different α chains came in 1975 when Daniels and Chou used pepsin to solubilize isolated bovine glomeruli and characterized the pepsin fragments so produced. This observation was later substantiated by a number of groups working on renal glomerulus (Tryggvason and Kivirikko, 1978; Dixit, 1979, 1980; Trüeb *et al.*, 1982), human placenta (Glanville *et al.*, 1979; Bailey *et al.*, 1979; Kresina and Miller, 1979; Sage *et al.*, 1979; MacWright *et al.*, 1983), lens capsule (Dixit, 1978; Gay and Miller, 1979), and chick muscle (Mayne and Zettergren, 1980).

An important step in the further characterization of type IV collagen was the discovery of a murine tumor that produced basement membrane components (Orkin *et al.*, 1977). When mice carrying this tumor were made lathyritic, it was possible to extract a small amount of type IV collagen from the tumor using dilute acetic acid (Timpl *et al.*, 1978). This material was later characterized using the rotary shadowing technique and shown to be various aggregates of intact type IV collagen molecules (Timpl *et al.*, 1981). Limited pepsin digestion of the tumor produced fragments P1 and P2 and, after bacterial collagenase digestion, the noncollagenous domains NC1 and NC2 (Timpl *et al.*, 1979a). These fragments were important in establishing the domain structure and primary structure of type IV collagen.

1. Helical Domain

Pepsin digestion of various basement membrane-containing tissue preparations from various species produces different-sized pepsin fragments. This led to some confusion in the early literature as to the

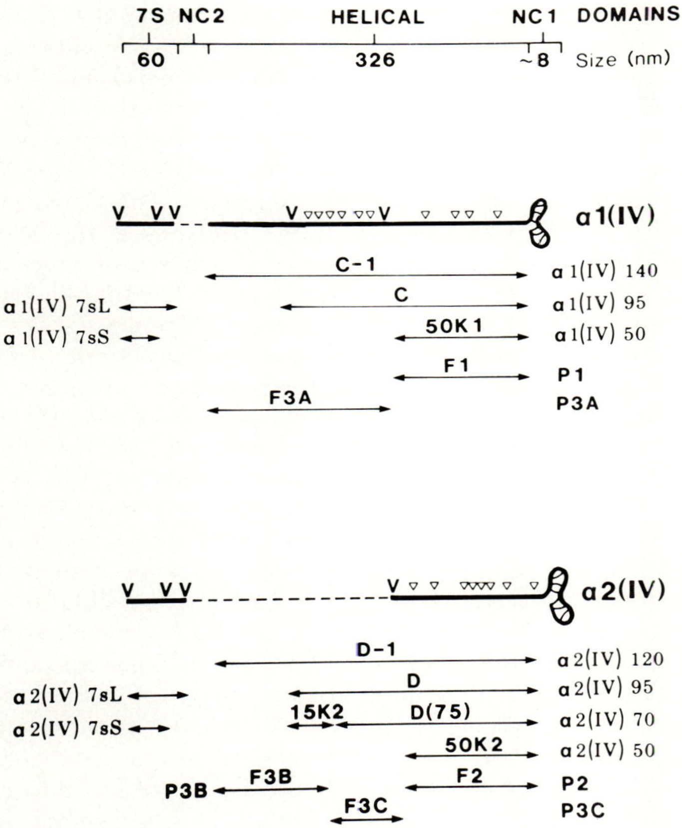

FIG. 4. Schematic diagram showing the positions and names of the major pepsin fragments of the α1 and α2 chains of type IV collagen. V, Major pepsin cleavage sites; ▽, nontriplet sequences; —, regions of known amino acid sequence; ---, regions of unknown amino acid sequence. C-1, C, D-1, D, 50K1, 50K2, 15K2 fragments from bovine type IV (Dixit and Kang, 1979); F1, F2, F3A, F3B, F3C fragments from chick type IV (Mayne and Zettergren, 1980); P1, P2, P3A, P3B, P3C fragments from mouse type IV (Timpl *et al.*, 1979a); α1(IV) 140, 95, 50, 7sL, 7sS and α2(IV) 120, 95, 70, 50, 7sL, 7sS fragments from human type IV (Glanville and Rauter, 1981; Glanville *et al.*, 1985).

origin of these fragments and their relationship to one another. Figure 4 is a schematic diagram of the type IV collagen α chains indicating the major structural features that have been described to date. Below each chain the position and name of the pepsin fragments that have been described are shown. Lens capsule, human placenta, and kidney appear to have similar fragmentation patterns, whereas mouse and chick are different.

The alignment of the pepsin fragments to either the α1 or α2 chain was possible, in most cases, by comparing their cyanogen bromide (CNBr) digest patterns on SDS electrophoresis gels. The initial assignment of one set of fragments to the α1 and the other to the α2 chain was arbitrary. CNBr peptides of the C-1 chain of bovine lens capsule (Dixit and Kang, 1979) and the porcine kidney D chain (Dixit and Kang, 1980) were isolated and characterized. The CNBr peptides of mouse P1 and P2 (Schuppan *et al.*, 1982; Schwartz *et al.*, 1986), human placenta (Babel and Glanville, 1984), and bovine lens capsule 50K (Schuppan *et al.*, 1984) have been sequenced. The P1 regions of mouse and human α1(IV) and part of the P2 region of mouse α2(IV) were also sequenced using cDNA sequencing methods (see Section III).

Amino acid sequencing studies of various pepsin fragments derived from mouse and human type IV collagen soon revealed that the continuous triplet sequence $(Gly\text{-}X\text{-}Y)_n$, so characteristic of all other sequenced collagens, was frequently interrupted in type IV collagen (Schuppan *et al.*, 1980; Glanville and Rauter, 1981). In fact, in the known sequence of the α1 chain there are 21 interruptions and in the α2 chain, for which only half as much sequence data is available, about 10 (see Sections IX and X). It has been possible to accurately align the sequences of the mouse P1 and P2 fragments because they both terminate at the carboxy-terminal noncollagenous domain. This alignment revealed that the nontriplet interruptions did not always coincide in the helix and they were usually of different lengths (Schwartz *et al.*, 1986). Presumably, the interruptions introduce flexibility into the arms of the type IV tetrameric structure as calculated from pictures of rotary shadowed molecules (see Section 11,A). The position and size of the interruptions has been conserved in different species as shown by comparing P1 regions of mouse, bovine, and human α1 type IV chains (Schuppan *et al.*, 1984; Babel and Glanville, 1984). It may be that at least some of these regions are binding sites for other basement membrane components (see Section VIII).

The positions and the sizes of the triplet interruptions within one chain do not appear to follow any regular pattern, and their compositions show no common characteristics. In several interruptions, hydroxyproline and hydroxylysine were found, indicating that the enzymes responsible for the hydroxylation of these amino acids do not require a continuous triplet sequence to function.

Comparing the P1 region of mouse α1(IV) with the corresponding region of human α1(IV) reveals a 90% identity of sequences (454 out of 501 residues). Of the 47 differences in sequences, 40 were conservative, that is, no change in polarity. Four of the conservative sub-

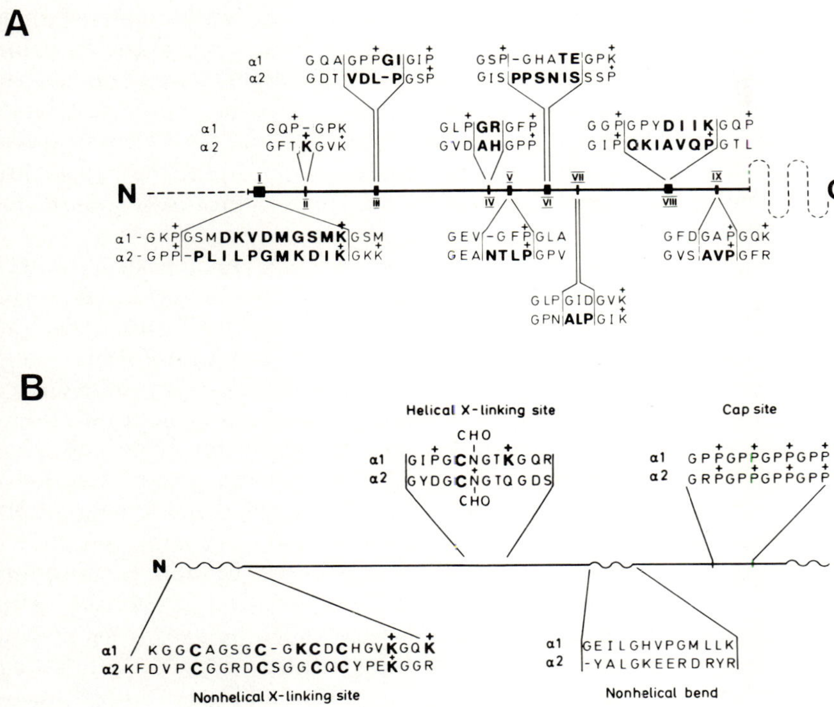

FIG. 5. (A) A schematic representation of the carboxy-terminal triple-helical region of type IV collagen (P1/P2 region). Nontriplet interruptions are indicated and numbered with Roman numerals. The amino acid sequences of both α1 and α2 chains in these regions are shown. Bold letters depict nontriplet sequences. Gaps, indicated by a broken line, are introduced to allow alignment of triplet sequences. Pluses mark the position of hydroxyproline and hydroxylysine residues. Note that in sections II, V, VII, and IX the tripeptide structure is interrupted only in α2. (B) A schematic diagram of the 7 S domain of type IV collagen indicating the positions of cross-linking sites, the bend region, and a particularly stable helical region termed the cap site. The amino acid sequences of both α1 and α2 chains are shown for these regions. Cysteine residues and lysine/hydroxylysine residues that may be involved in cross-linking are shown in bold type.

stitutions were in nontriplet sequences. This is in contrast to the interspecies homology in interstitial collagens where around 95% identity was found with only conservative substitutions being present.

If, however, position and size of the triplet sequence interruptions of P1 and P2 are compared, no similarities can be seen (Fig. 5A). P1 contains five interruptions whereas P2 has nine. Even in those positions where both chains have interruptions, they are of different lengths, the longer one usually being in the α2 chain.

There is also no significant homology between the triplet sequence of P1 and P2. Whether this low homology is restricted to the C-terminal half of the α chains or extends throughout the triple-helical arm region is not known. However, a considerably higher homology exists between the α1 and α2 chains within the 7 S and NC1 domains (see below). This suggests different constraints on different regions within type IV collagen during the evolutionary divergence of the genes for the α1(IV) and α2(IV) chains.

The distribution of amino acids between the X and Y positions of the repeating triplet, Gly-X-Y, is typical of any triple-helical molecule (Babel and Glanville, 1984). The large hydrophobic amino acids, leucine, tyrosine, phenylalanine, and tryptophan, are located exclusively in the X positions. Glutamic acid, aspartic acid, and serine also have a strong preference for X, while the small neutral amino acids are fairly evenly distributed. One difference between the interstitial collagens and type IV is the strong preference of lysine and proline for the Y position in type IV, whereas in the interstitial collagens they are more evenly distributed.

One small but interesting difference between human, bovine, and mouse type IV collagens is their content of 3-hydroxyproline. This amino acid is formed by a specific enzyme which hydroxylates proline in the amino acid sequence Gly-Pro-4-Hyp. Despite the frequent occurrence of this sequence in all collagens, only one 3-hydroxyproline residue was found in sequences of interstitial collagens, in bovine α1(I) CB6 (Fietzek *et al.*, 1972). Compositional data indicate the presence of two residues in α1(II) CB9,7 (Miller and Lunde, 1973). However, in both mouse and bovine P1 sequences there are two 3-hydroxyproline residues located in exactly the same positions. In human type IV the corresponding positions are not hydroxylated, and in mouse P2 these positions are not proline residues. The significance of these small amounts of this unusual amino acid remains unclear.

Another unique feature of type IV collagen is the presence of complex carbohydrate. These are attached through asparagine residues located in the helical regions. One was found on each of the α chains of the 7 S domain and the other on the P2 region of mouse type IV (Glanville *et al.*, 1985; Schwartz *et al.*, 1986). In addition to these complex sugars, every lysine residue occupying a Y position in a Gly-X-Y triplet is hydroxylated and glycosylated. Again, this is in contrast to the interstitial collagens in which very few of the lysine residues are hydroxylated and only some of these carry carbohydrates. A function for the carbohydrates still must be determined.

2. 7 S Domain

The amino-terminal of both chains isolated from the 7 S domain are blocked, probably by a pyroglutamic acid residue as is the case for all other sequenced collagen chains (Qian and Glanville, 1984). An amino-terminal sequence of the α1 chain, isolated from a trypsin-treated short form of 7 S, has been determined (Glanville *et al.*, 1985). The α1 chain has a 15-residue-long nontriplet sequence at its N-terminus which includes four cysteines and two lysines (Fig. 5B). All of these are capable of forming inter- and intramolecular cross-links. This is followed by 39 tripeptide units that include one cysteine residue in the sequence Gly-Cys-Asn-Gly-Thr-Hyl-Gly-Glu-Arg (Pos. 97–105), which is able to participate not only in a disulfide bridge but also in a lysine-derived cross-link. As this cysteine is present in a triple-helical portion of the molecule, the SH groups must project radially outwards from the helix and can, therefore, only form intermolecular disulfide bridges. It presumably reacts with one of the four cysteine residues in the N-terminal nonhelical region of an antiparallel-aligned neighboring molecule. It is not known which of the four cysteines react, but an arrangement in which the cysteine at position 14 is used would also allow for the formation of lysine-derived cross-links (positions 102 to 11) and would lead to an optimum interaction between the hydrophobic amino acids of two antiparallel-oriented type IV collagen molecules. How the second pair of molecules is oriented with respect to the first pair is unknown, and it is at present not clear how the disulfide bridges between the three remaining amino-terminal cysteine residues are formed. The position and number of cysteine residues in the α2 chain are the same as those in the α1, whereas lysine residues are fewer.

Considering the 7 S domain as a whole, there are 12 chains involved, each containing four cysteine residues. Therefore, as no free SH group has been detected in intact type IV collagen, there are 24 disulfide bridges to be located. Both inter- and intramolecular disulfide bonds are present. In the tissue form of type IV collagen, extensive lysine-derived cross-linking is evident, but the typical cross-linking site sequence, Arg-His-Gly-Hyl, that is present in all interstitial collagens, is not present in type IV. In pepsin-solubilized human placenta 7 S domain, only about 10% of the chains contained in the 7 S domain are not involved in lysine-derived cross-linking. Approximately equal proportions of dimers, trimers, tetramers, pentamers, and hexamers are present, all containing both α1 and α2 chains. A small proportion are involved in aggregates of 7–12 chains (Qian and Glanville, 1984). The structure of the cross-links have not been investigated in any great detail. The typical collagen difunctional cross-links that can be labeled using tritiated sodium borohydride are present in the 7 S domain (Bailey *et al.,* 1984). No hydroxypyridinium trifunctional cross-links have been found (Heathcote *et al.,* 1982), but there are indications of other, as yet uncharacterized, multifunctional cross-linking structures (Scott *et al.,* 1981, 1983). The high concentration of cross-links within the 7 S domain explains both its resistance to trypsin, pepsin, and collagenase digestion and its high thermal stability reflected in the denaturation temperature of 70°C (Risteli *et al.,* 1980a).

At the junction of the 7 S domain and the arms of the type IV tetramer, the molecules have a preferred angle of 40° (Hofmann *et al.,* 1984). This coincides with a long triple-helical sequence interruption in both the α1 and α2 chains, but the interruption is one residue shorter in the α2 than in the α1 (Fig. 5B). A similar situation is found in the complement factor C1q which contains three different polypeptide chains, A, B, and C, in a triple helix. At one position the helix makes a bend of 60° (Reid and Porter, 1976). Sequence studies have shown that at this position the A chain contains an extra threonine residue inserted between two triplets, the B chain has a continuous triplet sequence, and the C chain has a glycine substituted by an alanine residue (Reid, 1979). These interruptions of the triplet structure allow a bending of the molecule without disruption of the triple helix (Kilcherr *et al.,* 1985). However, as differences in the length of the adjacent nontriplet sequences in the α1 and α2 chains of type IV occur at several places along the molecule where no preferred angle was detected, there must be other functions for these regions.

3. NC1 Domain

The carboxy-terminal noncollagenous domain NC1 can be isolated from bacterial collagenase digests of basement membranes (Timpl *et al.,* 1981; Dean *et al.,* 1983; Weber *et al.,* 1984; Butkowski *et al.,* 1985; Oberbäumer *et al.,* 1985). When tissues are used as starting materials, the product is in fact a cross-linked dimer of the globular domains from two type IV molecules (Fig. 1D). About 50% of the human α1(IV) NC1 has been sequenced at the protein level, and complete sequences of both mouse and human α1(IV) NC1 have been determined via cDNA sequencing (Oberbäumer *et al.,* 1985; Brinker *et al.,* 1985a; Pihlajaniemi *et al.,* 1985). Species homology is very high with only seven conservative differences between the mouse and human sequences, which contain 229 amino acids (Section IX, positions 908–1136). This domain contains a high proportion of hydrophobic amino acids and 12 cysteine residues. Analysis of the sequence for internal repeats indicated that it consists of two homologous regions, each containing six cysteine residues (positions 908–1018 and 1019–1136). When the sequences are aligned to give maximum homology, 35% of the positions are identical and a further 21% are conservative substitutions. Therefore, the NC1 domain is probably formed from two spatially very similar substructures, each stabilized by disulfide bridges which may have arisen by duplication of a gene unit.

The function of the NC1 domain may be twofold. In analogy to the carboxy-terminal propeptides of the interstitial procollagens, this domain may be involved in chain selection and helix formation during the biosynthesis of type IV. Unlike interstitial procollagens, this domain is not removed and clearly plays an important role in the formation of the extracellular macromolecular structures. The NC1 domain of two molecules react to form a dimer which is covalently cross-linked by, as yet uncharacterized, nonreducible cross-links. According to one model of type IV superstructure, the NC1 domain can also interact specifically with certain regions of the triple-helical portion of the type IV molecule (see Section II,a).

C. Chain Composition of Molecules

Despite the vast amount of detailed structural data available for type IV collagen, it is still a matter of debate whether one molecule contains both α1 or α2 chains, i.e., $(\alpha1)_2\alpha2$(IV), whether two molecules exist as $\alpha1_3$(IV) and $\alpha2_3$(IV) homotrimers, or whether a combination of both possibilities is correct. There is considerable evidence to support a heterotrimeric molecule. This is based upon chromatographic behavior (Trüeb *et al.*, 1982; Mayne and Zettergren, 1980), the characterization of disulfide-bonded pepsin fragments (Fujiwara and Nagai, 1981; Dieringer *et al.*, 1985), cross-linked chains from the 7 S domain (Qian and Glanville, 1984), and studies with specific monoclonal antibodies (Odermatt *et al.*, 1984; Mayne *et al.*, 1984). Evidence for homotrimeric molecules originates from work on mouse tumor type IV pepsin fragments where $(P1)_3$ and $(P2)_3$ native fragments were characterized (Timpl *et al.*, 1979a). Support for this concept was provided by precipitation experiments in which the ratios of the α1(IV) to α2(IV) chains varied. However, in retrospect this was probably not valid as the preparations contained cross-linked type IV tetramers. The ratios of α1 to α2 would therefore depend upon the extent of involvement of these two chains in lysine-derived cross-links in the 7 S domain. More recent support for homotrimeric molecules is based upon cyanogen bromide peptide mapping and pepsin digestion of the chains produced by rat embryo-derived parietal yolk sac cells (Haralson *et al.*, 1985). However, this was not conclusive evidence because the cyanogen bromide peptide patterns of the α1(IV) and α2(IV) as visualized on SDS-PAGE are very similar (Robey and Martin, 1981). Also, characterization of pepsin-digested material may lead to erroneous conclusions, as it has been shown that the α2 chain is much more sensitive to pepsin digestion (Tryggvason *et al.*, 1980; Alitalo *et al.*, 1980; Robey and Martin, 1981) and may have been present as small fragments in the sample that was characterized. The existence of homotrimeric molecules therefore remains a matter for debate.

III. Molecular Biology

Applying genetic engineering technology to investigate the gene structure and control of genes of type IV collagen is in its infancy.

A. *Genomic DNA*

Genomic clones from the mouse α2(IV) collagen chain have been characterized and show some interesting differences in structure to the interstitial collagens gene structure (Kurkinen *et al.*, 1985). Three exons of the gene were sequenced which contained 64, 123, and 182 base pairs and coded triple-helical sequences from the amino-terminal region of P2 (Section X, positions 1–140). In the interstitial collagens, all exons are multiples of nine base pairs, and there is considerable evidence that the exons evolved by amplification of an ancestral 54-base pair exon unit (Yamada *et al.*, 1980). In type IV this is clearly not the case. Another difference is that interstitial collagen gene exons start with an intact glycine codon. In type IV, three out of the four exons characterized did not start with an intact glycine codon. The third base usage for glycine and proline also differs. In interstitial collagens it is most frequently U or C, but in the characterized α2(IV) exons it was not. These results suggest that the α2(IV) structure of the collagen gene is different and may have evolved independently from the fibrillar collagen genes.

Similar conclusions were drawn from studies of the human α1(IV) collagen gene (Soininen *et al.*, 1986). Here, seven exons of 134, 73, 72,

129, 99, 213, and 178 base pairs were characterized which coded for the carboxy-terminal region of the helical domain and about half of NC1 (Section IX, positions 970–1269). In addition it was shown that the introns were consistently large, the smallest being 960 and the largest more than 3500 base pairs.

B. cDNA

cDNA libraries from mouse parietal endoderm, human fibroblasts (Brinker *et al.,* 1985a), human tumor cell line HT-1080 (Pihlajaniemi *et al.,* 1985), mouse F9 cells (Wang and Gudas, 1983), mouse EHS tumor and F9 cells (Oberbäumer *et al.,* 1985), and mouse parietal endoderm cells (Kurkinen *et al.,* 1983) have been screened, using either antibodies or synthetic probes, to isolate clones containing α1(IV) and α2(IV) chain fragments. To date, only the carboxy-terminal one-third (i.e., P1 or P2 region) of the triple-helical domain and the complete globular domain (NC1) of both α1 and α2 chains have been isolated and sequenced (see Sections II,B,1 and II,B,3).

IV. Biosynthesis

The biosynthesis of type IV collagen has been studied in a large number of cells and tissues in culture (see Table 4 in Timpl and Dziadek, 1986). Early data summarized by Kefalides (1978) indicated the synthesis of a single polypeptide chain with a molecular weight in the range of 135,000–180,000 depending upon the system studied. More recent reports describe the synthesis of two polypeptide chains, α1(IV), with an apparent molecular weight of about 180,000, and α2(IV), about 165,000, in cell culture (Alitalo *et al.,* 1980; Crouch *et al.,* 1980; Tryggvason *et al.,* 1980; Fessler and Fessler, 1982; Foidart *et al.,* 1981) and in cell-free synthesis (Kurkinen *et al.,* 1982; Sawhney and Dixit, 1985). This is in good agreement with the size of material that can be directly extracted from mouse EHS tumor (Kleinman *et al.,* 1982; Karakashian *et al.,* 1982), indicating that the biosynthetic product is not enzymatically processed to a smaller tissue form as is the case for the interstitial collagens. Pulse–chase experiments have also failed to detect processing of type IV collagen (Minor *et al.,* 1976; Dehm and Kefalides, 1978; Crouch and Bornstein, 1979; Alitalo *et al.,* 1980; Heathcote *et al.,* 1978, 1980). There is some evidence from measurements of rotary shadowed molecules for the removal of small polypeptide extensions at both the amino-terminal (Bächinger *et al.,*

1982b) and carboxy-terminal (Oberbäumer *et al.,* 1982). However, a true type IV procollagen molecule has still to be described.

Synthesis of α1(IV) chains occurs in about 10 min (Fessler and Fessler, 1982). Hydroxylation of proline and lysine and glycosylation of hydroxylysine residues is complete in about 40 min. The completed type IV molecule is secreted after about 100 min (Pihlajaniemi *et al.,* 1981), which is considerably longer than the interstitial collagens which are secreted in about 20 min. The long synthesis time may reflect a slow association of α chains as the carboxy-terminal NC1 domains are not interchain disulfide bonded, slow folding of the triple helix because of the presence of many triplet sequence interruptions, or a long intracellular pathway.

Secreted type IV molecules contain interchain disulfide bridges. In cell cultures in the presence of α,α-dipyridyl, which inhibits proline and lysine hydroxylation, disulfide bridge formation is inhibited and random-coiled chains are secreted (Crouch and Bornstein, 1979; Fessler and Fessler, 1982). In contrast, in a tissue culture of parietal yolk sacs, it was reported that secreted type IV collagen was disulfide bonded but was nevertheless a random coil, as judged by its susceptibility to proteolysis (Karakashian and Kefalides, 1982).

The association of monomer type IV molecules into tetramers has been studied in a cell culture system in some detail (Duncan *et al.,* 1983). Most of the secreted molecules remain as monomers or tetramers in cell culture medium, although a small fraction of dimers is present (Bächinger *et al.,* 1982a; Oberbäumer *et al.,* 1982). The monomers associate into tetramers, initially through noncovalent interactions (Duncan *et al.,* 1983). This interaction is concentration dependent and reversible. Disulfide bonds are then formed between adjacent monomers, probably by disulfide exchange to form the 7 S domain. Dimer formation through interaction of the NC1 domains was rarely observed. In the cell layer, the tetramers are quickly incorporated into an insoluble matrix by formation of cross-links between NC1 domains and further lysine-derived cross-linking of the 7 S domain (Heathcote *et al.,* 1978, 1980; von der Mark *et al.,* 1985; Taylor and Grant, 1985) (see Section II,B,2).

The actual conformation of the secreted collagen molecule is not known. In outgrowths from embryonic explants of mouse lung, type IV collagen can be deposited in the form of a fibrillar network (Chen and Little, 1985), in contrast to the amorphous deposits of epithelial cells in culture. It was suggested that, during embryogenesis, components of basement membranes and of connective tissue extracellular matrices may not be as highly segregated as in later development. Developmental changes in basement membranes during embryogenesis have also been demonstrated in the chick lens capsule (Fitch and Linsenmayer, 1983; Fitch *et al.,* 1983), as has heterogeneity within one membrane. Microfibrillar structures have been observed in some mature basement membranes, for example the lung (Low, 1961), Reichert's membrane (Inoue *et al.,* 1983), and mouse EHS tumor (Inoue and Leblond, 1985), and in glomerular basement membrane (Farquhar *et al.,* 1961; Laurie *et al.,* 1984). It appears that the structure and composition of basement membranes may not be as constant as biochemical data suggest. Perhaps type IV collagen can assume different conformations as required, although the two models that have been suggested so far are not compatible with a fibrillar structure.

V. Immunochemistry

A. *Polyclonal Antibodies*

Antibodies to type IV collagen have been used extensively in the immunohistological analyses of basement membranes in normal and pathological tissues (see Table 3 in Timpl and Dziadek, 1986). Polyclonal antisera have been raised against the 7 S domain (Risteli *et al.,* 1981; Madri and Williams, 1983; Dixit *et al.,* 1984), NC1 (von der Mark *et al.,* 1985), various pepsin fragments (Timpl *et al.,* 1979c; Yaoita *et al.,* 1978; Garbisa *et al.,* 1981; Mayne *et al.,* 1982) and extracted type IV from the EHS tumor (Timpl *et al.,* 1978, 1979c; Madri and Furthmayr, 1980; Chen and Little, 1985), and lens capsule (Brinker *et al.,* 1985b). The 7 S and NC1 domains are strongly immunogenic, whereas the triple-helical domain is only weakly immunogenic.

In addition to immunohistochemical studies, these antibodies have been used in radioimmunoassays to detect the concentration of antigen in body fluids and cell culture media (Bowman *et al.,* 1980; Risteli *et al.,* 1981, 1985; Brocks *et al.,* 1985; von der Mark *et al.,* 1985; Schuppan *et al.,* 1985) as well as for a number of structural studies. It was possible in one case to characterize antibodies specific for the $\alpha 1$(IV) and $\alpha 2$(IV) chains (Risteli *et al.,* 1980b), again reflecting the dissimilarity of the two chains at the primary structural level as discussed earlier (Section II,B,1). Cross reactivity of an antibody with a purified 7 S domain preparation and a helical pepsin fragment, isolated from bovine kidney cortex, was used to demonstrate that the 7 S domain is composed of the amino-terminal ends of the type IV collagen molecule (Dixit *et al.,* 1983). Antibodies to type IV collagen were also instrumental in both understanding the pathophysiology of the tumor cell invasive process and the recognition of its presence in tissue sections (Barsky *et al.*, 1983).

B. *Monoclonal Antibodies*

Monoclonal antibodies have been prepared against the mouse NC1 domain (von der Mark *et al.*, 1985), human type IV pepsin fragments (Sakai *et al.*, 1982; Sundarraj and Wilson, 1982; Foellmer *et al.*, 1983; Odermatt *et al.*, 1984; Scheinman and Tsai, 1984), and chick type IV pepsin fragments (Fitch *et al.*, 1982; Mayne *et al.*, 1983, 1984). With one exception (see below), these antibodies are conformation specific. Mapping the antigenic determinants by rotary shadowing antigen antibody complexes (Fig. 6) revealed that there are three major determinants, one in the carboxy-terminal half of the triple helix (Mayne *et al.*, 1984; Odermatt *et al.*, 1984), one in the 7 S domain (Mayne *et al.*, 1984), and one approximately 55 nm from the 7 S domain (Foellmer *et al.*, 1983; Mayne *et al.*, 1984; Dieringer *et al.*, 1985). Although the position of epitopes is similar in different species, the antibodies do not always cross react, indicating that the structures of the epitopes are different.

Three monoclonal antibodies against chick type IV collagen were used in combination with rotary shadowing to align the native pepsin

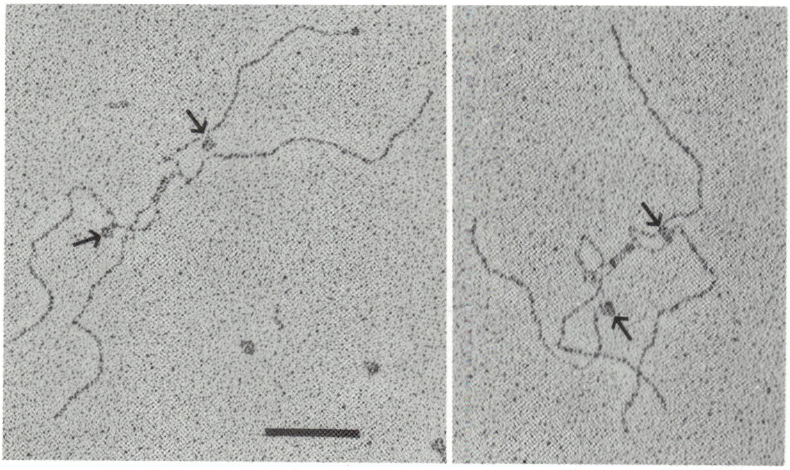

FIG. 6. Electron micrograph of a pepsin solubilized human type IV tetramer, with bound antibodies, prepared by the rotary-shadowing technique. The monoclonal antibodies, indicated by arrows, are bound to a site approximately 60 nm away from the 7 S domain. Bar, 100 nm.

fragments with the intact molecule (Mayne *et al.,* 1984). Two monoclonal antibodies, one against a native pepsin fragment of type IV collagen and one against the denatured α1(IV) chain, have been used to show that native type IV molecules are heterotrimers (Odermatt *et al.,* 1984) (see Section II,B,3). Finally, a combination of peptide chemistry and immunochemistry enabled an interchain disulfide bridge to be localized to a position close to the antigenic site (Dieringer *et al.,* 1985).

C. *Electron Microscopic Localization*

Immunoelectron microscopy has been used to demonstrate different compositions of the lamina rara and lamina densa. Type IV collagen was intially localized in glomerular basement membranes to either the lamina densa (Courtoy *et al.,* 1982) or throughout the basement membrane (Roll *et al.,* 1980; Mynderse *et al.,* 1983). Subsequently, studies on other basement membranes demonstrated a concentration of type IV collagen in the lamina densa but with projections into the lamina rara (Laurie *et al.,* 1982, 1984; Monaghan *et al.,* 1983; Fleischmajer *et al.,* 1985). These apparently conflicting results may be due to artifacts of fixation and antibody labeling, particularly when peroxidase is used (Courtoy *et al.,* 1983). In fact, it may not be possible to localize macromolecules to a specific region of basement membranes because their extended structure spans the entire width of the membrane. The apparent localization would then depend upon the position of the epitopes against which the antibodies are directed. There is also an apparent alteration in the distribution of type IV collagen during the development of diabetic microangiopathy in hyperglycemic rats (Bendayan, 1985). In normal animals it was localized to the lamina densa, but in diseased animals it was restricted to the subendothelial site of the lamina.

VI. Degradation

Proteases that specifically degrade type IV collagen have been difficult to characterize, presumably because of the presence of nontriplet sequences in the type IV collagen molecule which render it susceptible to attack by many nonspecific proteases. Therefore, the fact that a given enzyme preparation can solubilize type IV collagen is not a sufficient criteria to justify the designation type IV collagenase. For example, leukocyte elastase (Uitto *et al.,* 1980; Mainardi *et al.,* 1980) and bone metalloproteinases (Murphy *et al.,* 1981, 1985) both degrade not only type IV collagen but also laminin and other matrix proteins.

A specific type IV collagenase has been isolated from cell cultures derived from a malignant tumor (Liotta *et al.,* 1979) and the enzyme characterized (Salo *et al.,* 1983). It is a neutral metalloproteinase of 60,000–70,000 Da which is activated by trypsin. It has no action against the interstitial collagen types I, II, III, or V but cleaves both chains of type IV collagen at one specific locus. The fragments produced have been characterized (Fessler *et al.,* 1984) and the cleavage point located approximately 90 nm from the 7 S domain. This is not far from one of the antigenic determinants for monoclonal antibodies (Dieringer *et al.,* 1985; Fitch *et al.,* 1982), an interchain disulfide bridge (Dieringer *et al.,* 1985), and is a region of high flexibility (Hofmann *et al.,* 1984). As this region contains both triplet and nontriplet sequences (unpublished data), it is not known whether the enzyme cleaves the three chains in a triple-helical or nonhelical region of the molecule.

VII. Involvement in Disease Processes

Basement membranes are involved in a large variety of diseases either as the site of the primary lesion (as in autoimmune diseases such as Goodpasture syndrome, epidermolysis bullosa, bullous pemphigold), or in secondary complications (as in diabetes). Basement membranes are also degraded when cells, particularly tumor cells, invade tissues (for a review, see Liotta *et al.,* 1983; Martinez-Hernandez and Amenta, 1983; Timpl and Dziadek, 1986). However, a specific involvement of type IV collagen in these pathological conditions has, in most instances, not been shown.

Only in one pathological condition, Goodpasture syndrome, is there good evidence for a direct involvement of type IV collagen. Goodpasture syndrome, a rare form of glomerulonephritis, is an autoimmune disease in which the antigen resides in the basement membrane on a collagenase-resistant component (Marquardt *et al.,* 1973b; Fish *et al.,* 1984; Yoshioka *et al.,* 1985a,b; Wieslander *et al.,* 1984b), later shown to originate from the noncollagenous NC1 domain of type IV (Wieslander *et al.,* 1984a,b; Butkowski *et al.,* 1985). The epitope is concealed inside the hexameric structure of the NC1 domain in normal tissue (Wieslander *et al.,* 1985). It has also been possible to cause a Goodpasture-like syndrome by injecting mice with NC1 domain (von der Mark *et al.,* 1985).

Epithelial cell tumor invasion involves cells passing through various connective tissues and associated basement membranes (Liotta *et al.*, 1983). Tumor cells elaborate degradative enzymes, one of which is the type IV collagenase (see Section VI), which enable them to move through connective tissues. Benign tumors are surrounded by a continuous basement membrane, whereas most malignant tumors have very fragmented or no basement membranes (Barsky *et al.*, 1983). Collagenolytic activity of various tumor cell lines correlates well with their metastatic potential (Liotta *et al.*, 1980).

VIII. Interactions with Cells and Other Proteins

Now that the major components of basement membranes have been isolated and characterized, attention is turning to both how they interact with each other to form a membrane and which components within the membrane are responsible for interaction with cells. Although it is generally believed that epithelial cells do not attach directly to type IV collagen but utilize laminin as an attachment factor, a native type IV collagen binding protein called colligin has been isolated from the cell membrane of a variety of cell lines (Kurkinen *et al.*, 1984). It is a 47,000 Da glycoprotein which, it is speculated, either may be involved in the assembly of type IV collagen into a network or may be an intracellular component involved in the biosynthetic pathway that reaches the cell surface as a result of fusion of secretory vesicles from the Golgi with the cell membrane. Some cells do, however, attach readily and directly to type IV collagen, for example, aortic endothelial cells (Palotie *et al.*, 1983), hepatocytes (Rubin *et al.*, 1981), keratinocytes (Murray *et al.*, 1979), PF-HR9 cells (Engvall *et al.*, 1982), and metastatic melanoma cells (Dennis *et al.*, 1982). Asparagine-linked oligosaccharides on the cell surface have been implicated in this attachment (Dennis *et al.*, 1984).

Most attention has been paid to the interaction of type IV collagen with laminin and heparin sulfate proteoglycan. Laminin has a higher affinity for native type IV collagen than for denatured type IV or other collagen types, but it is nevertheless a weak affinity. Digesting the native type IV with pepsin, which removes the NC1 domain and about 100 nm of the helix, destroyed laminin binding, indicating that the binding site resides in NC1 or close to it (Rao *et al.*, 1985). Conversely, it was reported that isolated NC1 did not bind to laminin but pepsin-treated material did (Charonis *et al.*, 1985). Using thrombin or pepsin fragments of laminin, it was shown that the type IV binding site resides in the globular regions of the three short arms of laminin

(Woodley *et al.*, 1983; Charonis *et al.*, 1985). When binding studies are carried out in solution, a precipitate is formed which contains approximately equimolar amounts of type IV and laminin (Kleinman *et al.*, 1983).

Heparan sulfate proteoglycan also binds to both laminin and type IV collagen (Kleinman *et al.*, 1983; Fujiwara *et al.*, 1984; Woodley *et al.*, 1983). Precipitation studies using all three components resulted in a complex with a constant composition which contains type IV collagen, laminin, and heparin sulfate proteoglycan in molar ratios of 10:10:1, respectively, which are similar to those found in the EHS mouse tumor (Kleinman *et al.*, 1983; Grant *et al.*, 1985).

This is probably a gross simplification of the *in vivo* situation as interactions may be greatly modified by other minor components of basement membranes such as nidogen, entactin, or the unknown epidermolysis bullosa (Woodley *et al.*, 1984; Paller *et al.*, 1985) and bullous pemphigold (Stanley *et al.*, 1981) antigens. There is also no information available about either the order in which basement membrane components are laid down or their spatial relationship to one another within a membrane. Although biochemical investigations would indicate otherwise, there are some indications, based upon observation of *in vivo* systems, that at least some basement membranes exhibit thixotropic properties (Simpson, 1980, 1981). If this is indeed true, then interactions between various membrane components may be extremely weak, making the demonstration of specific interactions between the collagenous network and other components very difficult to define biochemically.

IX. Appendix A: α1(IV) Amino Acid Sequence

Amino acid sequence of a large cyanogen bromide peptide located at the amino-terminal of human α1(IV)140 (positions 1–255) (Glanville, unpublished data), human α1(IV)95 (positions 226–1162) from Babel and Glanville (1984), and human NC1 from Pihlajaniemi *et al.* (1985) and Brinker *et al.* (1985a), whose results were in complete agreement. The amino acids shown below the main sequence are substitutions found in the corresponding sequence of mouse P1 (Schuppan *et al.*, 1982) and NC1 (Oberbäumer *et al.*, 1985). P1 was corrected according to the results of Oberbäumer *et al.* (1985). (*, Indicates hydroxylated proline and lysine residues; +, indicates an asparagine-linked carbohydrate; and ⟨ ⟩, enclose nontriplet sequences.)

```
         5        10        15        20        25        30
 *         *         * *           *           *     *
 P<G V>G E K G E P G K P G P R G K P G K D G D K G E K G S P

         35        40        45        50        55        60
     *     *     *                       *           *     *
 G F P G E P G Y P G L I G R Q G P Q G E K G E A G P P G P P

         65        70        75        80        85        90
                       *           *     *           *     *
<G I V I G T>G P L G E K G E R G Y P G T P G P R G E P G P K

        95       100       105       110       115       120
           *     *     *                     *     *
 G F P G L P G Q P G P P<G L P V P>G Q A G A P G F P G E R G

       125       130       135       140       145       150
   *         *                 *     *     *     *     *
 F P<G T S L P>G P S G R D G L P G P P G S P G P P G Q P<G Y

       155       160       165       170       175       180
                       *           *     *     *
 T N G I V E C Q P>G P P G D Q G P P G I P G Q P G F I G E I

       185       190       195       200       205       210
     *     *                               *           *
 G E K G Q K<G E S C L I C D I D>G Y R G P P G P Q G P P G E

       215       220       225       230       235       240
       *     *     *           *                 *
 I G F P G Q P G A K G A R G L P G R D G V A G V P G P Q G T

       245       250       255       260       265       270
 *           *         *                   *     *
 P G L I G Q P<G A G Q P G E F Y F D L R L K>G D K G D P G F

       275       280       285       290       295       300
 *           *           *           *     *     *     *
 P G Q P G M P G R A G S P G R D G H P G L P G P K G S P G S

       305       310       315       320       325       330
       *           *           *                 *     *
 V G L K G E R G P P G G V G F P G S R G D T G P P G P P<G Y>

       335       340       345       350       355       360
                 *           *     *     *     *     *
 G P A G P I G D K G Q A G F P G G P G S P G L P G P K G E P

       365       370       375       380       385       390
             *     *           *     *     *
<G K I V P L P>G P P G A E G L P G S P G F P G P Q G D R G F
```

```
      395       400       405       410       415       420
*     *     *     *     *           *         *     *     *
P G T P G R P G L P G E K G A V G Q P<G I>G F P G P P G P K

      425       430       435       440       445       450
          *           *     *     *           *     *
G V D G L P G D M G P P G T P G R P G F D G L P G N P G V Q

      455       460       465       470       475       480
    *               *     *     *     *     *     *     *
G Q K G E P<G V>G L P G L K G L P G L P G I P G T P G E K G

      485       490       495       500       505       510
        *                       *                 *     *
S I G V P G V P G E H G A I G P P G L Q G I R G E P G P P G

      515       520       525       530       535       540
  *           *     *         *           *           *
L P G S V G S P G V P<G I>G P P G A R G P P G G Q G P P G L

      545       550       555       560       565       570
      *     *     *     *               *           *
S G P P G I K G E K G F P G F P<G L Y M P>G P P G D K G A Q

      575       580       585       590       595       600
                      *     *           *     *     *     *
G L P G I T G Q S G L P G L P G Q Q G A P G I P G F P G S K

      605       610       615       620       625       630
                *     *     *           *     *     *
G E M G V M G T P G Q P G S P G P V G A P G L P G Q K G D H

      635       640       645       650       655       660
    *     *           *     *     *           *     *
G F P G S K G P R G D P G L K G D K G D V G L P G K P<G S M

      665       670       675       680       685       690
                *     *           *                 *
D K V D M G S M K>G Q K G D Q G E K G Q I G P I G E K G S R
                                              T   D

      695       700       705       710       715       720
    *     *     *                 *           *
G D P G T P G V P G K D G Q A G Q P G Q P G P K G D P G I S
                                H                       L

      725       730       735       740       745       750
    *     *     *                       *     *     *
G T P G A P G L P G P P G S V G G M G L P G T P G E K G V P
        S K                                 S

      755       760       765       770       775       780
    *           *           *     *     *           *
G I P G P Q G S P G L P G D K G A K G E K G Q A G P P<G I>G
        S     V     S     E                       L
```

```
      785       790       795       800       805       810
  *           *                 *     *     *     *
I P G L R G E K G D Q G I A G F P G S P G E K G E K G S I G
        P   D           L                               A

      815       820       825       830       835       840
  *           *     *                 *     *           *
I P G M P G S P G L K G S P G S V G Y P G S P G L P G E K G
T                 P R         N I                     D E

      845       850       855       860       865       870
  *     *           *     *           *     *
D K G L P G L D G I P G V K G E A G L P G T P G P T G P A G
                  V

      875       880       885       890       895       900
  K                 *           *           *         *
Q K G E P G S D G I P G S A G E K G E P G L P<G R>G F P G F

      905       910       915       920       925       930
*     *     *     *                       *     *     *
P G A K G D K G S K G E V G F P G L A G S P G I P G S K G E
                                                    V Q

      935       940       945       950       955       960
            *                 *     *               *
Q G F M G P P G P Q G Q P G L P G S P<G H A T E>G P K G D R
                                  T       P V       P     Q

      965       970       975       980       985       990
          *     *     *           *     *           *     *
G P Q G Q P G L P G L P G P M G P P G L P G I D G V K G D K

      995      1000      1005      1010      1015      1020
          *     *     *     *                 *           *
G N P G W P G A P G V P G P K G D P G F Q G M P G I G G S P

     1025      1030      1035      1040      1045      1050
          *           *     *           *     *           *
G I T G S K G D M G P P G V P G F Q G P K G L P G L Q G I K
                    L                 Q                 V

     1055      1060      1065      1070      1075      1080
                *     *           *     *             *
G D Q G D Q G V P G A K G L P G P P G P P<G P Y D I I K>G E
                    P       Q                     V

     1085      1090      1095      1100      1105      1110
*     *                 *           *     *
P G L P G P E G P P G L K G L Q G L P G P K G Q Q G V T G L
                                  P                       S

     1115      1120      1125      1130      1135      1140
      *     *     *           *     *
V G I P G P P G I P G F D G A P G Q K G E M G P A G P T G P
    L           V                         T     F     P
```

```
      1145      1150      1155      1160      1165      1170
      *     *           *           *
R G F P G P P G P D G L P G S M G P P G T P/S V D H G F L V

      1175      1180      1185      1190      1195      1200
T R H S Q T I D D P Q C P S G T K I L Y H G Y S L L Y V Q G
            T       L     P

      1205      1210      1215      1220      1225      1230
N E R A H G Q D L G T A G S C L R K F S T M P F L F C N I N

      1235      1240      1245      1250      1255      1260
N V C N F A S R N D Y S Y W L S T P E P M P M S M A P I T G

      1265      1270      1275      1280      1285      1290
E N I R P F I S R C A V C E A P A M V M A V H S Q T I Q I P

      1295      1300      1305      1310      1315      1320
P C P S G W S S L W I G Y S F V M H T S A G A E G S G Q A L

      1325      1330      1335      1340      1345      1350
A S P G S C L E E F R S A P F I E C H G R G T C N Y Y A N A

      1355      1360      1365      1370      1375      1380
Y S F W L A T I E R S E M F K K P T P S T L K A G E L R T H

      1385      1390
V S R C Q V C M R R T
```

Amino acid sequence of the long form of human α1(IV) 7 S (from Glanville *et al.*, 1985).

```
         5        10        15        20        25        30
                                   *     *           *
<K G G C A G S G C G K C D C H>G V K G Q K G E R G L P G L Q

        35        40        45        50        55        60
           *                       *     *           *     *
 G V I G F P G M Q G P E G P Q G P P G Q K G D T G E P G L P

        65        70        75        80        85        90
     *           *           *     *     *                 *
 G T K G T R G P P G A S G Y P G N P G L P G I P G Q D G P P

        95       100       105       110       115       120
     *     *     +     *                 *     *           *
 G P P G I P G C N G T K G E R G P L G P P G L P G F A G N P

       125       130       135       140       145       150
     *     *     *     *               *         *
 G P P G L P G M K G D P<G E I L G H V P G M L L K>G E R G F
```

```
      155       160       165       170       175       180
*     *     *     *     *                 *           *
P G I P G T P G P P G L P G L Q G P V G P P G F T G P P G P

      185       190       195       200       205       210
*     *           *                     *     *
P G P P G P P G E K G Q M G L S F Q G P K G D K G D Q<G V K

      215

E P P K V P....>
```

X. Appendix B: α2(IV) Amino Acid Sequence

Amino acid sequence of mouse P2 (positions 1–511) (from Schwartz *et al.*, 1986) and NC1 (positions 512–738) (unpublished data kindly supplied by K. Kühn). Symbols as in Section IX.

```
             5        10        15        20        25        30
                                   *     *                 *
<... P P G P P P L I L P>G M K D I K G E K G D E G P M G L K G Y

            35        40        45        50        55        60
           *           *     *           *     *     *       *
     L G L K G I Q G M P G V P G V S G F P G L P G R P<G F I K>G

            65        70        75        80        85        90
       *           *     *     *     *           *           *
     V K G D I G V P G T P G L P G F P G V S G P P G I T G F P G

            95       100       105       110       115       120
                   *     *
     F T G S R G E K G T P G V A G V F G E T G P T G D F G D I<G

           125       130       135       140       145       150
                     *     *                 *     *           *
     D T V D L P>G S P G L K G E R G I T G I P G L K G F F G E K

           155       160       165       170       175       180
                     *                       *     *           *
     G A A G D I G F P G I T G M A G A Q G S P G L K G Q T G F P

           185       190       195       200       205       210
                           *           *     *           *     *
     G L T G L Q G P I G E P G R I G I P G D K G D F G W P G V P

           215       220       225       230       235       240
         *     *                       *     *           *
     G L P G F P G I R G I S G L H G L P G T K G F P G S P<G V D
```

```
      245       250       255       260       265       270
        *     *                                           *
A H>G D P G F P G P T G D R G D R<G E A N T L P>G P V G V P

      275       280       285       290       295       300
    *           *                 *           *
G Q K G E R G T P G E R G P A G S P G L Q G F P<G I S P P S

      305       310       315       320       325       330
          *           *                       *     *
N I S>G S P G D V G A P G I F G L Q G Y Q G P P G P P<G P N

      335       340       345       350       355       360
    *     *                                         *
A L P>G I K G D E G S S G A A G F P G Q K G W V G D P G P Q

      365       370       375       380       385       390
    *
G Q P G V L G L P G E K G P K G E Q G F M G N T G P S G A V

      395       400       405       410       415       420
                *           *     *           *     *     *
G D R G P K G P K G D Q G F P G A P G S M G S P G I P<G I P

      425       430       435       440       445       450
  *                                 *
Q K I A V Q P>G T L G P Q G R R G L P G A L G E I G P Q G P

      455       460       465       470       475       480
*     *           *                             *
P G D P G F R G A P G K A G P Q G R G<G V S A V P>G F R G D

      485       490       495       500       505       510
                              *     *     *           *
Q G P M G H Q G P V G Q E G E P G R P G S P G L P G M P G R

      515       520       525       530       535       540
S V S I G Y L L V K H S Q T D Q E P M C P V G M N K L W S G

      545       550       555       560       565       570
Y S L L Y F E G Q E K A H N Q D L G L A G S C L A R F S T M

      575       580       585       590       595       600
P F L Y C N P G D V C Y Y A S R N D K S Y W L S T T A P L P

      605       610       615       620       625       630
M M P L A E E E I K P Y I S R C S V C E A P A V A I A V H S

      635       640       645       650       655       660
Q D T S I P H C P A G W R S L W I G Y S F L M H T A A G D E

      665       670       675       680       685       690
G G G Q S L V S P G S C L E D F R A T P F I E C N G G R G T
```

```
       695       700       705       710       715       720
 C H Y F A N K Y S F W L T T I P E Q N F Q S T P S A D T L K

       725       730       735
 A G L I R T H I S R C Q V C M K N L
```

This amino acid sequence corresponds to the long form of human α2(IV) 7 S (Glanville, unpublished data).

```
         5        10        15        20        25        30
                                         *           *
<K F D V P C G G R D C S G G C Q C Y P E K>G G R G Q P G P V

        35        40        45        50        55        60
                 *           *                 *           *
 G P Q G Y N G P K G L Q G F P G L Q G R K G D K G E R G A P

        65        70        75        80        85        90
           *                       *           *     *
 G V T G P K G D V G A R G V S G F P G A D G I P G H P G Q G

        95       100       105       110       115       120
           *                                   *
 G P R G R P G Y D G C N G T Q G D S G P Q G P P G S E G F T

       125       130       135       140       145       150
     *                                                     *
 G P P G P Q G P P G Q Q G E P<Y A L G K E E R D R Y R>G E P

       155       160       165       170       175       180
                       *         *                   *     *
 G E P G L V G F Q G P P G R P G K V G Q M G P V G A P G R P

       185       190       195       200       205       210
     *     *                                 *     *
 G P P G P P G P P G Q Q G N R<G L>G F Y G V K G E K G D V G Q
```


Acknowledgments

The author would like to thank Professor Klaus Kühn, Hanna Wiedemann, Hans Peter Bächinger, Greg Lunstrum, Lynn Sakai, and Doug Keene for supplying electron micrographs of rotary shadowed type IV collagen molecules. Special thanks are due to Professor Klaus Kühn for making unpublished data available and providing Fig. 5A and 5B, and to Hans Peter Bächinger for critically reading the manuscript.

References

Alitalo, K., Vaheri, A., Krieg, T., and Timpl, R. (1980). *Eur. J. Biochem.* **109,** 247–255.

Babel, W., and Glanville, R. W. (1984). *Eur. J. Biochem.* **143,** 545–556.

Bächinger, H. P., Doege, K. J., Petschek, J. P., Fessler, L. I., and Fessler, J. H. (1982a). *J. Biol. Chem.* **257,** 14590–14592.

Bächinger, H. P., Fessler, L. I., and Fessler, J. H. (1982b). *J. Biol. Chem.* **257,** 9796–9803.

Bailey, A. J., Sims, T. S., Duance, V. C., and Light, N. (1979). *FEBS Lett.* **99,** 361–366.

Bailey, A. J., Sims, T. S., and Light, N. (1984). *Biochem. J.* **218,** 713–723.

Barsky, S. H., Siegal, G., Janotta, F., and Liotta, L. A. (1983). *Lab. Invest.* **49,** 140–147.

Bendayan, M. (1985). *Diabetologia* **28,** 373–378.

Bowman, B. H., Schneider, L., Barnett, D. R., Kurosky, A., and Goldblum, R. M. (1980). *J. Biol. Chem.* **255,** 9484–9489.

Brinker, J. M., Gudas, L. J., Loidl, H. R., Wang, S. Y., Rosenbloom, J., Kefalides, N. A., and Myers, J. C. (1985a). *Proc. Natl. Acad. Sci. U.S.A.* **82,** 3649–3654.

Brinker, J. M., Pegg, M. T., Howard, P. S., and Kefalides, N. A. (1985b). *Collagen Relat. Res.* **5,** 233–244.

Brocks, D. G., Neubauer, H. P., and Strecker, H. (1985). *Diabetologia* **28,** 928–932.

Butkowski, R. J., Wieslander, J., Wisdom, B. J., Barr, J. F., Noelken, M. E., and Hudson, B. G. (1985). *J. Biol. Chem.* **260,** 3739–3747.

Carlson, E. C., Meezan, E., Brendel, K., and Kenney, M. C. (1981). *Anat. Rec.* **200,** 421–436.

Charonis, A. S., Tsilibary, E. C., Yurchenco, P. D., and Furthmayr, H. (1985). *J. Cell Biol.* **100,** 1848–1853.

Chen, J. M., and Little, C. D. (1985). *J. Cell Biol.* **101,** 1175–1181.

Courtoy, P. J., Timpl, R., and Farquhar, M. G. (1982). *J. Histochem. Cytochem.* **30,** 874–886.

Courtoy, P. J., Picton, D. H., and Farquhar, M. G. (1983). *J. Histochem. Cytochem.* **31,** 945–951.

Crouch, E., and Bornstein, P. (1979). *J. Biol. Chem.* **254,** 4197–4204.

Crouch, E., Sage, H., and Bornstein, P. (1980). *Proc. Natl. Acad. Sci. U.S.A.* **77,** 745–749.

Daniels, J. R., and Chou, H. C. (1975). *J. Biol. Chem.* **250,** 3531–3537.

Dean, C. D., Barr, J. F., Freyteg, J. W., and Hudson, B. G. (1983). *J. Biol. Chem.* **258,** 590–596.

Dehm, P., and Kefalides, N. A. (1978). *J. Biol. Chem.* **19,** 6680–6686.

Dennis, J., Waller, C., Timpl, R., and Schirrmacher, V. (1982). *Nature (London)* **300,** 274–276.

Dennis, J., Waller, C., and Schirrmacher, V. (1984). *J. Cell Biol.* **99,** 1416–1423.

Dieringer, H., Hollister, D. W., Glanville, R. W., Sakai, L. Y., and Kühn, K. (1985). *Biochem. J.* **227,** 217–222.

Dixit, S. N. (1978). *FEBS Lett.* **85,** 153–157.

Dixit, S. N. (1979). *FEBS Lett.* **106,** 379–384.

Dixit, S. N. (1980). *Eur. J. Biochem.* **106,** 563–570.

Dixit, S. N., and Kang, A. H. (1979). *Biochemistry* **18,** 5686–5692.

Dixit, S. N., and Kang, A. H. (1980). *Biochemistry* **19,** 2692–2696.

Dixit, S. N., Stuart, J. M., Seyer, J. M., Risteli, J., Timpl, R., and Kang, A. H. (1981). *Collagen Relat. Res.* **1,** 549–556.

Dixit, S. N., Mainardi, C. L., Beachey, E. H., and Kang, A. H. (1983). *Collagen Relat. Res.* **3,** 263–270.

Dixit, S. N., Robinson, W., Dixit, P., and Kang, A. H. (1984). *Connect. Tissue Res.* **12,** 297–306.

Duncan, K. G., Fessler, L. I., Bächinger, H. P., and Fessler, J. F. (1983). *J. Biol. Chem.* **258,** 5869–5877.

Engvall, E., Bell, M. L., Carlsson, R. N. K., Miller, E. J., and Ruoslahti, E. (1982). *Cell* **29,** 475–482.

Farquhar, M. G., Wissig, S. L., and Palade, G. E. (1961). *J. Exp. Med.* **113,** 47–66.

Fessler, L. I., and Fessler, J. H. (1982). *J. Biol. Chem.* **257,** 9804–9810.

Fessler, L. I., Duncan, K. G., Fessler, J. H., Salo, T., and Tryggvason, K. (1984). *J. Biol. Chem.* **259,** 9783–9789.

Fietzek, P. P., Rexrodt, F. W., Wendt, P., Stark, M., and Kühn, K. (1972). *Eur. J. Biochem.* **30,** 163–168.

Fish, A. J., Lockwood, M. C., Wong, M., and Price, R. G. (1984). *Clin. Exp. Immunol.* **55,** 58–66.

Fitch, J. M., and Linsenmayer, T. F. (1983). *Dev. Biol.* **95,** 137–153.

Fitch, J. M., Gibney, E., Sanderson, R. D., Mayne, R., and Linsenmayer, T. F. (1982). *J. Cell Biol.* **95,** 641–647.

Fitch, J. M., Mayne, R., and Linsenmayer, T. F. (1983). *J. Cell Biol.* **97,** 940–943.

Fleischmajer, R., Timpl, R., Dziadek, M., and Lebwohl, M. D. (1985). *J. Invest. Dermatol.* **85,** 54–59.

Foellmer, H. G., Madri, J. A., and Furthmayr, H. (1983). *Lab. Invest.* **48,** 639–649.

Foidart, J. M., Tryggvason, K., Gehron-Robey, P., Liotta, L. A., and Martin, G. R. (1981). *Collagen Relat. Res.* **1,** 137–150.

Fujiwara, S., and Nagai, Y. (1981). *Collagen Relat. Res.* **1,** 491–504.

Fujiwara, S., Wiedemann, H., Timpl, R., Lustig, A., and Engel, J. (1984). *Eur. J. Biochem.* **143,** 145–157.

Furthmayr, H., and Madri, J. A. (1982). *Collagen Relat. Res.* **2,** 349–363.

Garbisa, S., Liotta, L. A., Tryggvason, K., and Siegal, G. P. (1981). *FEBS Lett.* **127,** 257–262.

Gay, S., and Miller, E. J. (1979). *Arch. Biochem. Biophys.* **198,** 370–378.

Glanville, R. W., and Rauter, A. (1981). *Hoppe-Seyler's Z. Physiol. Chem.* **362,** 943–951.

Glanville, R. W., Rauter, A., and Fietzek, P. P. (1979). *Eur. J. Biochem.* **95,** 383–389.

Glanville, R. W., Voss, T., and Kühn, K. (1982). *In* "New Trends in Basement Membrane Research" (K. Kühn, H. H. Schoene, and R. Timpl, eds.), pp. 69–77. Raven, New York.

Glanville, R. W., Qian, R. Q., Siebold, B., Risteli, J., and Kühn, K. (1985). *Eur. J. Biochem.* **152,** 213–219.

Grant, D. S., Kleinman, H. K., Leblond, C. P., Inoue, S., Chung, A. E., and Martin, G. R. (1985). *Am. J. Anat.* **174,** 387–398.

Haralson, M. A., Federspeil, S. J., Martinez-Hernandez, A., Rhodes, R. K., and Miller, E. J. (1985). *Biochemistry* **24,** 5792–5797.

Heathcote, J. G., Sear, C. H., and Grant, M. E. (1978). *Biochem. J.* **176,** 283–294.

Heathcote, J. G., Bailey, A. J., and Grant, M. E. (1980). *Biochem. J.* **190,** 229–237.

Heathcote, J. G., Eyre, D. R., and Gross, J. (1982). *Biochem. Biophys. Res. Commun.* **108,** 1588–1594.

Hofmann, H., Voss, T., Kühn, K., and Engel, J. (1984). *J. Mol. Biol.* **172,** 325–343.

Hudson, B. G., and Spiro, R. G. (1972). *J. Biol. Chem.* **247,** 4229–4238.

Inoue, S., and Leblond, C. P. (1985). *Am. J. Anat.* **174,** 373–386.

Inoue, S., Leblond, C. P., and Laurie, G. W. (1983). *J. Cell Biol.* **97,** 1524–1537.

Karakashian, M. W., and Kefalides, N. A. (1982). *Connect. Tissue Res.* **10,** 247–262.

Karakashian, M. W., Dehm, P., Gramling, T. S., and LeRoy, E. C. (1982). *Collagen Relat. Res.* **2,** 3–17.

Kefalides, N. A. (1968). *Biochemistry* **7,** 3103–3112.

Kefalides, N. A. (1973). *Int. Rev. Connect. Tissue Res.* **6,** 63–104.

Kefalides, N. A. (1978). "Biology and Chemistry of Basement Membranes." Academic Press, New York.

Kefalides, N. A., and Winzler, R. J. (1966). *Biochemistry* **5,** 702–713.

Kirchherr, E., Hofmann, H., Steigemann, W., and Engel, J. (1985). *J. Mol. Biol.* **186,** 403–415.

Kleinman, H. K., McGarvey, M. L., Liotta, L. A., Gehron-Robey, P., Tryggvason, K., and Martin, G. R. (1982). *Biochemistry* **2,** 6188–6193.

Kleinman, H. K., McGarvey, M. L., Hassell, J. R., and Martin, G. R. (1983). *Biochemistry* **22,** 4969–4974.

Kresina, T. F., and Miller, E. J. (1979). *Biochemistry* **18,** 3089–3097.

Kühn, K. (1982). *Collagen Relat. Res.* **2,** 61–80.

Kühn, K., Wiedemann, H., Timpl, R., Risteli, J., Dieringer, H., Voss, T., and Glanville, R. W. (1981). *FEBS Lett.* **125,** 123–128.

Kühn, K., Schoene, H. H., and Timpl, R. (1982). "New Trends in Basement Membrane Research." Raven, New York.

Kurkinen, M., Foster, L., Barlow, D. P., and Hogan, B. L. M. (1982). *J. Biol. Chem.* **257,** 15151–15155.

Kurkinen, M., Barlow, D. P., Helfman, D. M., Williams, J. G., and Hogan, B. L. M. (1983). *Nucleic Acids Res.* **11,** 6199–6209.

Kurkinen, M., Taylor, A., Garrels, J. I., and Hogan, B. L. M. (1984). *J. Biol. Chem.* **259,** 5915–5922.

Kurkinen, M., Bernard, M. P., Barlow, D. P., and Chou, L. T., (1985). *Nature (London)* **317,** 177–179.

Laurie, G. W., Leblond, C. P., and Martin, G. R. (1982). *J. Cell Biol.* **95,** 340–350.

Laurie, G. W., Leblond, C. P., Inoue, S., Martin, G. R., and Chung, A. (1984). *Am. J. Anat.* **169,** 463–481.

Levine, M. J., and Spiro, R. G. (1979). *J. Biol. Chem.* **254,** 8121–8124.

Liotta, L. A., Abe, S., Gehron-Robey, P., and Martin, G. R. (1979). *Proc. Natl. Acad. Sci. U.S.A.* **76,** 2268–2272.

Liotta, L. A., Tryggvason, K., Garbisa, S., Hart, I., Foltz, C. M., and Shafie, S. (1980). *Nature (London)* **284,** 67–68.

Liotta, L. A., Rao, C. N., and Barsky, S. H. (1983). *Lab Invest.* **49,** 636–649.

Low, F. N. (1961). *Anat. Rec.* **139,** 105–123.

MacWright, R. S., Benson, V. A., Lovello, K. T., van der Rest, M., and Fietzek, P. P. (1983). *Biochemistry* **22,** 4940–4948.

Madri, J. A., and Furthmayr, H. (1980). *Hum. Pathol.* **11,** 353–366.

Madri, J. A., and Williams, S. K. (1983). *J. Cell Biol.* **97,** 153–165.

Madri, J. A., Foellmer, H. G., and Furthmayr, H. (1983). *Biochemistry* **22,** 2797–2804.

Mainardi, C. L., Dixit, S. N., and Kang, A. H. (1980). *J. Biol. Chem.* **255,** 5435–5451.

Marquardt, H., Wilson, C. B., and Dixon, F. J. (1973a). *Biochemistry* **12,** 3260–3266.

Marquardt, H., Wilson, C. B., and Dixon, F. J. (1973b). *Kidney Int.* **3,** 57–65.

Martinez-Hernandez, A., and Amenta, P. S. (1983). *Lab Invest.* **48,** 656–677.

Mayne, R., and Zettergren, J. G. (1980). *Biochemistry* **19,** 4065–4072.

Mayne, R., Wiedemann, H., Dessau, W., von der Mark, K., and Bruckner, P. (1982). *Eur. J. Biochem.* **126,** 417–423.

Mayne, R., Sanderson, R. D., Wiedemann, H., Fitch, J. M., and Linsenmayer, T. F. (1983). *J. Biol. Chem.* **258,** 5794–5797.
Mayne, R., Wiedemann, H., Irwin, M. H., Sanderson, R. D., Fitch, J. M., Linsenmayer, T. F., and Kühn, K. (1984). *J. Cell Biol.* **98,** 1637–1644.
Miller, E. J., and Lunde, L. G. (1973). *Biochemistry* **12,** 3153–3159.
Minor, R. R., Clark, C. C., Strause, E. L., Koszalka, T. R., Brent, R. L., and Kefalides, N. A. (1976). *J. Biol. Chem.* **251,** 1789–1794.
Monaghan, P., Warburton, M. J., Perisinghe, N., and Rudland, P. S. (1983). *Proc. Natl. Acad. Sci. U.S.A.* **80,** 3344–3348.
Murphy, G., Cawston, D. E., Galloway, W. A., Barnes, M. J., Bunning, R. A. D., Mercer, E., Reynolds, J. J., and Burgeson, R. E. (1981). *Biochem. J.* **199,** 807–811.
Murphy, G., McAlpine, C. G., Poll, C. T., and Reynolds, J. J. (1985). *Biochim. Biophys. Acta* **831,** 49–58.
Murray, J. C., Stingl, G., Kleinman, H. K., Martin, G. R., and Katz, S. I. (1979). *J. Cell Biol.* **80,** 197–202.
Mynderse, L. A., Hassell, J. R., Kleinman, H., Martin, G. R., and Martinez-Hernandez, A. (1983). *Lab. Invest.* **48,** 292–302.
Oberbäumer, I., Wiedemann, H., Timpl, R., and Kühn, K. (1982). *EMBO J.* **1,** 805–810.
Oberbäumer, I., Laurent, M., Schwarz, U., Sakurai, Y., Yamada, Y., Vogeli, G., Voss, T., Siebold, B., Glanville, R. W., and Kühn, K. (1985). *Eur. J. Biochem.* **147,** 217–224.
Odermatt, B. F., Lang, A. B., Ruttner, J. R., Winterhalter, K. H., and Trüeb, B. (1984). *Proc. Natl. Acad. Sci. U.S.A.* **81,** 7343–7347.
Orkin, R. W., Gehron, P., McGoodwin, E. B., Martin, G. R., Valentine, T., and Swarm, R. (1977). *J. Exp. Med.* **145,** 204–220.
Paller, A. S., Queen, L. L., Woodley, D. T., Gammon, W. R., O'Keefe, E. J., and Briggaman, M. D. (1985). *J. Invest. Dermatol.* **84,** 215–217.
Palotie, A., Tryggvason, K., Peltonen, L., and Seppa, H. (1983). *Lab. Invest.* **49,** 362–370.
Pihlajaniemi, T., Myllylä, R., Alitalo, K., Vaheri, A., and Kivirikko, K. (1981). *Biochemistry* **20,** 7409–7415.
Pihlajaniemi, T., Tryggvason, K., Myers, J. C., Kurkinen, M., Lebo, R., Cheung, M. C., Prockop, D. J., and Boyd, C. D. (1985). *J. Biol. Chem.* **260,** 7681–7687.
Pirie, A. (1951). *Biochemistry* **48,** 368–371.
Qian, R. G., and Glanville, R. W. (1984). *Biochem. J.* **222,** 447–452.
Rao, C. N., Margulies, I. M. K., and Liotta, L. A. (1985). *Biochem. Biophys. Res. Commun.* **128,** 45–52.
Reid, K. B. H. (1979). *Biochem. J.* **179,** 367–371.
Reid, K. B. H., and Porter, R. R. (1976). *Biochem. J.* **155,** 19–23.
Risteli, J., Bächinger, H. P., Engel, J., Furthmayr, H., and Timpl, R. (1980a). *Eur. J. Biochem.* **108,** 239–250.
Risteli, L., von Koskull, H., Autio-Harmainen, H., and Risteli, J. (1985). *Clin. Chim. Acta,* **147,** 283–290.
Risteli, J., Rohde, H., and Timpl, R. (1981). *Anal. Biochem.* **113,** 372–378.
Risteli, L., von Koskull, H., Autio-Harmainen, H., and Risteli, J. (1985). *Clin. Chim. Acta,* in press.
Robert, A. M., Boniface, R., and Robert, L. (1979). "Frontiers of Matrix Biology," Vol. 7. Karger, Basel.
Robey, P., and Martin, G. R. (1981). *Collagen Relat. Res.* **1,** 27–36.
Roll, F. J., Madri, J. A., Albert, J., and Furthmayr, H. (1980). *J. Cell Biol.* **95,** 597–616.
Rubin, K., Höök, M., Öbrink, B., and Timpl, R. (1981). *Cell* **24,** 463–470.

Sage, H., Woodbury, R. G., and Bornstein, P. (1979). *J. Biol. Chem.* **254,** 9893–9900.
Sakai, L. Y., Engvall, E., Hollister, D. W., and Burgeson, R. E. (1982). *Am. J. Pathol.* **108,** 310–318.
Salo, T., Liotta, L. A., and Tryggvason, K. (1983). *J. Biol. Chem.* **258,** 3058–3063.
Sato, T., and Spiro, R. G. (1976). *J. Biol. Chem.* **251,** 4062–4070.
Sawada, H. (1982). *Cell Tissue Res.* **226,** 241–255.
Sawhney, R. S., and Dixit, S. N. (1985). *Eur. J. Biochem.* **151,** 11–16.
Scheinman, J. I., and Tsai, C. (1984). *Lab. Invest.* **50,** 101–112.
Schuppan, D., Timpl, R., and Glanville, R. W. (1980). *FEBS Lett.* **115,** 297–300.
Schuppan, D., Glanville, R. W., and Timpl, R. (1982). *Eur. J. Biochem.* **123,** 505–512.
Schuppan, D., Glanville, R. W., Timpl, R., Dixit, S. N., and Kang, A. H. (1984). *Biochem. J.* **220,** 227–233.
Schuppan, D., Ruhlmann, T., and Hahn, E. G. (1985). *Anal. Biochem.* **149,** 238–247.
Schwartz, D., and Veis, A. (1978). *FEBS Lett.* **85,** 326–332.
Schwartz, D., and Veis, A. (1979). *Biopolymers* **18,** 2363–2367.
Schwartz, U., Schuppan, D., Oberbäumer, I., Glanville, R. W., Deutzmann, R., Timpl, R., and Kühn, K. (1986). *Eur. J. Biochem.* **157,** 49–56.
Scott, J. E., Hughes, E. W., and Shuttleworth, A. (1981). *Biosci. Rep.* **1,** 611–618.
Scott, J. E., Qian, R. G., Henkel, W., and Glanville, R. W. (1983). *Biochem. J.* **209,** 263–264.
Simpson, L. O. (1980). *Pathology* **12,** 377–389.
Simpson, L. O. (1981). *Nephron* **27,** 105–112.
Soininen, R., Tikka, L., Chow, L., Pihlajaniemi, T., Kurkinen, M., Prockop, D. J., Boyd, C. D., and Tryggvason, K. (1986). *Proc. Natl. Acad. Sci. U.S.A.* **83,** 1568–1572.
Stanley, J. R., Hawley-Nelson, P., Yuspa, S. H., Shevach, E. M., and Katz, S. I. (1981). *Cell* **24,** 897–903.
Sundarraj, N., and Wilson, J. (1982). *Immunology* **47,** 133–140.
Taylor, C. M., and Grant, M. E. (1985). *Biochem. J.* **226,** 527–536.
Timpl, R., and Dziadek, M. (1986). *Int. Rev. Exp. Pathol.* **29,** 1–112.
Timpl, R., and Martin, G. R. (1982). *In* "Immunochemistry of the Extracellular Matrix" (H. Furthmayr, ed.), pp. 119–150. CRC Press, Boca Raton, Florida.
Timpl, R., Martin, G. R., Bruckner, P., Wick, G., and Wiedemann, H. (1978). *Eur. J. Biochem.* **84,** 43–52.
Timpl, R., Bruckner, P., and Fietzek, P. (1979a). *Eur. J. Biochem.* **95,** 255–263.
Timpl, R., Risteli, J., and Bächlinger, H. P. (1979b). *FEBS Lett.* **101,** 265–268.
Timpl, R., Glanville, R. W. Wick, G., and Martin, G. R. (1979c). *Immunology* **38,** 109–116.
Timpl, R., Wiedemann, H., van Delden, V., Furthmayer, H., and Kühn, K. (1981). *Eur. J. Biochem.* **120,** 203–211.
Trüeb, B., Grobli, B., Spiess, M., Odermatt, B. F., and Winterhalter, K. H. (1982). *J. Biol. Chem.* **257,** 5239–5245.
Tryggvason, K., and Kivirikko, K. I. (1978). *Nephron* **21,** 230–235.
Tryggvason, K., Gehron-Robey, P., and Martin, G. R. (1980). *Biochemistry* **19,** 1284–1289.
Uitto, V. J., Schwartz, D., and Veis, A. (1980). *Eur. J. Biochem.* **105,** 409–417.
von der Mark, H., Oberbäumer, I., Timpl, R., Kemler, R., and Wick, G. (1985). *Eur. J. Biochem.* **146,** 555–562.
Wang, S. Y., and Gudas, L. J. (1983). *Proc. Natl. Acad. Sci. U.S.A.* **80,** 5880–5884.
Weber, S., Engel, J., Wiedemann, H., Glanville, R. W., and Timpl, R. (1984). *Eur. J. Biochem.* **139,** 401–410.

Wieslander, J., Bygren, P., and Heinegård, D. (1984a). *Proc. Natl. Acad. Sci. U.S.A.* **81,** 1544–1548.

Wieslander, J., Barr, J. F., Butkowski, R. J., Edwards, S. J., Bygren, P., Heinegård, D., and Hudson, B. G. (1984b). *Proc. Natl. Acad. Sci. U.S.A.* **81,** 3838–3842.

Wieslander, J., Langeveld, J., Butkowski, R., Jodlowski, M., Noelken, M., and Hudson, B. G. (1985). *J. Biol. Chem.* **260,** 8564–8570.

Woodley, D. T., Rao, C. N., Hassell, J. R., Liotta, L. A., Martin, G. R., and Kleinman, H. K. (1983). *Biochim. Biophys. Acta* **761,** 278–283.

Woodley, D. T., Briggaman, R. A., O'Keefe, E. J., Inman, A. O., Queen, L. L., and Gammon, W. R. (1984). *N. Engl. J. Med.* **310,** 1007–1013.

Yamada, Y., Avvedimento, V. E., Mudryj, M., Ohkubo, H., Vogeli, G., Irani, M., Pastan, I., and de Crombrugghe, B. (1980). *Cell* **22,** 887–892.

Yaoita, H., Foidart, J. M., and Katz, S. I. (1978). *J. Invest. Dermatol.* **70,** 191–193.

Yoshioka, K., Kleppel, M., and Fish, A. J. (1985a). *J. Immunol.* **134,** 3831–3837.

Yoshioka, K., Michael, A. F., Velosa, J., and Fish, A. J. (1985b). *Am. J. Pathol.* **121,** 156–165.

Yurchenco, P. D., and Furthmayr, H. (1984). *Biochemistry* **23,** 1839–1850.

Type V Collagen

John H. Fessler and Liselotte I. Fessler

Department of Biology and the Molecular Biology Institute, University of California, Los Angeles, Los Angeles, California 90024

I. Introduction

This review is focused on the biosynthesis of collagen V, the structure of the procollagens V, and their processing. In addition, an attempt is made to summarize current information regarding the fiber form of collagens V, their occurrence in various connective tissues, and their susceptibility to degradative enzymes. Originally, this collagen type was reported as AB-collagen, which had been extracted from tissues after pepsin digestion (Burgeson *et al.,* 1976; Chung *et al.,* 1976). This material was then named type V collagen (Bornstein and Sage, 1980). As pepsin digestion removes significant noncollagenous peptide regions, we refer to such preparations as collagen V (pepsin), in contrast to the term collagen V which implies the end stage of physiologically processed material. In most of the literature this distinction is not made. Collagen V is also a generic term for all variants of this collagen type. Reports of amino acid analyses and sequences are mostly on pepsin-extracted material (Bornstein and Sage, 1980). Although these analyses showed distinct differences from the fibrous collagens I, II, and III, which are also evident in other ways, type V collagen is frequently associated with collagens I and III.

II. Compositions of Collagens V

Four chains have been described for collagen V: α1 and α1′, α2, and α3 (Bornstein and Sage, 1980). Two related collagen chains were isolated from cartilage and named 1α and 2α (Burgeson and Hollister, 1979). They are regarded as distinct gene products from α1(V) and α2(V) chains and are structurally different, though they have substantial similarities in amino acid and carbohydrate composition, solubility properties (Reese and Mayne, 1981; Burgeson *et al.,* 1982), and susceptibility to proteases (Liotta *et al.,* 1982), and show immunologi-

STRUCTURE AND FUNCTION OF
COLLAGEN TYPES

cal cross-reactivity (von der Mark and Öcalan, 1982; Gay *et al.*, 1981a). The α(V) chains are defined by the pattern of peptides obtained after cyanogen bromide cleavage of the pepsin-derived chains as well as by other properties (Bornstein and Sage, 1980). The principal, widely distributed chains are α1(V) and α2(V). Several methods have been developed for the identification and quantitation of these chains in different tissues, based on the characteristic peptides formed by cleavage with cyanogen bromide and proteases (Rhodes and Miller, 1979, 1981; Rhodes *et al.*, 1981; Miller *et al.*, 1983; von der Mark and von der Mark, 1979; Tanaka *et al.*, 1981; Cole and Chan, 1981; Kelly *et al.*, 1981; Sage *et al.*, 1981a; Reiser and Last, 1983; Bellon, 1985).

A homotrimeric molecule of α1(V) chains was found in a hamster lung cell line (Haralson *et al.*, 1980, 1984; Haralson and Mitchell, 1981) and is said to exist in liver (Biempica *et al.*, 1980). *In vitro* reconstitution experiments of pepsin-derived chains indicated that a corresponding homotrimer of α2(V) chains would be unstable (Bentz *et al.*, 1978). Initially, various combinations of α1(V) and α2(V) chains were proposed according to the varied recovery of these chains from tissues, though the heterotrimer $(\alpha 1)_2\alpha 2$ was particularly likely (Burgeson and Hollister, 1979; Rhodes and Miller, 1978, 1981; Deyl *et al.*, 1979; Sage *et al.*, 1981b; Trelstad *et al.*, 1981; Madri *et al.*, 1982; Abedin *et al.*, 1982). A disulfide-linked form of the corresponding three precursor chains, $(\text{pro}\alpha 1)_2\text{pro}\alpha 2$(V), was found in small amounts (Kumamoto and Fessler, 1980). From tissues in which the additional α3(V) chain occurs, the heterotrimers $(\alpha 1)_2\alpha 2$(V) and α1α2α3(V) have been separated and the stoichiometry of their components demonstrated (Rhodes and Miller, 1981; Abedin *et al.*, 1982; Niyibizi *et al.*, 1984; van der Rest *et al.*, 1985). These investigators failed to find the homotrimer $(\alpha 3)_3$(V), which has been reported (Madri *et al.*, 1982). Experiments assessing the susceptibility of these molecules to protease digestion also confirm the existence of heterotrimers (Sage *et al.*, 1981b). The different chain compositions impart slightly different stabilities and flexibilities to the molecules. For example, while the $(\alpha 1)_2\alpha 2$(V) heterotrimer melts at 39°C, the α1α2α3 heterotrimer dissociates at 36°C (Niyibizi *et al.*, 1984), and the latter combination displays a greater flexibility as determined by analysis of rotary shadowed molecules (J. Rauterberg *et al.*, personal communication). In different tissues, type V molecules with different compositions may be better suited for their particular functions.

An additional chain was obtained from chick embryo tendon and initially named α4(V) (Fessler *et al.*, 1985b), and then α1′(V) because of the similarity of its properties to the α1(V) chain (Fessler *et al.*,

1985a). An equivalent chain has been described in human epitheloid lung cells (Bloch *et al.,* 1986). The precursor forms of the α1(V) and α1′(V) chains and their physiologically derived pα- and α-chains differ distinctly from each other in the substantial noncollagenous peptide regions that are retained at one end of the chains. In the α1′(V) series, this region is smaller and differs in peptide fragments which carry sulfated tyrosine residues and in glycosylation (Fessler *et al.,* 1985a, 1986b). The noncollagenous peptide regions at the other end of the main collagen domain are of similar size but differ in their propensity to form disulfide cross-links to companion chains in the trimeric procollagen molecules. After pepsin treatment, it has not been possible to separate the α1(V) (pepsin) and α1′(V) (pepsin) chains. The cyanogen bromide peptide pattern of such mixed α(V) (pepsin) chains did not show clear differences. We suggest that the proα1′(V) and proα1(V) chains are either the products of two similar genes, which differ primarily in the noncollagenous peptide regions, or that they arise from one gene by different splicing of RNA. The proα1′(V) chains associate into a homotrimer and into the heterotrimer (proα1 proα1′ proα2)(V).

III. Nucleotide Sequences of Procollagens V

Recombinant human cDNA sequences were first correlated with the CB9 cyanogen bromide peptide in the carboxyl-end collagen region of α2(V) (pepsin) (Myers *et al.,* 1985a). The pattern of charged groups found in the α1(I), α2(I), and α1(III) chains is strongly conserved in this region, and the results showed that these three interstitial collagen chains are more similar to each other than to α2(V). The region 918–944 lacks any prolines, a finding that had also been made on α1(I). The telopeptide lacks lysine residues and therefore no lysine-derived cross-links could occur in this region of this chain. The DNA sequences were then extended in the 3′ direction to cover all of the carboxyl propeptide (Myers *et al.,* 1985b). The carboxyl propeptide of α2(V) has a greater degree of similarity to the corresponding domains of the interstitial procollagens than to the basement membrane procollagen IV. The carboxyl propeptide has the same number of cysteine residues as in proα2(I), and most are at the same sites. There is also a carbohydrate attachment site at the same position as in the interstitial procollagens, and there may be an additional carbohydrate attachment site within the hypervariable N-terminal region of this carboxyl propeptide. With this cloned DNA, the human proα2(V) gene was located on chromosome 2 at q24.3 to q31, a site where the gene for proα1(III) is also located (Emanuel *et al.,* 1985).

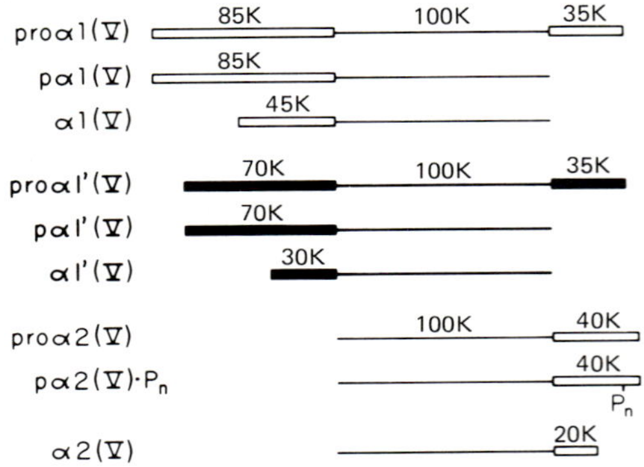

FIG. 1. Diagram of structure of proα(V) chains and their derivatives in chick embryo. The central collagenous sequence is indicated as a thin line and is assigned a nominal molecular size of 100 kDa. The nominal molecular sizes of the flanking, noncollagenous sequences are indicated. The proα2(V) chain may have an additional, small amino-terminal sequence that would have been missed by the analytical techniques that were used. By analogy with the interstitial procollagens, there could also be short, terminal collagenous sequences separated from the main collagen helix by telopeptide regions. The carboxyl end of each peptide is presumed to be at the right side of the diagram, though this is only known for the proα2 chain. Upon conversion of proα2(V) to pα2(V) two or more peptides, called P-peptides, become disulfide linked to the noncollagenous region of pα2(V). The chains occur in various trimeric combinations with different degrees of interchain disulfide linkage, as given in the text. These covalent connections between the main polypeptide chains are lost on conversion to p-collagens (V).

IV. Structure of Procollagens V

The domain structure of procollagens V is equivalent to that of the interstitial collagens (Fig. 1). A central collagen helix of practically the same length as that of type I has knoblike projections of noncollagenous propeptides at each end, as shown by electron microscopy of sprayed, rotary shadowed molecules (Bächinger *et al.*, 1982; Fessler *et al.*, 1982). The two knobs have different diameters. After physiological processing, a small knob is still seen at one end of the molecule (Elstow and Weiss, 1983; Broek *et al.*, 1985), and presumably this corresponds to a residual portion of the original amino propeptide. However, at present the only definitive structural location of the type V propeptides is the above recombinant DNA sequence of the carboxyl propeptide of proα2(V). This sequence corresponds to the noncollagenous peptide of approximately 40 kDa, as judged by SDS–polyacrylamide gel electrophoresis, that is formed from proα2(V) chains upon digestion with bacterial collagenase (Kumamoto and Fessler, 1981). Slightly smaller sized, equivalent noncollagenous peptides are obtained from proα1(V) and proα1′(V), with nominal molecular size of 35 kDa (Kumamoto and Fessler, 1981; Fessler *et al.,* 1985a). (These authors rounded off all nominal molecular sizes to the nearest 5 kDa.) This suggests that a trimeric carboxyl propeptide simlar to that of the interstitial procollagens exists. However, there are also some differences. The diameter of the knoblike carboxyl propeptides of sprayed, rotary shadowed procollagen I was intermediate between the diameters of the two procollagen V projections (Bächinger *et al.,* 1982). The interchain disulfide linking of the carboxyl V propeptides is usually incomplete, and the peptide connection to the collagen helical portion of the proα1(V) and proα1′(V) chains is cleaved much earlier than that of proα2(V) (Kumamoto and Fessler, 1981). It is assumed that in the type V collagen helix the amino-to-carboxyl sense of the constituent peptides is parallel.

The molecular masses of the procollagens V assigned from electrophoretic mobilities of these hybrid proteins are subject to error, as the collagen and noncollagen peptide regions contribute differently to the net mobilities. Glycosylation and internal disulfide links cause additional anomalies. However, within a precursor-to-product related series of peptides, defined by pulse–chase experiments, there is good correlation of electrophoretic mobilities and ultracentrifuge sedimentation velocities, supporting the concept of sequential proteolytic processing of precursors (Fessler *et al.,* 1981a). There are a number of reports of large, recently formed potential precursors of collagen V (pepsin) chains (e.g., Linsenmayer and Little, 1978; Jimenez *et al.,* 1978; Kumamoto and Fessler, 1980, 1981; Foidart *et al.,* 1981; Alitalo *et al.,* 1982; Narayanan and Page, 1983; Kao *et al.,* 1983; Elstow and Weiss, 1983). The nominal size of the human proα1(V) chain is 220–240 kDa and of the proα2(V) chain is 150–170 kDa (Alitalo *et al.,* 1982; Narayanan and Page, 1983). Differences may only be due to experimental conditions and the molecular-weight standards that were used. The summed values for the sizes of the collagen helical portions and the propeptides in chick embryo are about 220 kDa for the proα1(V) chain and 140 kDa for proα2(V) (Kumamoto and Fessler, 1981), and the relationships between these values agree with the relative sedimentation coefficients of the denatured chains (Fessler *et al.,* 1981a). The proα1(V) chain of hamster lung cells is the same size, 220 kDa (Fessler *et al.,* 1981b). These determinations were made on chains that were identified by peptide mapping to correspond with the previously isolated α1(V) and α2(V) (pepsin) chains. When hydroxylation and the accompanying O-linked glycosylation were prevented by synthesis in the presence of 2,2′-dipyridyl, the apparent size of the proα1(V) chain was decreased from 220 to 200 kDa, and the proα2(V) chain from 150 to 145 kDa (Alitalo *et al.*, 1982). A separate small decrease of apparent size occurred when attachment of N-linked carbohydrate was inhibited by incubation in the presence of tunicamycin (Fessler *et al.*, 1985a). The chick proα1′(V) chain is about 15 kDa smaller than the proα1(V) chain, and this difference is ascribed to a decrease in the size of the presumed amino propeptide. Digestion with bacterial collagenase showed that the chick proα1(V) and proα1′(V) chains have large noncollagenous peptides at both ends of the collagen region (Kumamoto and Fessler, 1981; Fessler *et al.*, 1985a). Assignment of the 35 kDa noncollagenous peptides as carboxyl propeptides, by analogy with the proα2(V) chain, leaves 85 and 70 kDa noncollagenous peptides as amino propeptides of, respectively, proα1(V) and proα1′(V). In chick, the proα2(V) chain lacks a large noncollagenous peptide at the amino end such as could be detected by the polyacrylamide gels used in these investigations. However, either a short collagenous region, as in proα2(I), or some additional, smaller sequences of amino acid residues might, of course, occur.

V. Biosynthesis

Embryonic chick tendons synthesise proα1(V) and proα1′(V) chains in about 10–12 min at 37°C (Fessler *et al.*, 1986b) (Fig. 2). This is consistent with the synthesis times of proα1(I) and proα1(IV) and their relative electrophoretically determined sizes. Within each propeptide, intrachain disulfide links are formed first, as judged by changes in electrophoretic mobility on reduction. Assembly of trimeric molecules is relatively slow (Kumamoto and Fessler, 1981; Fessler *et al.*, 1981a; Alitalo *et al.*, 1982). In human rhabdomyosarcoma A-204 cells, the first interchain disulfide-linked forms appeared 30 min after a 5-min pulse. Eventually, all of the proα2(V) chains formed disulfide-linked oligo/multimers of proα1(V) and proα2(V) chains which were not further characterized. Some proα1(V) chains were not disulfide linked, and the proα1(V) was present in excess of 2:1 over the proα2(V) chain (Alitalo *et al.*, 1982). Human gingival fibroblast cultures also formed procollagen V molecules with interchain disulfide links, but the degree of linkage was not determined (Narayanan and Page, 1983). The $(\text{pro}\alpha 1)_3$(V) made by hamster lung cells lacked any interchain disulfide bonds (Fessler *et al.*, 1981b). In contrast, the homotrimer $(\text{pro}\alpha 1')_3$(V) made by chick embryo tendon is fully interchain disulfide linked (Fessler *et al.*, 1985a,b). Mostly incomplete disulfide linkage

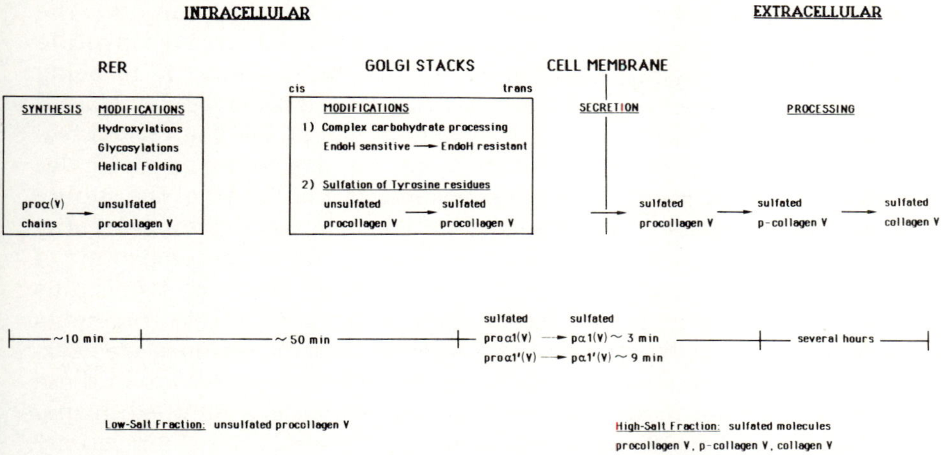

FIG. 2. Temporal relationships of biosynthesis and modifications of procollagens V in chick embryo tissues. Formation of individual proα (V) chains occurs in the rough endoplasmic reticulum (RER) and takes about 10 min. The indicated modifications of these chains and their triple-helical assembly are also assigned to the RER by analogy with other procollagens. Chick tendon procollagen V molecules are various trimeric combinations of proα2(V) and the similar, but different, proα1(V) and proα1′(V) chains. The procollagen V molecules with N-linked carbohydrate that is degraded by endoglycosidase H (EndoH sensitive) are assigned to RER and the cis Golgi compartment. Sulfation of tyrosine residues is assigned to some other Golgi compartment, possibly the trans stack, and occurs about 50 min after completion of peptide synthesis. The sulfated procollagen V molecules are secreted by cell cultures. Extracellularly, the carboxyl propeptides are considered to be cleaved off first, and the resulting sulfated pα1(V) chains are formed more rapidly than their pα1′(V) counterparts. Several hours later, distal portions of the remaining, large noncollagenous domains are removed, but most of the sulfated tyrosine residues are retained in the residual noncollagenous domains of collagens (V). Further processing may occur on residence in the tissues. The unsulfated procollagen V molecules elute in a low-salt fraction from DEAE-cellulose, while all their derivatives elute later in a high-salt fraction.

was found in the trimeric procollagen molecules made by various chick tissues. Attempts to increase the time of intracellular residence of these procollagen molecules did not increase the degree of disulfide linkage (Kumamoto and Fessler, 1981; Fessler *et al.*, 1981a). Although reversal of existing disulfide bridges by disulfide exchange during extraction cannot be excluded, this was guarded against, and the procollagens I and III that were coextracted were normally disulfide linked.

The procollagen V molecules move through the cell in the sequential order in which they are synthesized and undergo further modifications in the Golgi organelle, probably in the trans-Golgi, 50 min after the completion of polypeptide synthesis. The N-linked carbohydrate changes from having been sensitive to endoglycosidase H to being resistant to this enzyme, though it is still cleaved by endoglycosidase F (Fessler *et al.,* 1986b). Several tyrosine residues become O-sulfated, and this is accompanied by changes in the charge of the molecules (Fessler *et al.,* 1986a,b). Within experimental limits, all of the sulfate bound to procollagens V is attached in this form and is not on carbohydrate. About 40% of the tyrosine residues of the amino propeptide of proα1(V) and proα1′(V) chains become sulfated, with the two chains giving different maps of tyrosine sulfate-labeled peptide fragments after cleavage with proteases. The sulfated tyrosine residues are nearly all in the presumed amino propeptides, and most or all are in those portions of these noncollagenous regions which are retained during physiological processing to collagens V. The proα2(V) chain is only weakly tyrosine sulfated, and this is consistent with its lack of a large amino propeptide. The tyrosine sulfation proceeds independently of the attachment of N-linked carbohydrate, i.e., it continues when this has been suppressed with the drug tunicamycin. A cell-free system of Golgi-enriched membranes, together with the sulfate donor 3′-phosphoadenosine 5′-phosphosulfate (PAPS), sulfated tyrosine residues both in endogenous and in added unsulfated procollagens V. The enzyme is presumably membrane bound, and detergent disruption of the vesicles is required for added procollagen to be sulfated. All of the procollagen V becomes sulfated, but only shortly before secretion and cleavage to p-collagens V. Therefore, most of the extracted procollagen V is not sulfated. This tyrosine sulfation of procollagens V occurs in all tested chick embryo tissues, in neonatal rat, and in cultured human cells. The antimicrotubular drug Nocodazole speeds up tyrosine sulfation of procollagens V without influencing the final degree of sulfation (Fessler *et al.,* 1986a,b).

VI. Secretion of Procollagens V and Processing

All of the procollagens V and their derivatives found in culture media of human and chick cells were tyrosine sulfated. Secretion and cleavage of procollagens V rapidly follow sulfation (Fessler *et al.,* 1986a,b). As with the interstitial procollagens, secretion is inhibited when hydroxylation of the procollagens V has been prevented by synthesis in the presence of 2,2′-dipyridyl (Alitalo *et al.,* 1982). Also, as with the interstitial procollagens, addition of arginine to the culture medium slows the cleavage conversion to p-collagens V (Kumamoto and Fessler, 1981; Fessler *et al.*, 1981b). The kinetics of secretion from corneal cells have been studied (Kao *et al.*, 1983), and a secretion time of about 100 min was estimated for rhabdomyosarcoma cells (Alitalo *et al.*, 1982). Presumably, cleavage to p-collagens V occurs at the cell surface or extracellularly, and cell culture fluids contain enzymes that cleave procollagens V (Fessler *et al.*, 1981b; Alitalo *et al.*, 1982; Narayanan and Page, 1983).

In whole chick tendons, conversion to p-collagens V occurs about 60 min after completion of polypeptide synthesis (Fessler *et al.*, 1986b). The secreted procollagens V contain trimeric molecules of different chain compositions, e.g., homotrimers or heterotrimers consisting of two or three different proα(V) chains. During the conversion to p-collagen V, the constituent chains are cut at different rates (Kumamoto and Fessler, 1981; Fessler *et al.*, 1981b, 1986b), and overall the proα1(V) chains are cut about three times faster than the proα1′(V) chains. Thus, several forms of trimeric molecules will exist in which one or more polypeptide strands are cut without loss of cleavage products from the original procollagen V molecule. Equivalent intermediates have been described in the conversion of procollagen I to p-collagen I (Davidson *et al.*, 1977; Morris *et al.*, 1979). However, there are significant differences in the processing of procollagens V and procollagen I, and additional changes must occur. We shall first describe the changes observed for each proα(V) chain, and then discuss the possible three-chained molecular forms.

The smaller, 35 kDa noncollagenous peptide is split off the proα1(V) and proα1′(V) chains. Presumably, this is the carboxyl propeptide. The resulting pα1(V) and pα1′(V) chains are not disulfide linked to any other chains. This processing of proα1(V) occurs both in chick tissues (Kumamoto and Fessler, 1981; Fessler *et al.*, 1985a) and in human cell cultures (Alitalo *et al.*, 1982; Narayanan and Page, 1983; Bloch *et al.*, 1986). The conversion of proα2(V) to pα2(V) may be slightly different in human and chick. In human gingival cell cultures, three chains of the α2(V) series were found: 174, 157, and 132 kDa. Presumably, these correspond to the proα2(V), pα2(V), and α2(V) chains, though this sequential relationship was not established (Narayanan and Page, 1983). In chick, the reduced forms of the proα2(V) and pα2(V) chains have the same electrophoretic mobility and give rise to the same noncollagenous peptide upon digestion with bacterial collagenase. Thus, the peptide structure of chick proα2(V) is presumably unchanged, and the carboxyl propeptide is retained in pα2(V). However, both electrophoretic and velocity sedimentation analysis of the pα2(V) chain without reduction show it to be larger than the original proα2(V) chain. This is because more than one copy of an approximately 30 kDa peptide has become disulfide linked to the carboxyl propeptide of pα2(V). Reduction releases these peptides, which were called P-peptides, and yields the pα2(V) chain. Collagenase digestion of the nonreduced pα2(V)–P-peptide complex gives the carboxyl–α2(V) peptide disulfide linked to the P-peptides (Kumamoto and Fessler, 1981; Fessler et al., 1981a).

The most likely sources of the P-peptides are the cleaved-off carboxyl propeptides of proα1(V) and proα1′(V) chains. The P-peptides electrophorese slightly faster than the presumed carboxyl propeptides prepared from proα1(V) chains by digestion with bacterial collagenase. This is consistent with an expected difference between the likely physiological site of cleavage, in a telopeptide region, and the place where bacterial collagenase will stop digestion when it reaches the end of the collagenous sequence. Both carboxyl propeptides and P-peptides carry N-linked carbohydrate, and both become labeled during incubation with radioactive amino acids. However, if the P-peptides are derived from the carboxyl propeptides of adjacent proα1(V) and proα1′(V) chains, then an increase of interchain disulfide bonding must accompany the procollagen V to p-collagen V transition. The P-peptides could be carboxyl propeptides which are unable to undergo disulfide linkage to neighboring carboxyl propeptides as long as they retain continuity with the main collagenous peptide portion and, only after cleavage, could undergo some rearrangement needed for disulfide link formation. However, when the proα1(V) chains of isolated procollagen V were cleaved to pα1(V) with the enzymes of fibroblast culture media, none of the resulting pα2(V) chains had P-peptides attached (Fessler *et al.*, 1981b). It is not necessary that all P-peptides originally belonged to the set of chains of one trimeric molecule. They could be derived from an adjacent, juxtaposed molecule. These considerations apply equally if the proα1(V) and proα2(V) chains within one molecule had an antiparallel arrangement.

The conversion of p-collagens V to collagens V only occurs slowly, over a number of hours, in chick tissues incubated *in vitro* (Fessler *et al.*, 1981a,b, 1985a). This conversion is strikingly different from the corresponding change in the interstitial p-collagens I, II, and III. Only the distal portion of the carboxyl propeptide of pα2(V) is cut off, so that an approximately 20 kDa noncollagenous peptide remains attached to the main collagen region. With this change, the P-peptides are lost, and they are, presumably, disulfide linked to the distal portion of the carboxyl propeptide. The presumed amino propeptides of pα1(V) and pα1′(V) also have their distal portions removed, leaving substantial peptides of, respectively, 45 and 30 kDa that are resistant to bacterial collagenase and remain continuous with the main collagen regions. Most or all of the sulfated tyrosine residues are retained during this conversion. While native procollagen V and p-collagen V molecules could not be distinguished by velocity sedimentation, the collagen V molecules formed by the above changes sedimented more slowly. This implies that while the conversion from procollagen V to p-collagen V may primarily represent nicks in peptide strands and changes in disulfide linkage of the originally secreted triple-chained molecule, the further change to collagen V removes some peripheral, noncollagenous portions at both ends of this molecule. At the time that this change occurs, these molecules have been a part of the extracellular matrix for a relatively long time.

Further, slow processing may occur at times longer than the 20 hr of organ explants maintenance. Extracts of chick embryo tendons show two corresponding smaller electrophoretic bands which chromatograph with collagen V and stain with Coomassie Blue but do not incorporate radioactive amino acids during 20 hr of incubation. The nature of these materials has not yet been established. From lathyritic chick tendon (Jimenez *et al.*, 1978; Broek *et al.*, 1985), bone, and crop (Broek *et al.*, 1985), collagen V has been extracted that probably corresponds to this material. The lathyritic α1(V) and α2(V) chains contain, respectively, 18 and 29 kDa noncollagenous regions. This lathyritic collagen V was then associated into segment long spacing (SLS) crystallites that had the same 300-nm-long banded region as that made from collagen V (pepsin) and had an additional 25-nm domain. Comparison with SLS crystallites of collagen I indicated that the additional domain is at the amino-terminus. In rotary shadowed molecules, the 304-nm collagen thread had a knob at one end, and several collagen molecules were connected by one large knob (Broek *et al.*, 1985). Similar collagen V fractions were isolated from calf skin (Elstow and Weiss, 1983) and from lightly pepsin-digested placenta (Adachi and Hayashi, 1985). An unresolved paradox is that the single amino domain of collagen V molecules seen in the electron microscope is presumed to contain the noncollagenous regions of both the α1(V) and the α2(V) chains. But the single, substantial noncollagenous domain of chick proα2(V) (Kumamoto and Fessler, 1981; Fessler *et al.*, 1981a) is assigned as a carboxyl propeptide because of the existence of a carboxyl propeptide of this size in a human cDNA sequence (Myers *et al.*, 1985b). Thus, this noncollagenous region of proα2(V) and its processed derivative in α2(V) are assigned to opposite ends of one molecule. It is not established what the end form of type V collagen molecules is, and there could be more than one form in different tissues.

VII. Location of Collagens V in the Extracellular Matrix

Immunofluorescence microscopy shows, in a wide variety of tissues, that collagen V occurred both in association with basement membranes and, quite separately, in fibrous stroma such as tendon and cornea. Recently, the basement membrane localization has been questioned, and identification of collagen V as an interstitial fibrous collagen has been suggested (Schuppan *et al.,* 1986). Investigation of the relationship of collagen V to basement membranes requires resolution with the electron microscope. Initial immunoelectron microscopy by the ferritin–ultrathin frozen section technique showed colocalization of collagens V and IV in three different basement membranes of the mouse kidney, and their occurrence in parts of the mesangium as well (Roll *et al.,* 1980; Furthmayr *et al.,* 1982). Different results were obtained using immunoelectron microscopy with cross-adsorbed antibodies and the peroxidase technique. The results suggested that type V collagen is present in the rat renal interstitium in different forms: in close apposition to interstitial collagen fibers, in the stromal aspect of vascular basement membranes, and in a particulate material not bound to other structures. Type V collagen was specifically noted as being absent from epithelial and glomerular basement membranes. It was sometimes found on only the stromal aspect of vascular smooth muscle as well as in between collagen bundles as part of the matrix. A binding or connecting function was postulated for collagen V (Martinez-Hernandez *et al.,* 1982). In a third study, application of the immunogold-labeling technique to human amnion showed type V collagen in the immediate vicinity of basement membranes but distinctly separate from the locations of laminin and collagen IV. Unbanded fibrils of 12 nm were stained. These were separate from the striated, 67-nm fibers that were labeled by antibodies to collagen I, though these different fibrils were enmeshed. An anchoring function between membranes and stromal matrix was postulated for these 12-nm collagen V fibrils (Modesti *et al.,* 1984).

The relationship of collagen V to muscle cells is of particular interest. During formation of myotubes *in vitro,* myoblasts make small amounts of collagen V, but pronounced expression of collagen V occurs when they have fused (Bailey *et al.,* 1979; Sasse *et al.,* 1981; von der

Mark and Öcalan, 1982; Nusgens *et al.*, 1986). Collagen V is primarily found in the endomysium of skeletal muscle (Duance *et al.*, 1977; Bailey *et al.*, 1979) and was undetected within the synaptic regions of muscle basement membranes (Sanes and Cheney, 1982). In contrast, cultures of smooth muscle cells synthesize and secrete collagen V over the entire plasma membrane (Gay and Miller, 1978), but within tissues collagen V tends to be associated with only the stromal aspects of the cell surface (Gay *et al.*, 1981b; Miller *et al.*, 1982).

Collagen V has also been reported peripherally adjacent to capillaries, at the surface of blood vessels denuded of their endothelial lining (Keremyi *et al.*, 1985), and in atherosclerotic plaques (McCullagh *et al.*, 1980; Ooshima, 1981; Shekhonin *et al.*, 1985). The overall impression and interpretation of such findings is that, in these locations, collagen V has one or more functions as a connector between basement membrane and stroma.

Partly in contrast to such functions is the concept of collagen V as a component of extended collagen fibers. Purified collagens V (pepsin) can be precipitated as thin fibrils (Chiang *et al.*, 1980; Broek *et al.*, 1985), even with 67-nm periodicities (Barnes *et al.*, 1980; Adachi and Hayashi, 1985). The central collagen domain has the same 300-nm length in collagens I and V (Kühn, 1982; Bächinger *et al.*, 1982; Elstow and Weiss, 1983; Broek *et al.*, 1985). Monoclonal antibodies against collagens V (pepsin) have been made (von der Mark and Öcalan, 1982; Linsenmayer *et al.*, 1983a; Mayne *et al.*, 1984b), and it was found that, at the light-microscopic level, the epitopes appeared to be masked in tissues by association with collagen I. They were unmasked by acetic acid swelling of the fibers, low temperature dissolution of lathyritic material, and cleavage of collagen I with vertebrate collagenase (Linsenmayer *et al.*, 1983a, 1984, 1985; Fitch *et al.*, 1984). These data suggested two separate models for fibers containing more than one collagen type: in one model the different components are intimately mixed throughout the whole cross section of a fiber, and in the other there are regional condensations of the separate types. In the latter case the type V collagen contributions would be located more centrally within a fiber. Either combination is likely to imbue a fiber with special properties (Linsenmayer *et al.*, 1985). In addition to biomechanical properties such as tensile strength and flexibility of a fiber, such combinations could also influence final fibril diameter and the processes of fibrillogenesis (Adachi and Hayashi, 1986). Fibers which contain collagen V in the chick embryo cornea only appear during a second stage of fiber formation; therefore, in this tissue the collagen V does not seem to be involved in the additional biological problem of how the first lead fibers are laid down. Polyclonal antibodies against collagen V (pepsin) indicated that, at the electron-microscope level, this material was on or close to collagen fibrils of the stroma of human cornea and in Bowman's layer (Nakayasu *et al.,* 1986).

In chick embryo tendon, the proα1′(V) chain is only made within the tendon proper and not by the cells of the tendon sheath, which do, however, make proα1(V) and proα2(V) chains (Fessler *et al.,* 1985a). The proα3(V) chain is primarily found in placenta and in some other tissues (Brown *et al.,* 1978; Sage and Bornstein, 1979; Brown and Weiss, 1979; Abedin *et al.,* 1981; Morton and Barnes, 1982; Leushner, 1983; Leushner and Haust, 1984). These occurrences could be further indications of specialization of collagen V fiber formation for particular regions.

The potentially intimate molecular association of collagens V with collagen I is a property reminiscent of collagen III, which probably also is rarely, if ever, found alone. The presence of amino propeptides on pN-collagens I and III clearly influences the size of the fibers deposited (Fleischmajer *et al.,* 1981, 1985; Hulmes, 1983), and a similar role for pN-collagen V may be postulated. Both procollagens V and III are also only slowly processed at their amino ends (Fessler *et al.,* 1981c), and both have sulfated tyrosine residues (Fessler *et al.,* 1986a,b; Niemela *et al.,* 1985; Jukkola *et al.,* 1986). There is an impression that both collagen types are also associated with fibers of thinner diameter, but it is probably premature to generalize.

VIII. Other Properties of Collagens V

In addition to biomechanical contributions, collagen V (pepsin) has been tested for its interactive properties with cells and other components of the extracellular matrix. It has been reported to promote cell attachment and migration (Stenn *et al.,* 1979; Grotendorst *et al.,* 1981), and specifically cyanogen bromide peptide CB6 of the α2(V) chain serves as a binding site to some plasma membrane glycoproteins (Leushner and Haust, 1985). Type V collagen interacts preferentially, relative to other collagens (pepsin), with thrombospondin (Mumby *et al.,* 1984). The extensive literature on the interaction of collagens with platelets, which includes interaction with collagens V, (e.g., Chiang *et al.,* 1980; Barnes *et al.,* 1980), and the role of collagens in blood clot formation will not be reviewed here.

The collagen helix of collagens V is resistant to the vertebrate collagenases which cleave collagens I and III in a characteristic manner (Harper, 1980). Other enzymes, with various degrees of specificity for collagens V (pepsin), have been described (Mainardi *et al.,* 1980, 1984; Liotta *et al.,* 1981; Murphy *et al.,* 1981, 1982, 1985; Hibbs *et al.,* 1982; Rich *et al.,* 1983; Haralson *et al.,* 1984). A class of these neutral metalloproteinases, termed "gelatinases," specifically degrade collagen V in addition to denatured collagens (Murphy *et al.,* 1981, 1982; Mainardi *et al.,* 1983; Hibbs *et al.,* 1985), and gelatinases are often present during matrix degradation together with the interstitial collagenases (Seltzer *et al.,* 1981; Harris and Krane, 1972; Sopata and Dancewiez, 1974), suggesting that they may act synergistically with the classical collagenases during matrix remodeling. The regions flanking the collagen V helix must be important for maintaining the state of insolubility of collagens V in tissues from which they are extracted after digestion with pepsin. It is likely that some of the pathologically degradative enzymes might act at such sites. There may be turnover of collagens V associated with inflammatory conditions and wound healing (Hering *et al.,* 1983; Kurita *et al.,* 1985). The synthesis of collagen V by gingival cells is increased during inflammation, but this condition does not become an inherent property of the cells that is maintained in culture (Narayanan and Page, 1983, 1985).

IX. Occurrence of Collagens V

We hope that the following list of locations in which collagens V have been found or investigated in normal adult and embryonic tissues and in some pathological conditions will help the reader, and we apologize to the authors of any additional publications that we may have omitted.

Vascular tissues and cell cultures: Chung *et al.* (1976), Sage *et al.* (1979, 1981c), Sage (1982), Alitalo *et al.* (1980), McCullagh *et al.* (1980), Roll *et al.* (1980), Fessler *et al.* (1981a), Sankey *et al.* (1981, Sankey and Barnes (1984), Borg *et al.* (1982), Tseng *et al.* (1982b), Morton and Barnes (1982, 1983), Bartholomew and Anderson (1983), Bellon *et al.* (1985), Keremyi *et al.* (1985), and Jerdan and Glaser (1986).

Kidney: Roll *et al.* (1980), Martinez-Hernandez *et al.* (1982), Bartholomew and Anderson (1983), Foidart *et al.* (1983), and Bonadio *et al.* (1984).

Lung: Madri and Furthmayr (1979, 1980), Haralson *et al.* (1980), Reiser and Last (1983), Konomi *et al.* (1984).

Liver and hepatocyte cell cultures: Biempica *et al.* (1980), Hatahara and Seyer (1982), Rojkind *et al.* (1983), Tseng *et al.* (1983), Murata *et al.* (1984), Yamamoto *et al.* (1984), Hata *et al.* (1985), and Schuppan *et al.* (1986).

Pancreas: Ingber *et al.* (1985).

Placenta and uterus: Burgeson *et al.* (1976), Sage and Bornstein (1979), Hong *et al.* (1979), Foidart *et al.* (1981), Rhodes and Miller (1981), Abedin *et al.* (1982), Madri *et al.* (1982), Mandell and Sodek (1982), Modesti *et al.* (1984), and Niyibizi *et al.* (1984).

Amniocytes: Alitalo *et al.* (1980).

Chinese hamster ovary cells: Limeback *et al.* (1982).

Mammary gland: Warburton *et al.* (1983).

Gingiva, oral mucosa, periodontal tissue and tooth development: Sodek and Mandell (1982), Sodek *et al.* (1982), Narayanan and Page (1983, 1985), Narayanan *et al.* (1983), Limeback *et al.* (1983), Silver *et al.* (1984), Kurita *et al.* (1985), Dabbous *et al.* (1981), and Schuppan *et al.* (1986).

Tendon, skin, and bone: Chung *et al.* (1976), Jimenez *et al.* (1978), Laurain *et al.* (1980), Herrmann *et al.* (1980), Fessler *et al.* (1981b, 1985a), Aubin *et al.* (1982), Linsenmayer *et al.* (1983a), von der Mark and Öcalan (1982), Elstow and Weiss (1983), Kenney *et al.* (1983), Woodley *et al.* (1983), Weber *et al.* (1984), Konomi *et al.* (1984), Broek *et al* (1985), and Smith *et al.* (1986).

Cornea: Linsenmayer *et al.* (1985), Davison *et al.* (1979), Yue *et al.* (1979, 1983), Conrad *et al.* (1980), Pöschl and von der Mark (1980), Welsh *et al.* (1980), Kao and Foreman (1980), Kao *et al.* (1982), Tseng *et al.* (1981, 1982a), Sankey *et al.* (1981), Sankey and Barnes (1984), Newsome *et al.* (1982), von der Mark and Öcalan (1982), and Nakayasu *et al.* (1986).

Keratinoctyes: Birkedal-Hansen *et al.* (1982) and Prunieras *et al.* (1983).

Nervous system: Shellswell *et al.* (1979), Maxwell *et al.* (1984), and Salonen *et al.* (1985a).

Smooth muscle and smooth-muscle cell cultures: Chung *et al.* (1976), Mayne *et al.* (1978), McCullagh *et al.* (1980), Fessler *et al.* (1981a), Bartholomew and Anderson (1983), Leushner and Haust (1984), Bellon *et al.* (1985), and Schuppan *et al.* (1986).

Skeletal muscle and cell cultures: Duance *et al.* (1977, 1980), Bailey *et al.* (1979), Shellswell *et al.* (1980), John and Lawson (1980), Sasse *et al.* (1981), von der Mark and Öcalan (1982), Linsenmayer *et al.* (1983b), Sanes and Cheney (1982), Stephens *et al.* (1982), Salonen *et al.* (1985b), DeMichele *et al.* (1985), and Nusgens *et al.* (1986).

Tumors:

Breast carcinoma: Barsky *et al.* (1982), Gordenne *et al.* (1982), and Van Cauwenberge *et al.* (1983).

Osteosarcoma: Shapiro and Eyre (1982) and Yamagata and Yamagata (1984).

Neurofibroma: Peltonen *et al.* (1981, 1984) and Shinkai (1984).

Rhabdomyosarcoma: Krieg *et al.* (1979), and Alitalo *et al.* (1982).

Basal cell carcinoma: Van Cauwenberge *et al.* (1983).

Mesothelial cells: Harvey and Amlot (1983).

Dedifferentiated chondrocyte cultures: Benya *et al.* (1977, 1978) and Mayne *et al.* (1984a).

Invertebrate (jellyfish): A mesoglial collagen was suggested to be like type V, (Miura and Kimura, 1985).

X. Conclusions

The presence of collagen V in the vicinity of basement membranes and in collagen fibers suggests that it can act as a linker and also contribute to fiber structure. This collagen has several α chains in different trimeric combinations. It is not known whether these various type V molecules have different physiological functions. The existence of the related 1α and 2α chains in cartilage (see Eyre and Wu, this volume) and of the $\alpha 1'(V)$ chain in tendon suggests that there could be tissue-specific adaptation. Significant noncollagenous regions are maintained from the procollagens V and are only slowly cut off. These noncollagenous peptides must influence the interactions of the newly made p-collagens V with other components of the extracellular matrix.

Acknowledgments

We thank the Fogarty International Center for a Senior International Fellowship (JHF), the Biozentrum of the University of Basel for hospitality, and the US Public Health Service for Grant AM-13748.

References

Abedin, M. Z., Ayad, S., and Weiss, J. B. (1981). *Biochem. Biophys. Res. Commun.* **102,** 1237–1245.

Abedin, M. Z., Ayad, S., and Weiss, J. B. (1982). *Biosci. Rep.* **2,** 493–502.

Adachi, E., and Hayashi, T. (1985). *Collagen Relat. Res.* **5,** 225–232.

Adachi, E., and Hayashi, T. (1986). *Connect. Tissue Res.* **14,** 257–266.

Alitalo, K., Kurkinen, M., Vaheri, A., Krieg, T., and Timpl, R. (1980). *Cell* **19,** 1053–1062.

Alitalo, K., Myllylä, R., Sage, H., Pritzl, P., Vaheri, A., and Bornstein, P. (1982). *J. Biol. Chem.* **257,** 9016–9024.

Aubin, J. E., Heersche, J. N. M., Merrilees, M. J., and Sodek, J. (1982). *J. Cell Biol.* **92,** 452–461.

Bächinger, H. P., Doege, K. J., Petschek, J. P., Fessler, L. I., and Fessler, J. H. (1982). *J. Biol. Chem.* **257,** 14590–14592.

Bailey, A. J., Shellswell, G. B., and Duance, V. C. (1979). *Nature (London)* **278,** 67–69.

Barnes, M. J., Bailey, A. J., Gordon, J. L., and MacIntyre, D. E. (1980). *Thromb. Res.* **18,** 375–388.

Barsky, S. H., Rao, C. N., Grotendorst, G. R., and Liotta, L. A. (1982). *Am. J. Pathol.* **108,** 276–283.

Bartholomew, J. S., and Anderson, J. C. (1983). *Histochem. J.* **15,** 1177–1190.

Bellon, G. (1985). *Anal. Biochem.* **150,** 188–202.

Bellon, G., Caulet, T., Cam, Y., Pluot, M., Poulin, G., Pytlinska, M., and Bernard, M. H. (1985). *Acta Neuropathol.* **66,** 245–252.

Bentz, H., Bächinger, H. P., Glanville, R., and Kühn, K. (1978). *Eur. J. Biochem.* **92,** 563–567.

Benya, P. P., Padilla, S. R., and Nimni, M. E. (1977). *Biochemistry* **16,** 865–872.

Benya, P. P., Padilla, S. R., and Nimni, M. E. (1978). *Cell* **15,** 1313–1321.

Biempica, L., Morecki, R., Wu, C. H., Giambrone, M. A., and Rojkind, M. (1980). *Am. J. Pathol.* **98,** 591–602.

Birkedal-Hansen, H., Munksgaard, E. C., Hansen, I. L., Nellemann, K., and Gay, R. (1982). *Collagen Relat. Res.* **2,** 287–300.

Bloch, J. M., Stone, K., and Crouch, E. (1986). *Fed. Proc., Fed. Am. Soc. Exp. Biol.* **45,** 814.

Bonadio, J. F., Sage, H., Cheng, F., Bernstein, J., and Striker, G. E. (1984). *Am. J. Pathol.* **116,** 289–296.

Borg, T. K., Gay, R. E., and Johnson, L. D. (1982). *Collagen Relat. Res.* **2,** 211–218.

Bornstein, P., and Sage, H. (1980). *Annu. Rev. Biochem.* **49,** 957–1003.

Brodek, D. L., Madri, J., Eikenberry, E. F., and Brodsky, B. (1985). *J. Biol. Chem.* **260,** 555–562.

Brown, R. A., and Weiss, J. B. (1979). *FEBS Lett.* **106,** 71–75.

Brown, R. A., Shuttleworth, C. A., and Weiss, J. B. (1978). *Biochem. Biophys. Res. Commun.* **80,** 866–872.

Burgeson, R. E., and Hollister, D. W. (1979). *Biochem. Biophys. Res. Commun.* **87,** 1124–1131.

Burgeson, R. E., El Adli, F. A., Kaitila, I. I., and Hollister, D. W. (1976). *Proc. Natl. Acad. Sci. U.S.A.* **73,** 2579–2583.

Burgeson, R. E., Hebda, P. A., Morris, N. P., and Hollister, D. W. (1982). *J. Biol. Chem.* **257,** 7852–7856.

Chiang, T. M., Mainardi, C. L., Seyer, J. M., and Kang, A. H. (1980). *J. Lab. Clin. Med.* **95,** 99–107.

Chung, E., Rhodes, R. K., and Miller, E. J. (1976). *Biochem. Biophys. Res. Commun.* **71,** 1167–1174.

Cole, W. G., and Chan, D. (1981). *Biochem. J.* **197,** 377–383.

Conrad, G. W., Dessau, W., and von der Mark, K. (1980). *J. Cell Biol.* **84,** 501–512.

Dabbous, M. K., Hammouda, O., and Brinkley, B. (1981). *Mol. Cell. Biochem.* **34,** 86–93.

Davidson, J. M., McEneany, L. S. G., and Bornstein, P. (1977). *Eur. J. Biochem. 81,* 349–355.

Davison, P. F., Hong B.-S., and Cannon, D. J. (1979). *Exp. Eye Res.* **29,** 97–107.

DeMichele, S. J., Brown, R. G., Krasin, B. W., and Sweeny, P. R. (1985). *Comp. Biochem. Physiol.* **81,** 149–157.

Deyl, Z., Macek, K., and Adam, M. (1979). *Biochem. Biophys. Res. Commun.* **89,** 627–634.

Duance, V. C., Restall, D. J., Beard, H., Bourne, F. J., and Bailey, A. J. (1977). *FEBS Lett.* **79,** 248–252.

Duance, V. C., Black, C. M., Dubowitz, B., Hughes, G. R. V., and Bailey, A. J. (1980). *Muscle Nerve* **39,** 487–490.

Elstow, S. F., and Weiss, J. B. (1983). *Collagen Relat. Res.* **3,** 181–193.

Emanuel, B. S., Cannizzaro, L. A., Seyer, J. M., and Myers, J. C. (1985). *Proc. Natl. Acad. Sci. U.S.A.* **82,** 3385–3389.

Fessler, L. I., Kumamoto, C. A., Meis, M. E., and Fessler, J. H. (1981a). *J. Biol. Chem.* **256,** 9640–9645.

Fessler, L. I., Robinson, W. J., and Fessler, J. H. (1981b). *J. Biol. Chem.* **256,** 9646–9651.

Fessler, L. I., Timpl, R., and Fessler, J. H. (1981c). *J. Biol. Chem.* **256,** 2531–2537.

Fessler, J. H., Bächinger, H. P., Lunstrum, G., and Fessler, L. I. (1982). *In* "New Trends in Basement Membrane Research" (K. Kühn, H. H. Schoene, and R. Timpl, eds.), pp. 145–154. Raven, New York.

Fessler, L. I., Shigaki, N., and Fessler, J. H. (1985a). *J. Biol. Chem.* **260,** 13286–13293.

Fessler, J. H., Shigaki, N., and Fessler, L. I. (1985b). *Ann. N.Y. Acad. Sci.* **460,** 181–186.

Fessler, L. I., Brosh, S., Chapin, S., and Fessler, J. H. (1986a). *J. Biol. Chem.* **261,** 5034–5040.

Fessler, L. I., Chapin, S., Brosh, S., and Fessler, J. H. (1986b). *Eur. J. Biochem.* **158,** 511–518.

Fitch, J. M., Gross, J., Mayne, R., Johnson-Wint, B., and Linsenmayer, T. F. (1984). *Proc. Natl. Acad. Sci. U.S.A.* **81,** 2791–2795.

Fleischmajer, R., Timpl, R., Tuderman, L., Raisher, L., Wiestner, M., Perlish, J. S., and Graves, P. N. (1981). *Proc. Natl. Acad. Sci. U.S.A.* **78,** 7360–7364.

Fleischmajer, R., Perlish, J. S., and Timpl, R. (1985). *Ann. N.Y. Acad. Sci.* **460,** 246–257.

Foidart, J. M., Tryggvason, K., Gehron-Robey, P., Liotta, L. A., and Martin, G. R. (1981). *Collagen Relat. Res.* **1,** 137–150.

Foidart, J. B., Foidart, J. M., Hassell, J., and Mahieu, P. (1983). *Renal Physiol.* **6,** 163–170.

Furthmayr, H., Roll, F. J., Madri, J. A., and Foellmer, H. (1982). *In* "New Trends in Basement Membrane Research" (K. Kühn, H. H. Schoene, and R. Timpl, eds.), pp. 31–48. Raven, New York.

Gay, S., and Miller, E. J. (1978). *In* "Collagen in the Physiology and Pathology of Connective Tissue," pp. 30–31. Fischer, Stuttgart.

Gay, S., Rhodes, R. K., Gay, R. E., and Miller, E. J. (1981a). *Collagen Relat. Res.* **1,** 53–58.

Gay, S., Martinez-Hernandez, A., Rhodes, R. K., and Miller, E. J. (1981b). *Collagen Relat. Res.* **1,** 377–384.

Gordenne, W., Foidart, J. M., and Lapiere, C. M. (1982). *J. Gynecol. Obstet. Biol. Reprod.* **11,** 549–554.

Grotendorst, G. R., Seppa, H. E., Kleinman, H. K., and Martin, G. R. (1981). *Proc. Natl. Acad. Sci. U.S.A.* **78,** 3669–3672.

Haralson, M. A., and Mitchell, W. M. (1981). *Collagen Relat. Res.* **1,** 309–325.

Haralson, M. A., Mitchell, W. M., Rhodes, R. K., Kresina, T. F., Gay, R., and Miller, E. J. (1980). *Proc. Natl. Acad. Sci. U.S.A.* **77,** 5206–5210.

Haralson, M. A., Mitchell, W. M., Rhodes, R. K., and Miller, E. J. (1984). *Arch. Biochem. Biophys.* **229,** 509–518.

Harper, E. (1980). *Annu. Rev. Biochem.* **49,** 1063–1078.
Harris, E. D., Jr., and Krane, S. M. (1972). *Biochim. Biophys. Acta* **258,** 566–576.
Harvey, W., and Amlot, P. L. (1983). *J. Pathol.* **139,** 337–347.
Hata, R., Ninomiya, Y., Sano, J., Konomi, H., Hori, H., Sunada, H., Tanaka, S., Kabuki, K., Nagai, Y., and Tsukada, Y. (1985). *J. Cell. Physiol.* **122,** 333–342.
Hatahara, T., and Seyer, J. M. (1982). *Biochim. Biophys. Acta* **716,** 431–438.
Hering, T. M., Marchant, R. E., and Anderson, J. M. (1983). *Exp. Mol. Pathol.* **39,** 219–229.
Herrmann, H., Dessau, W., Fessler, L. I., and von der Mark, K. (1980). *Eur. J. Biochem.* **105,** 63–74.
Hibbs, M. S., Mainardi, C. L., and Kang, A. H. (1982). *Biochem. J.* **207,** 621–624.
Hibbs, M. S., Hasty, K. A., Seyer, J. M., Kang, A. H., and Mainardi, C. L. (1985). *J. Biol. Chem.* **260,** 2493–2500.
Hong, B.-S., Davison, P. F., and Cannon, D. J. (1979). *Biochemistry* **18,** 4278–4282.
Hulmes, D. (1983). *Collagen Relat. Res.* **3,** 317–322.
Ingber, D. E., Madri, J. A., and Jamieson, J. D. (1985). *Am. J. Pathol.* **121,** 248–260.
Jerdan, J. A., and Glaser, B. M. (1986). *Invest. Ophthalmol. Visual Sci.* **27,** 194–203.
Jimenez, S. A., Yankowski, R., and Bashey, R. I. (1978). *Biochem. Biophys. Res. Commun.* **81,** 1298–1306.
John, H. A., and Lawson, H. (1980). *Cell Biol. Int. Rep.* **4,** 841–850.
Jukkola, A., Risteli, J., Niemela, O., and Risteli, L. (1986). *Eur. J. Biochem.* **154,** 219–224.
Kao, W. W. Y., and Foreman, C. A. (1980). *Eur. J. Biochem.* **106,** 41–48.
Kao, W. W. Y., Mai, S. H., and Chou, K. L. L. (1982). *Invest. Opthalmol. Visual Sci.* **23,** 787–795.
Kao, W. W. Y., Mai, S. H., Chou, K. L. L., and Ebert, J. (1983). *J. Biol. Chem.* **258,** 7779–7787.
Kelly, P. T., von der Mark, K., and Conrad, G. W. (1981). *Anal. Biochem.* **112,** 105–116.
Kenney, M. C., Cheung, C. C., and Smith, R. E. (1983). *Exp. Eye Res.* **36,** 35–46.
Keremyi, T., Voss, B., Rauterberg, J., Fromme, H. G., Jellinek, H., and Hauss, W. H. (1985). *Exp. Mol. Pathol.* **43,** 151–161.
Konomi, H., Hayashi, T., Nakayasu, K., and Arima, M. (1984). *Am. J. Pathol.* **116,** 417–426.
Krieg, T., Timpl, R., Alitalo, K., Kurkinen, M., and Vaheri, A. (1979). *FEBS Lett.* **104,** 405–410.
Kühn, K. (1982). *Collagen Relat. Res.* **2,** 61–80.
Kumamoto, C. A., and Fessler, J. H. (1980). *Proc. Natl. Acad. Sci. U.S.A.* **77,** 6434–6438.
Kumamoto, C. A., and Fessler, J. H. (1981). *J. Biol. Chem.* **256,** 7053–7058.
Kurita, K., Hashimoto, Y., Takei, T., Kawai, T., and Hayakawa, T. (1985). *J. Dent. Res.* **64,** 28–32.
Laurain, G., Delvincourt, T., and Szymanowicz, A. G. (1980). *FEBS Lett.* **120,**44–48.
Leushner, J. R. (1983). *Can. J. Biochem. Cell Biol.* **61,** 1012–1017.
Leushner, J. R., and Haust, M. D. (1984). *Atherosclerosis* **50,** 11–27.
Leushner, J. R., and Haust, M. D. (1985). *Can. J. Biochem. Cell Biol.* **63,** 1176–1182.
Limeback, H., Sodek, J., and Aubin, J. E. (1982). *Biochemistry* **21,** 4720–4729.
Limeback, H., Sodek, J., and Aubin, J. E. (1983). *J. Periodontal Res.* **18,** 242–248.
Linsenmayer, T. F., and Little, C. D. (1978). *Proc. Natl. Acad. Sci. U.S.A.* **75,** 3235–3239.
Linsenmayer, T. F., Fitch, J. M., Schmid, T. M., Zak, N. B., Gibney, E., Sanderson, R. D., and Mayne, R. (1983a). *J. Cell Biol.* **96,** 124–132.
Linsenmayer, T. F., Fitch, J. M., and Mayne, R. (1983b). *Prog. Clin. Biol. Res.* **110b,** 369–378.

Linsenmayer, T. F., Fitch, J. M., and Mayne, R. (1984). *Invest. Ophthalmol. Visual Sci.* **25,** 41–47.

Linsenmayer, T. F., Fitch, J. M., Gross, J., and Mayne, R. (1985). *Ann. N.Y. Acad. Sci.* **460,** 232–245.

Liotta, L. A., Lanzer, W. L., and Garbisa, S. (1981). *Biochem. Biophys. Res. Commun.* **98,** 184–190.

Liotta, L. A., Kalebic, T., Reese, C. A., and Mayne, R. (1982). *Biochem. Biophys. Res. Commun.* **104,** 500–506.

McCullagh, K. G., Duance, V. C., and Bishop, K. A. (1980). *J. Pathol.* **130,** 45–55.

Madri, J. A., and Furthmayr, H. (1979). *Am. J. Pathol.* **94,** 323–331.

Madri, J. A., and Furthmayr, H. (1980). *Hum. Pathol.* **11,** 353–366.

Madri, J. A., Foellmer, H. G., and Furthmayr, H. (1982). *Collagen Relat. Res.* **2,** 19–29.

Mainardi, C. L., Seyer, J. M., and Kang, A. H. (1980). *Biochem. Biophys. Res. Commun.* **97,** 1108–1115.

Mainardi, C. L., Hibbs, M. S., and Seyer, J. M. (1983). *Clin. Res.* **31,** 652A.

Mainardi, C. L., Hibbs, M. S., Hasty, K. A., and Seyer, J. M. (1984). *Collagen Relat. Res.* **4,** 479–492.

Mandell, M. S., and Sodek, J. (1982). *J. Biol. Chem.* **257,** 5268–5273.

Martinez-Hernandez, A., Gay, S., and Miller, E. J. (1982). *J. Cell Biol.* **92,** 343–349.

Maxwell, W. L., Duance, V. C., Lehto, M., Ashurst, D. E., and Berry, M. (1984). *Histochem. J.* **16,** 1215–1229.

Mayne, R., Vail, M. S., and Miller, E. J. (1978). *Biochemistry* **17,** 446–452.

Mayne, R., Elrod, B. W., Mayne, P. M., Sanderson, R. D., and Linsenmayer, T. F. (1984a). *Exp. Cell Res.* **151,** 171–182.

Mayne, R., Wiedemann, H., Irwin, M. H., Sanderson, R. D., Fitch, J. M., Linsenmayer, T. F., and Kühn, K. (1984b). *J. Cell Biol.* **98,** 1637–1644.

Miller, E. J., Gay, S., Haralson, M. A., Martinez-Hernandez, A., and Rhodes, R. K. (1982). *In* "New Trends in Basement Membrane Research" (K. Kühn, H. H. Schoene, and R. Timpl, eds.), pp. 99–106. Raven, New York.

Miller, E. J., Rhodes, R. K., and Furuto, D. K. (1983). *Collagen Relat. Res.* **3,** 79–87.

Miura, S., and Kimura, S. (1985). *J. Biol. Chem.* **260,** 15352–15356.

Modesti, A., Kalebic, T., Scarpa, S., Togo, S., Grotendorst, G., Liotta, L. A., and Triche, T. J. (1984). *Eur. J. Cell Biol.* **35,** 246–255.

Morris, N. P., Fessler, L. I., and Fessler, J. H. (1979). *J. Biol. Chem.* **254,** 11024–11032.

Morton, L. F., and Barnes, M. J. (1982). *Atherosclerosis* **42,** 41–51.

Morton, L. F., and Barnes, M. J. (1983). *Artery* **11,** 361–383.

Mumby, S. M., Raugi, G. J., and Bornstein, P. (1984). *J. Cell Biol.* **98,** 646–652.

Murata, K., Kudo, M., Onuma, F., and Motoyama, T. (1984). *Hepatogastroenterology* **31,** 158–161.

Murphy, G., Cawston, T. E., Galloway, W. A., Barnes, M. J., Bunning, R. A. D., Mercer, E., Reynolds, J. J., and Burgeson, R. E. (1981). *Biochem. J.* **199,** 807–811.

Murphy, G., Reynolds, J. J., Bretz, U., and Baggiolini, M. (1982). *Biochem. J.* **203,** 209–221.

Murphy, G., McAlpine, C. G., Poll, C. T., and Reynolds, J. J. (1985). *Biochim. Biophys. Acta* **831,** 49–58.

Myers, J. C., Loidl, H. R., Stolle, C. A., and Seyer, J. M. (1985a). *J. Biol. Chem.* **260,** 5533–5541.

Myers, J. C., Loidl, H. R., Seyer, J. M., and Dion, A. S. (1985b). *J. Biol. Chem.* **260,** 11216–11222.

Nakayasu, K., Tanaka, M., Konomi, H., and Hayashi, T. (1986). *Ophthalmic Res.* **18,** 1–10.

Narayanan, A. S., and Page, R. C. (1983). *J. Biol. Chem.* **258,** 11694–11699.

Narayanan, A. S., and Page, R. C. (1985). *Collagen Relat. Res.* **5,** 297–304.

Narayanan, A. S., Engel, L. D., and Page, R. C. (1983). *Collagen Relat. Res.* **3,** 323–334.

Newsome, D. A., Gross, J., and Hassell, J. R. (1982). *Invest. Ophthalmol. Visual Sci.* **22,** 376–381.

Niemela, O., Risteli, L., Parkkinen, J., and Risteli, J. (1985). *Biochem. J.* **232,** 145–150.

Niyibizi, C., Fietzek, P. P., and van der Rest, M. (1984). *J. Biol. Chem.* **259,** 14170–14174.

Nusgens, B., Delain, D., Senechal, H., Winand, R., Lapierre, C. M., and Wahrmann, J. P. (1986). *Exp. Cell Res.* **162,** 51–62.

Ooshima, A. (1981). *Science* **213,** 666–668.

Peltonen, J., Marttala, T., Vihersaari, T., Renvall, S., and Penttinen, R. (1981). *Acta Neuropathol.* **55,** 183–187.

Peltonen, J., Foidart, J. M., and Aho, H. J. (1984). *Virchows Arch. B* **47,** 291–301.

Pöschl, A., and von der Mark, K. (1980). *FEBS Lett.* **115,** 100–104.

Prunieras, M., Regnier, M., Fougere, S., and Woodley, D. (1983). *J. Invest. Dermatol.* **81s,** 74–81.

Reese, C. A., and Mayne, R. (1981). *Biochemistry* **20,** 5443–5448.

Reiser, K. M., and Last, J. A. (1983). *J. Biol. Chem.* **258,** 269–275.

Rhodes, R. K., and Miller, E. J. (1978). *Biochemistry* **17,** 3442–3448.

Rhodes, R. K., and Miller, E. J. (1979). *J. Biol. Chem.* **254,** 12084–12087.

Rhodes, R. K., and Miller, E. J. (1981). *Collagen Relat. Res.* **1,** 337–343.

Rhodes, R. K., Gibson, K. D., and Miller, E. J. (1981). *Biochemistry* **20,** 3117–3121.

Rich, E. A., Seyer, J. M., Kang, A. H., and Mainardi, C. L. (1983). *Am. Rev. Respir. Dis.* **128,** 166–169.

Rojkind, M., Rojkind, M. H., and Cordero-Hernandez, J. (1983). *Collagen Relat. Res.* **3,** 335–347.

Roll, F. J., Madri, J. A., Albert, J., and Furthmayr, H. (1980). *J. Cell Biol.* **85,** 597–616.

Sage, H. (1982). *J. Invest. Dermatol.* **79,** 51–59.

Sage, H., and Bornstein, P. (1979). *Biochemistry* **18,** 3815–3822.

Sage, H., Crouch, E., and Bornstein, P. (1979). *Biochemistry* **18,** 5433–5442.

Sage, H., Pritzl, P., and Bornstein, P. (1981a). *Collagen Relat. Res.* **1,** 3–15.

Sage, H., Pritzl, P., and Bornstein, P. (1981b). *Biochemistry* **20,** 3778–3784.

Sage, H., Pritzl, P., and Bornstein, P. (1981c). *Arteriosclerosis* **1,** 427–442.

Salonen, V., Lehto, M., Vaheri, A., Aro, H., and Peltonen, J. (1985a). *Acta Neuropathol.* **67,** 315–321.

Salonen, V., Lehto, M., Kalimo, M., Penttinen, R., and Aro, H. (1985b). *Muscle Nerve* **8,** 125–131.

Sanes, J. R., and Cheney, J. M. (1982). *J. Cell Biol.* **93,** 442–451.

Sankey, E. A., and Barnes, M. J. (1984). *Biochem. J.* **218,** 11–18.

Sankey, E. A., Brown, F. E., Morton, L. F., Scott, D. M., and Barnes, M. J. (1981). *Biochem. J.* **198,** 707–710.

Sasse, J., von der Mark, H., Kühl, U., Dessau, W., and von der Mark, K. (1981). *Dev. Biol.* **83,** 79–89.

Schuppan, D., Becker, J., Boehm, H., and Hahn, E. G. (1986). *Cell Tissue Res.* **243,** 535–543.

Seltzer, J. L., Adams, S. A., Grant, G. A., and Eizen, A. Z. (1981). *J. Biol. Chem.* **256,** 4662–4668.

Shapiro, F. D., and Eyre, D. R. (1982). *J. Natl. Cancer Inst. (U.S.)* **69,** 1009–1016.

Shekhonin, B. V., Domogatsky, S. P., Muzykantov, V. R., Idelson, G. L., and Rukosuev, V. S. (1985). *Collagen. Relat. Res.* **5,** 355–368.

Shellswell, G. B., Restall, D. J., Duance, V. C., and Bailey, A. J. (1979). *FEBS Lett.* **106,** 305–308.

Shellswell, G. B., Bailey, A. J., Duance, V. C., and Restall, D. J. (1980). *J. Embryol. Exp. Morphol.* **60,** 245–254.

Shinkai, H. (1984). *J. Dermatol.* **11,** 57–62.

Silver, M. H., Murray, J. C., and Pratt, R. M. (1984). *Differentiation* **27,** 205–208.

Smith, L. T., Holbrook, K. A., and Madri, J. A. (1986). *Am. J. Anat.* **175,** 507–521.

Sodek, J., and Mandell, S. M. (1982). *Biochemistry* **21,** 2011–2015.

Sodek, J., Feng, J., Yen, E. H. K., and Melcher, A. H. (1982). *Calcif. Tissue Int.* **34,** 408–415.

Sopata, I., and Dancewicz, A. M. (1974). *Biochim. Biophys. Acta* **370,** 510–523.

Stenn, K. S., Madri, J. A., and Roll, F. J. (1979). *Nature (London)* **277,** 229–232.

Stephens, H. R., Duance, V. C., Dunn, M. J., Bailey, A. J., and Dubovitz, V. (1982). *J. Neurol. Sci.* **53,** 45–62.

Tanaka, S., Hata, R., and Nagai, Y. (1981). *Collagen Relat. Res.* **1,** 445–452.

Trelstad, R. L., Lawley, K. R., Hayashi, K., Ehrlich, H. P., and Silver, F. H. (1981). *Collagen Relat. Res.* **1,** 39–52.

Tseng, S. C. G., Savion, N., Gospodarowicz, D., and Stern, R. (1981). *J. Biol. Chem.* **256,** 3361–3365.

Tseng, S. C. G., Smucker, D., and Stern, R. (1982a). *J. Biol. Chem.* **257,** 2627–2633.

Tseng, S. C. G., Savion, N., Stern, R., and Gospodarowicz, D. (1982b). *Eur. J. Biochem.* **122,** 355–360.

Tseng, S. C. G., Smuckler, E. A., and Stern, R. (1983). *Hepatology* **3,** 955–963.

Van Cuawenberge, D., Pierard, G. E., Foidart, J. M., and Lapiere, C. M. (1983). *Br. J. Dermatol.* **108,** 163–170.

van der Rest, M., Niyibizi, C., and Fietzek, P. P. (1985). *Ann. N.Y. Acad. Sci.* **460,** 517–519.

von der Mark, K., and Öcalan, M. (1982). *Collagen Relat. Res.* **2,** 541–555.

von der Mark, H., and von der Mark, K. (1979). *FEBS Lett.* **99,** 101–105.

Warburton, J. J., Monaghan, P., Ferns, S. A., Hughes, C. M., and Rudland, P. S. (1983). *J. Histochem. Cytochem.* **31,** 1265–1273.

Weber, L., Kirsch, E., Müller, P., and Krieg, T. (1984). *J. Invest. Dermatol.* **82,** 156–160.

Welsh, C., Gay, S., Rhodes, R. K., Pfister, R., and Miller, E. J. (1980). *Biochim. Biophys. Acta* **625,** 78–88.

Woodley, D., Sauder, D., Talley, M. J., Silver, M., Grotendorst, G., and Qwarnstrom, E. (1983). *J. Invest. Dermatol.* **81,** 149–153.

Yamagata, S., and Yamagata, T. (1984). *J. Biochem. (Tokyo)* **96,** 17–26.

Yamamoto, M., Sumiyoshi, H., Nakagami, K., and Tahara, E. (1984). *Acta Pathol. Jpn.* **34,** 77–86.

Yue, B. Y. J. T., Baum, J. L., and Smith, B. D. (1979). *Biochem. Biophys. Res. Commun.* **86,** 465–472.

Yue, B. Y., Baum, J. L., and Smith, B. D. (1983). *Biochim. Biophys. Acta* **755,** 318–325.

Type VI Collagen

Rupert Timpl

Max-Planck-Institut für Biochemie, D-8033 Martinsried bei München, Federal Republic of Germany

and

Jürgen Engel

Biozentrum der Universität, Abteilung Biophysikalische Chemie, CH-4056 Basel, Switzerland

I. Introduction

A new protein now known to be a fragment of collagen VI (Engel *et al.*, 1985; Rauterberg *et al.*, 1986) was first described about 10 years ago. It soon became clear that collagen VI possesses several unique properties which distinguish this protein from any other type of collagen known so far. One unusual characteristic is the rather short triple-helical domain which accounts for less than half of the total mass of the protein. Another unique feature is the assembly of collagen VI monomers into well-defined oligomers which are the building blocks of microfibrils found in tissues and cell cultures and are now the best-characterized microfibrillar structures existing in the extracellular matrix (for a review on microfibrils, see Cleary and Gibson, 1983). The present article will review the data which led to the understanding of the structure and macromolecular organization of collagen VI and, in addition, the data related to its biosynthesis, immunology, and tissue distribution.

Lines of Discovery

Chung *et al.* (1976) were the first to report on the isolation of a large, disulfide-bonded collagen complex from a pepsin digest of the intimal layer of human blood vessels. This material had a unique amino acid composition, was rich in carbohydrate, and could be split into peptide

STRUCTURE AND FUNCTION OF
COLLAGEN TYPES

chains with a M_r of about 50,000 by reduction. Similar or identical protein complexes were subsequently prepared from cirrhotic liver (Rojkind *et al.,* 1979), human and bovine placenta (Furuto and Miller, 1980; Risteli *et al.,* 1980; Jander *et al.,* 1981), calf skin (Laurain *et al.,* 1980), and uterus (Abedin *et al.,* 1982). This material was originally referred to as high-molecular-weight aggregates, short chain collagen, or intima collagen. Peptide map and N-terminal amino acid sequence analyses (Jander *et al.,* 1983) as well as electron microscopy (Furthmayr *et al.,* 1983) demonstrated no relationship of the protein to the collagens I–V and led to the proposal to name this new protein collagen type VI.

It is now known that pepsin solubilization of collagen VI causes the loss of about 50% of its original mass. Larger and presumably intact forms of collagen VI chains were also studied at the same time, but their relationship to pepsin-solubilized collagen VI was not immediately recognized. Sear *et al.* (1978, 1981a) had characterized a M_r = 150,000 polypeptide chain (termed MFP I), from nuchal ligament cell cultures, which was considered to be a subunit of the microfibrils associated with elastin. A similar, disulfide-bonded polypeptide (GP-140) with a M_r of 140,000 was identified in the detergent-insoluble matrix of cultured skin and lung fibroblasts (Carter and Hakomori, 1981; Carter, 1982a). Both polypeptides contained significant amounts of hydroxyproline and hydroxylysine, indicating that they originated from a collagenous protein. Glycoproteins of similar size and composition were also found in urea or guanidine–HCl extracts of muscle (Marton and Arnason, 1982) and aorta and ligamentum nuchae (Gibson and Cleary, 1982) and were, after reduction of disulfide bonds, obtained in purified form. A possible relationship between these collagen-like glycoproteins and pepsin-solubilized collagen VI was indicated because of certain similarities in compositional features (Gibson and Cleary, 1982; Cleary and Gibson, 1983).

Convincing evidence for this relationship came both from immunological studies of tissue and cell culture-derived glycoprotein using antibodies against either pepsin-solubilized collagen VI (Jander *et al.,* 1984; von der Mark *et al.,* 1984; Hessle and Engvall, 1984; Trüeb and Bornstein, 1984) or the larger, reduced polypeptide chains (Heller-Harrison and Carter, 1984; Gibson and Cleary, 1985), and by studies of the degradation of the intact form with pepsin (Knight *et al.,* 1984; Ayad *et al.,* 1985). These findings were strongly supported by electron microscopy of rotary shadowed specimens of nondegraded collagen VI (von der Mark *et al.,* 1984; Jander *et al.,* 1984) which demonstrated the same monomer–oligomer structures previously observed in preparations of pepsin-solubilized collagen VI (Furthmayr *et al.*, 1983). In the following sections we will refer to the material consisting of M_r 110,000–140,000 polypeptides as intact collagen VI. There are a few more reports on similarly sized polypeptides from lung tissue (Lafuma *et al.*, 1982) and cultured fibroblasts (Lehto *et al.*, 1981; Lehto, 1983), but the definite establishment as collagen VI structures or fibronectin receptors (see Section III) is less clear.

II. Protein Chemistry and Macromolecular Organization

A. *Sources of Collagen VI*

Pepsin digests prepared from various tissues in 0.1–0.5 *M* acetic or formic acid are still the most convenient source for obtaining collagen VI (although not the intact molecule). Most purification procedures then use differential precipitation with increasing NaCl concentrations to separate collagen VI from other collagens (Furuto and Miller, 1980). Most of the collagen VI, precipitating between 1.2 and 1.8 *M* NaCl, is redissolved in 0.1 *M* acetic acid or 2 *M* guanidine HCl, and precipitated again by dialysis against 0.02 *M* Na_2HPO_4. The latter step removes efficiently the more soluble 7 S collagen which is a cross-linked fragment of collagen IV (Risteli *et al.*, 1980; Abedin *et al.*, 1982). Further purification by chromatography on carboxymethylcellulose (CM-cellulose) or a molecular sieve column may be necessary to obtain a more homogenous preparation of pepsin-solubilized collagen VI (Furuto and Miller, 1980; Jander *et al.*, 1981; Odermatt *et al.*, 1983).

The complete purification of intact collagen VI has been only accomplished with its constituent polypeptide chains (Gibson and Cleary, 1982, 1985; Marton and Arnason, 1982; Carter, 1982a; Jander *et al.*, 1984; Knight *et al.*, 1984). Here, tissues or cell cultures are extracted with 6 *M* guanidine HCl or 6–8 *M* urea in the absence or presence of a reducing agent, and the extracted proteins are subsequently reduced and/or alkylated. Further purification steps usually include ion exchange and molecular sieve chromatography or preparative gel electrophoresis. Carbohydrate affinity chromatography on phenylboronate–agarose was also used with success (Gibson and Cleary, 1985). The purification of intact collagen VI in a nonreduced triple-helical form has been attempted several times (Carter, 1982a; Jander *et al.*, 1984; von der Mark *et al.*, 1984; Knight *et al.*, 1984; Gibson and Cleary, 1985), but the final products were not pure as judged by electrophoresis. They were nevertheless useful in electron microscopic and biological studies. Electrophoretically pure material was recently obtained from bovine uterus after molecular sieve chromatography in SDS (Trüeb and Winterhalter, 1986).

B. *Collagen VI Monomers and Constituent Chains*

Intact collagen VI monomers appear in electron micrographs after rotary shadowing as dumbbell-shaped molecules with two globular domains of about equal size connected by a 105-nm-long triple-helical rod (Figs. 1 and 2). These monomers are rarely seen in tissue extracts (von der Mark *et al.*, 1984) but are efficiently generated from oligomers by reduction of disulfide bonds under nondenaturing conditions (Jander *et al.*, 1984). Similar structures have been produced from pepsin-solubilized collagen VI by partial reduction (Furthmayr *et al.*, 1983), but these monomers appear to have lost almost completely one

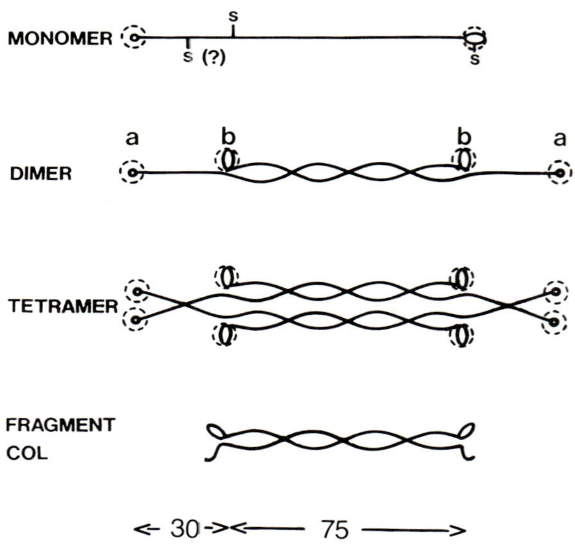

FIG. 1. Models of collagen VI monomers, dimers, and tetramers and of a dimeric fragment COL produced by bacterial collagenase. The relative size of the globular domains in intact collagen VI are denoted by dotted circles. They are degraded by pepsin to smaller structures indicated by full lines. Oligomer formation allows the distinction between outer (a) and inner (b) globular domains. The triple-helical domain (represented by a straight or twisted full line) was found to be the same in intact and pepsin-solubilized collagen VI. The relative positions of cysteine residues essential for intermolecular cross-linking are denoted by S. Dimensions between arrows are given in nanometers.

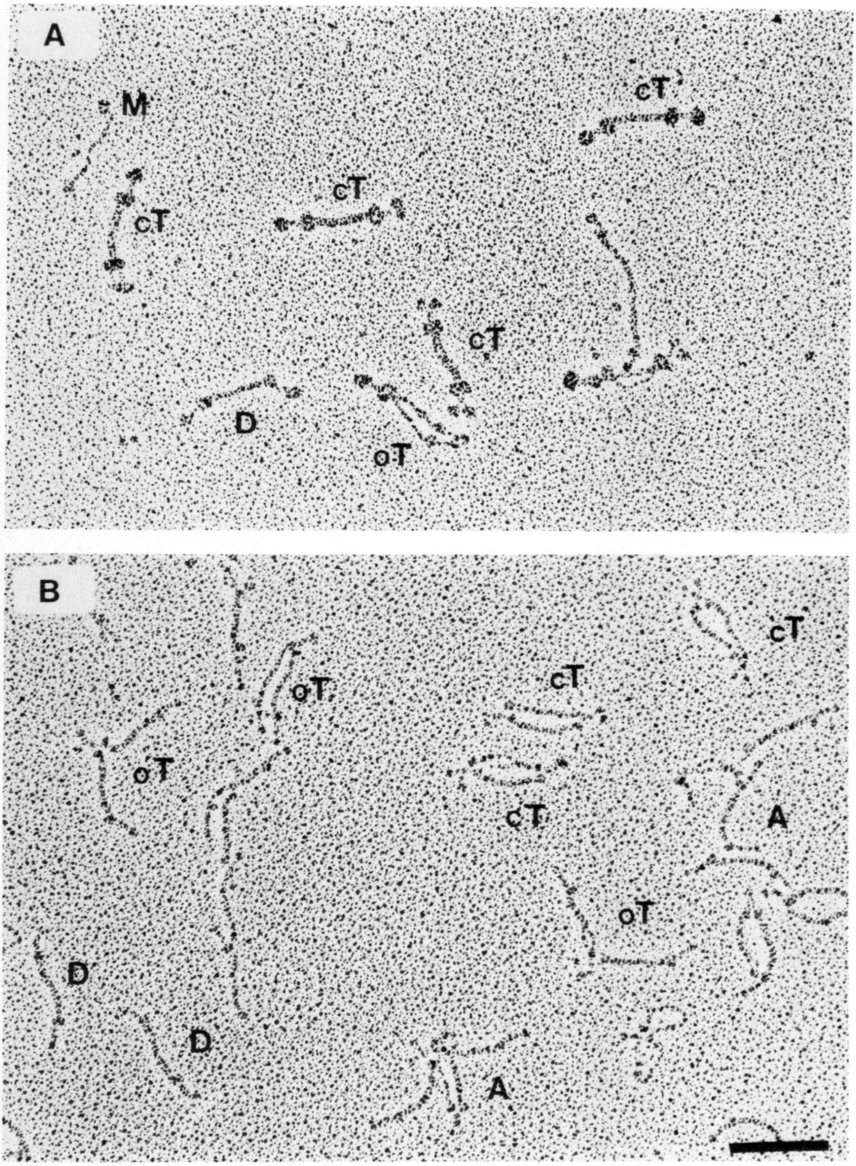

FIG. 2. Rotary-shadowing images of intact collagen VI (A) obtained from a NaCl extract of neurofibroma and of pepsin-solubilized collagen VI (B) obtained from human placenta. Selected monomers (M), dimers (D), open (oT) and closed (cT) tetramers and larger aggregates (A) are indicated for better identification. Bar, 100 nm. Reproduced with permission from Engel *et al.* (1985). The original studies were described by von der Mark *et al.* (1984) and Furthmayr *et al.* (1983).

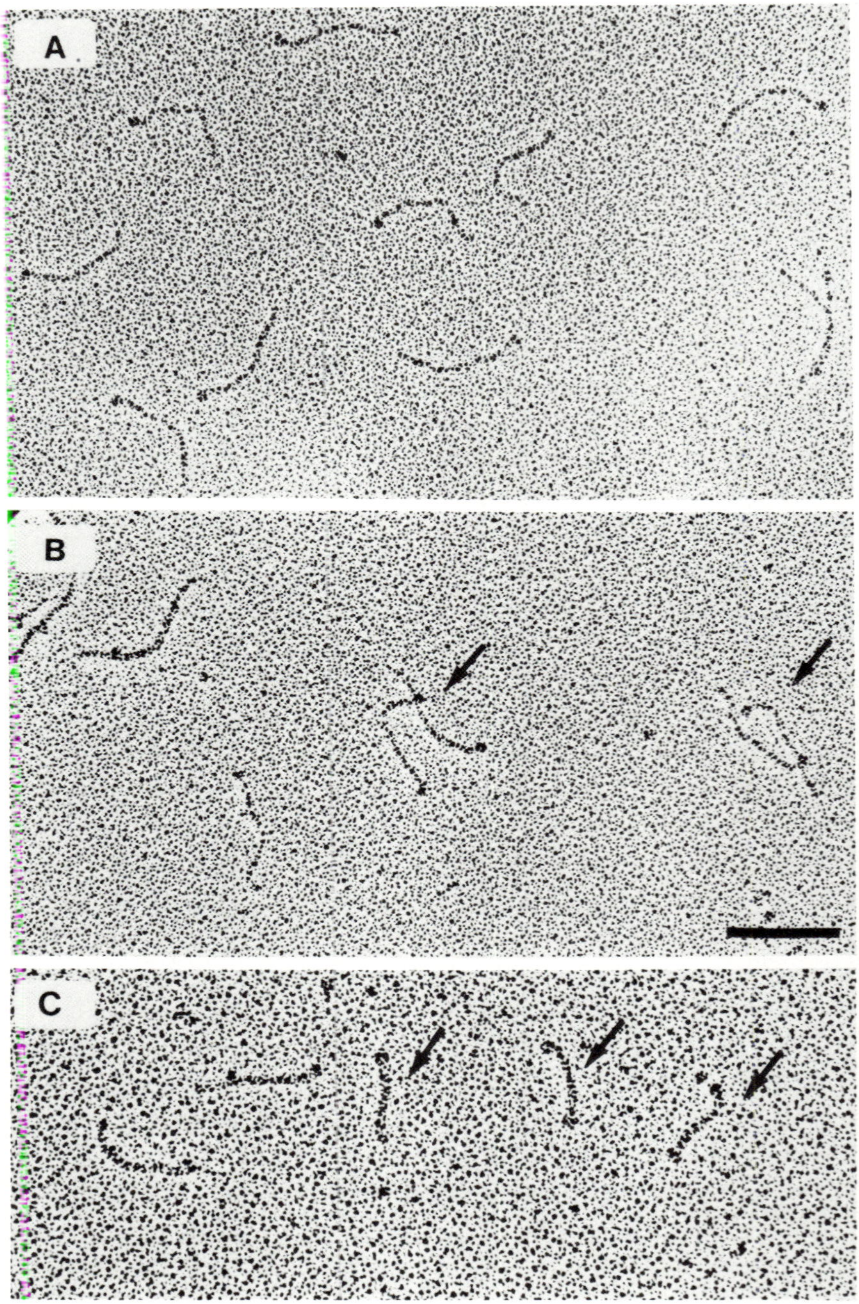
A
B
C

globular domain and significant portions of the other (Fig. 3). The molecular mass of intact collagen VI monomers has not yet been established by ultracentrifugation, but estimates based on the size of its chains (see below) indicates a M_r in the range 330,000–420,000. Values of 340,000 and 380,000 have been reported based on domain analysis (Aumailley *et al.*, 1985) and electrophoretic mobility (Gibson and Cleary, 1985) of partially reduced collagen VI monomers. Monomers obtained from pepsin-solubilized collagen VI when reduced show a M_r of 170,000 by ultracentrifugation (Furthmayr *et al.*, 1983). Together with data on the conformation of the protein and chain size (see below), it appears that collagen VI, like other collagens, is a triple-stranded molecule in which each constituent chain very likely spans the length of the entire molecule in a single direction.

Total reduction of disulfide bonds in intact collagen VI followed by denaturation produces unfolded polypeptide chains which migrate as a single band or occasionally as a close doublet of bands in SDS electrophoresis. These polypeptides are now referred to as α(VI) chains and show, after purification, a unique amino acid composition (Table I) characterized by 15–30 residues per 1000 of 4-hydroxyproline, hydroxylysine, tyrosine, and half cystine, and by a glycine content distinctly less than 330 residues per 1000. Molecular mass estimates were made from electrophoresis runs and consistently showed M_r values of 110,000 (using collagenous proteins) and 140,000 (using globular proteins as standards) (Gibson and Cleary, 1982; Carter, 1982a; Knight *et al.*, 1984; Jander *et al.*, 1984; von der Mark *et al.*, 1984). The presence of two or three distinct α(VI) chains in these materials can be anticipated from studies with pepsin-solubilized collagen VI, but their separation has not yet been reported. The splitting of the M_r 140,000 band observed in immunoblotting (Trüeb and Bornstein, 1984) was interpreted as indicating the separation of α1(VI) and α3(VI) chains.

More recent studies provided, however, convincing evidence that collagen VI from uterus consists of stoichiometric amounts of α1(VI) and α2(VI) chains with a M_r of 140,000, and a distinctly larger α3(VI) chain with a M_r of 200,000 (Trüeb and Winterhalter, 1986). These chains differ in amino acid composition and patterns of collagenase-

---

FIG. 3. Molecular images of collagen VI monomers (A), half-open dimers (B), and fragment COL (C) as visualized by rotary shadowing. Representative particles in B and C are indicated by arrows. The materials were prepared from pepsin-solubilized collagen VI of human placenta either by partial reduction under nondenaturing conditions (A and B) or by treatment with bacterial collagenase (C) (Odermatt *et al.*, 1983; Furthmayr *et al.*, 1983). Bar, 100 nm. Reproduced with permission from Engel *et al.* (1985).

TABLE I

AMINO ACID COMPOSITION OF α(VI) CHAINS OBTAINED FROM INTACT AND PEPSIN-SOLUBILIZED COLLAGEN VI[a]

	Intact α(VI) chains				Pepsin-treated α(VI) chains			
	Aorta[b]	Nuchal ligament[b]	Muscle[c]	Fibroblasts[d]	Total[e]	α1(VI)p[f]	α2(VI)p[f]	α3(VI)p[f]
4-Hydroxyproline	24	22	19	14	65	71	69	45
Aspartic acid	102	98	99	103	78	84	74	81
Threonine	37	36	46	70	16	15	14	20
Serine	45	40	109	71	27	27	28	30
Glutamic acid	117	105	144	116	104	117	88	91
Proline	64	64	67	69	92	82	97	91
Glycine	179	194	158	134	318	307	325	331
Alanine	69	67	75	59	41	52	43	34
Half cystine	18	16	8	25	20	11	6	15
Valine	57	65	38	60	21	22	15	22
Methionine	9	10	14	11	8	7	5	3
Isoleucine	35	38	26	44	20	19	19	32
Leucine	61	62	52	79	24	23	27	26
Tyrosine	19	18	15	26	18	17	13	15
Phenylalanine	30	34	27	32	15	24	14	14
Histidine	14	13	17	16	3	4	2	2
Hydroxylysine	25	23	11	13	48	50	74	45
Lysine	38	35	25	34	18	11	14	22
Arginine	56	59	45	24	64	57	73	81

[a]Expressed as residues per 1000. Intact α1(VI) chains with a M_r of 140,000.
[b]From Gibson and Cleary (1982).
[c]From Marton and Arnason (1982).
[d]From Carter (1982a).
[e]From vessel intima (Chung *et al.*, 1976).
[f]Triple-helical domains after two treatments with pepsin (Jander *et al.*, 1983).

resistant peptides. Similar results were also obtained for collagen VI from human lung (H. von der Mark, unpublished). The large form of the α3(VI) chain may have escaped detection in previous studies due to it lability against proteolysis or to the failure of specific antibody reactions in immunoblotting. The presence of a M_r 200,000 component has been, in fact, noted in earlier studies of intact collagen VI (Jander *et al.*, 1984; Gibson and Cleary, 1985) but was not interpreted at this time to represent a constituent chain.

More complex chain patterns with two to more than four electrophoretic bands with M_r of 40,000–70,000 were observed after total reduction of pepsin-solubilized collagen VI (Furuto and Miller, 1980; Rojkind *et al.*, 1979; Laurain *et al.*, 1980; Abedin *et al.*, 1982; Jander *et al.*, 1981, 1983; Odermatt *et al.*, 1983; Trüeb and Bornstein, 1984). Several of the major chain fragments were purified, indicating the presence of a single acidic chain and two more basic chains in the molecule. These chain fragments were named SC1 (acidic) and SC2 and SC3 (basic); these were equated with chain fragments originating from α1(VI), α2(VI) and α3(VI), respectively (Jander *et al.*, 1981, 1983). Since the SC fragments can usually be separated by SDS electrophoresis, it indicates that the individual α(VI) chains are degraded to a different degree by pepsin. Further limited heterogeneity may arise by nonuniform cleavage within single chains (Furuto and Miller, 1980; Odermatt *et al.*, 1983) and can make the interpretation of chain patterns difficult. Ultracentrifugal and molecular sieve analyses indicated, however, that the major chain fragments have a molecular mass in the range 43,000–53,000 (Furuto and Miller, 1980; Odermatt *et al.*, 1983). Their composition is characterized by a strong increase in glycine, hydroxyproline, and hydroxylysine content (cf. Table I) which indicates, in agreement with electron microscopic data, extensive but not complete loss of globular domain segments. Reduction of pepsin-solubilized collagen VI produces a small amount (5–10%) of other peptides which are smaller than the major chain fragments and show a noncollagenous composition. Their origin from the globular domains or from other proteins is not clear.

The complete degradation to the triple helical domain of collagen VI requires reduction of disulfide bonds under nondenaturing conditions followed by a second pepsin treatment (Jander *et al.*, 1981; 1983; Odermatt *et al.*, 1983). This procedure produced some further shortening of the chain fragments which now appear more similar in electrophoresis. The data also indicated a M_r of about 120,000 for the triple helix, a value compatible with its length of 105 nm observed by electron microscopy (Furthmayr *et al.*, 1983; Rauterberg *et al.*, 1986).

α1(VI)p:	Val-Xaa-Xaa-Arg-Phe-Glu-Gly-Gln-Pro-$^{\text{Gly}}_{\text{Ala}}$-Leu-Gly-Pro
α2(VI)p:	Leu-Glu-Ile-Pro-Gly-Pro-Arg-Gly-Pro-Xaa-Gly-Tyr-Xaa-Gly
α3(VI)p:	Ser-Ala-Ile-Ala-Gly-Val-Ala-Gly-Val-Gly

FIG. 4. Amino-terminal sequences of homologous fragments obtained from three different α(VI) chains of collagen VI which constitute the triple-helical domain (Jander *et al*, 1983). The fragments were prepared from pepsin digest of human placenta which were then partially reduced to allow a second, limited degradation by pepsin to obtain α(VI)p chains. Xaa denotes unidentified residues.

Three different chain fragments could be separated after denaturation of the triple-helical domain [hereafter referred to as α1(VI)p, α2(VI)p, and α3(VI)p] which all showed a collagenous amino acid composition (Table I) but differed distinctly in their N-terminal amino acid sequences (Fig. 4) and in peptide maps produced with cyanogen bromide or proteases (Jander *et al.*, 1983). The nearly equal ratio between the three different α(VI)p chains was interpreted to indicate a heterotrimer composition, α1α2α3, for intact collagen VI (Jander *et al.*, 1983). Direct proof for this composition, i.e., by chain cross-linking experiments, is still lacking.

Amino acid sequence data are so far only available for short N-terminal segments (see Fig. 4) of α(VI)p chains from human and bovine placenta (Jander *et al.*, 1983) and demonstrated the presence of Gly-X-Y repeats which are characteristic for the collagen triple helix. Four cyanogen bromide peptides (27–194 amino acid residues) have been isolated from the acidic chain of pepsin-solubilized collagen VI, and all were found to contain less than one-third glycine (Furuto and Miller, 1980). This could indicate the presence of short interruptions of the Gly-X-Y repeats in the triple helix of collagen VI similar to those found in collagens IV and IX. Sequence data for cDNA corresponding to mRNA of collagen VI are not yet available, but work is in progress in several laboratories.

Physical properties of the triple helix were studied with pepsin-solubilized collagen VI which contains up to 60–80% collagenous sequences. This material shows a strong negative optical rotation [$(\alpha)_{313} = -1130°$] (Furuto and Miller, 1981) and characteristic signals in its CD spectrum ($\theta_{200nm} = -43{,}000$ and $\theta_{222nm} = 3700° \cdot cm^2/dmol$) (Odermatt *et al.*, 1983) which are comparable in their magnitude to those of other collagenous proteins. Monomers prepared from this fragment by reduction under nondenaturing conditions show a sharp melting profile of the triple-helix with T_m = 32–35°C (Jander *et al.*, 1981; Furuto and Miller, 1981; Odermatt *et al.*, 1983). These observations, together with electron microscopic measurements (Hofmann *et al.*, 1984), demonstrated that collagen VI possesses a typical triple helix with a comparable rigidity (persistence length) to those of interstitial collagens. Based on reduction data, Furuto and Miller (1981) proposed a different structure where only two constituent chains were involved in triple-helical folding while the third chain was associated in a random-coiled fashion. This model could not explain the high protease resistance found for pepsin-solubilized collagen VI and was not supported by subsequent studies (Odermatt *et al.*, 1983; Jander *et al.*, 1983).

When compared to the triple-helical domain less is known about the structure of the globular domains of collagen VI. A M_r of 150,000 was estimated for each globule based on electron microscopic comparison with other globular structures (Engel *et al.*, 1985) which is in reasonable agreement with mass estimates based on collagenase-resistant fragments (Aumailley *et al.*, 1985). The globular domains are very likely to be different as indicated by differences in pepsin resistance (see Fig. 3) and topological orientation in collagen VI dimers (inner and outer globules) (Section II,C). It is also very likely that these globular domains are folded from either C- or N-terminal segments within the α(VI) chains, but this assignment has not yet been made. Controversial data exist about the size and number of noncollagenous segments resistant to bacterial collagenase which apparently constitute the globular domains of intact collagen VI. Studies with fibroblast-derived material demonstrated either several bands with a M_r of 30,000–50,000 (Sear *et al.*, 1981a,b; Aumailley *et al.*, 1985), a particularly large fragment of M_r about 120,000 (Trüeb and Bornstein, 1984), or failure to detect large fragments (Heller-Harrison and Carter, 1984). The latter two results are not compatible with the general structural model and may be caused by either incomplete cleavage due to incomplete reduction (Aumailley *et al.*, 1985) or the failure to detect such fragments by immunoblotting.

Three different fragments were consistently found in collagenase digests of reduced α(VI) chain (M_r = 140,000 species) prepared from tissues with M_rs of about 35,000, 55,000, and 70,000 (Marton and Arnason, 1982; Jander *et al.*, 1984; Gibson and Cleary, 1985). These fragments were separated and found to lack hydroxyproline and to differ in amino acid composition and cyanogen bromide patterns (Gibson and Cleary, 1985). It is not clear whether they represent homogeneous peptide segments since some six different peptides would be expected to arise after collagenase cleavage of the three α(VI) chains.

They appear to originate from α1(VI) and α2(VI) chains, while a M_r 160,000 fragment has been reported for the α3(VI) chain (Trüeb and Winterhalter, 1986). A heterogeneous mixture of noncollagenous peptides was also found after collagenase cleavage of pepsin-solubilized collagen VI (Odermatt *et al.*, 1983) which were all smaller than those obtained from the intact protein chains.

Collagen VI is characterized by an unusually high number (20–30) of cysteine residues which may be present in each α(VI) chain (Table I). Only 2–3 of these residues are presumably utilized to form intermolecular disulfide bonds to stabilize the oligomer forms (Section II,C). Most are, therefore, very likely used for intramolecular or intrachain bonds or exist with free sulfhydryls. The presence of intramolecular bonds is indicated by the high conformational stability of collagen VI against denaturing agents as revealed by electron microscopy (Jander *et al.*, 1984; von der Mark *et al.*, 1984) and by protease resistance (see below). Intramolecular bonds may be crucially involved in the stabilization of globular domains since they disappear after complete reduction under nondenaturing conditions of disulfide bonds (Furthmayr *et al.*, 1983). Intrachain disulfide bonds very likely exist at least in the triple-helical domain since α(VI)p chains contain 3–6 cysteine residues (Jander *et al.*, 1983; see Table I). These residues are not regularly distributed as indicated by cleavage with 2-nitro-5-thiocyanobenzoic acid (Trüeb and Bornstein, 1984).

The high degree of disulfide-bonding found in collagen VI endows the protein with a high resistance to proteolytic attack. When extracted from tissues in intact form with denaturing agents but without reduction, collagen VI resists digestion by bacterial collagenase but not digestion by pepsin. Reduction under nondenaturing or denaturing conditions resulted in a considerable increase in protease sensitivity (Gibson and Cleary, 1982, 1985; Jander *et al.*, 1984). Collagen VI isolated from cell cultures of both medium and matrix forms was also stable against neutral proteases such as trypsin and thrombin and was a poor substrate for collagenase (Carter and Hakomori, 1981; Carter, 1982b; Heller-Harrison and Carter, 1984; Bruns, 1984; Aumailley *et al.*, 1985). A distinct cleavage by collagenase was, however, reported for nuchal ligament collagen VI (Sear *et al.*, 1981a; Knight *et al.*, 1984). Protease stability is also a characteristic feature of pepsin-solubilized collagen VI which, after isolation, cannot be cleaved further by additional exposure to pepsin, various neutral proteases, or bacterial collagenase (Furuto and Miller, 1981; Odermatt *et al.*, 1983). Here again, reduction of disulfide bonds increases protease sensitivity (see above). No reports exist so far on the cleavage of collagen VI by known tissue collagenases which are able to split a single peptide bond in either collagens I–III, IV, or V.

Collagen VI is also subjected to several other posttranslational modifications besides disulfide bond formation, such as hydroxylation of 30–50% of its proline residues within the triple-helical segment to form 4-hydroxyproline and of more than 70–90% of its lysine residues to form hydroxylysine (see Table I). More than 90% of the hydroxylysine residues are then substituted with glucosyl–galactosyl or, to a lesser degree, with galactosyl groups (Chung *et al.,* 1976; Furuto and Miller, 1981; Jander *et al.,* 1981, 1983). These studies also demonstrated substantial amounts of glucosamine but no galactosamine in pepsin-solubilized collagen VI, indicating the presence of N-linked oligosaccharide even within the triple-helical segment of the protein. From the reported data it can be calculated that pepsin-solubilized collagen VI contains 20–25% carbohydrate. Lower values, however, have been reported for intact α(VI) chains isolated from fibroblasts (2.7%; Carter, 1982a) or from nuchal ligament (6.1%; Knight *et al.,* 1984). The latter study demonstrated the presence of galactose, glucose, mannose, glucosamine, and fucose in a ratio of 33:24:25:13:5 but only trace amounts of sialic acids. As expected from these data, α(VI) chains usually show strong staining with the periodic acid–Schiff reagent and also react with lectins including concanavalin A, wheat germ agglutinin and ricin (Carter, 1982b; Gibson and Cleary, 1982; Marton and Arnason, 1982). These reactions have been frequently used to monitor purification and fragmentation of collagen VI. Carbohydrate incorporation was also demonstrated in biosynthetic and cell surface labeling studies (see Section III).

C. *Oligomeric Forms of Collagen VI*

The most unique feature of collagen VI is its assembly into well-defined dimers and tetramers which, in varying proportions, are the major particles in preparations of pepsin-solubilized collagen VI from human placenta visualized by rotary shadowing and negative staining (Fig. 2). Molecular masses and sedimentation coefficients calculated for models of these oligomers (Fig. 1) are in good agreement with ultracentrifugal data demonstrating that the dimers and tetramers seen by electron microscopy are indeed the most prominent components in these preparations (Furthmayr *et al.,* 1983). Identical oligomers have been observed in subsequent studies using pepsin-solubilized collagen VI from a variety of sources (Jander *et al.,* 1984; Hessle and Engvall, 1984; Fleischmajer *et al.,* 1985; Schuppan *et al.,* 1985; Linsenmayer *et al.,* 1986).

Dimers of collagen VI are formed by a staggered, antiparallel alignment of monomers with a 75-nm overlap in their triple-helical domains (Fig. 1). This arrangement gives rise to 30-nm-long single triple-helical segments at each end of the dimer and allows the distinction between outer and inner globular domains differing in size due to a different susceptibility to pepsin. Other structural features as observed in negatively stained dimers (Furthmayr *et al.,* 1983) include a supertwisting (four to five turns) within the overlapping triple-helical segments and an egg-shaped contour of the inner globules pointing toward one side of the particles. Strong evidence for this antiparallel, staggered assembly came from partial reduction experiments (Odermatt *et al.,* 1983; Furthmayr *et al.,* 1983) which, after opening of a single disulfide bond, resulted in the appearance of asymmetric, half-open dimers (Fig. 3) as well as more recently from an asymmetric distribution of monoclonal antibodies along the inner triple helix of dimers (Linsenmayer *et al.,* 1986) as demonstrated by rotary shadowing (see Fig. 10).

Tetramers of collagen VI are formed by the lateral association of dimers with their ends in register (Fig. 1). Stabilization of this assembly apparently occurs by a crossing-over of the outer triple-helical segments into a scissors-like structure at one or both ends producing both open and closed tetramers (Fig. 2). The frequent separation of the two inner triple helical domains indicates little interaction in this region of the tetramer. Major contributions to the stabilization of the tetramer (and dimer) structures are provided by 3–4 disulfide bonds located in different portions of the monomers (Fig. 1). Two disulfide bonds connect the inner globular domains to the opposite triple-helical domains in the dimers, and further disulfide bonds may be involved in stabilizing the scissors-like connections in the tetramers (Odermatt *et al.,* 1983; Furthmayr *et al.,* 1983). Noncovalent interactions between central regions of tetramers and dimers (supercoiled triple helix) may also exist, but they are apparently weak and lost after partial reduction.

The basic features of pepsin-solubilized collagen VI oligomers were confirmed in electron microscopic studies of intact collagen VI obtained from neutral salt extracts of neurofibroma (von der Mark *et al.,* 1984) and guanidine–HCl extracts of nuchal ligament (Jander *et al.,* 1984). The materials consisted either mainly of tetramers together with some dimers and monomers (Fig. 2) or, in the latter case, mainly of dimers mixed with some open tetramers and end-to-end linked dimers. These structures possessed the 105-nm-long triple-helical domain observed in pepsin-solubilized collagen VI plus two larger globular domains which, due to their size, often obscure the scissors-like connection at the ends of tetramers. Similar structures were recently found for collagen VI isolated from cell cultures with monoclonal antibodies (Engvall *et al.,* 1986). These intact oligomers are also held together by disulfide bonds as shown by the release of monomers after reduction under nondenaturing condtions (Jander *et al.,* 1984). The number of disulfide bonds involved is not known and could be larger than those determined by partial reduction of pepsin-solubilized collagen VI (Odermatt *et al.,* 1983).

Oligomer formation increases both thermal stability and protease resistance of collagen VI. Pepsin-solubilized collagen VI showed an increased thermal stability compared to collagen VI monomers with either a clear biphasic melting profile with T_m of 42 and 62°C (Jander *et al.,* 1981; Odermatt *et al.,* 1983) or a more complex profile within the same temperature range (Furuto and Miller, 1981). This thermal stability could be due to triple-helical contacts (superhelix) and/or stabilizing effects by disulfide bonds. The biphasic melting indicates differences in stability within a single triple-helical domain which was supported by isolating a defined fragment Col after treatment of heat-denatured (55°C), pepsin-solubilized collagen VI with bacterial collagenase (Odermatt *et al.,* 1983; Furthmayr *et al.,* 1983). This fragment (M_r of 240,000) corresponded to the inner triple-helical domain of collagen VI dimers (Figs. 1 and 3) and had a T_m of 42°C. Quite often, it showed an unraveling of the two supercoiled triple-helical segments which, however, always remained connected at both ends due to disulfide bonds. Preparations of fragment Col did not contain any tetrameric structures or larger aggregates, demonstrating that the outer segments in the dimers are exclusively responsible for connecting tetramers and microfibrils.

The data on the structure and degradation of oligomers are highly suggestive for a single pathway in the assembly process which very likely starts with the formation of staggered dimers followed by open and closed forms of tetramers. Some support for this suggestion is obtained from biosynthetic studies (Section III), but more data are still needed for a full understanding of these processes.

D. *Microfibrillar Assembly*

The tissue organization of collagen VI into structures larger than the oligomers described above was suggested by studies which failed to extract substantial amounts of the protein from muscle or placenta

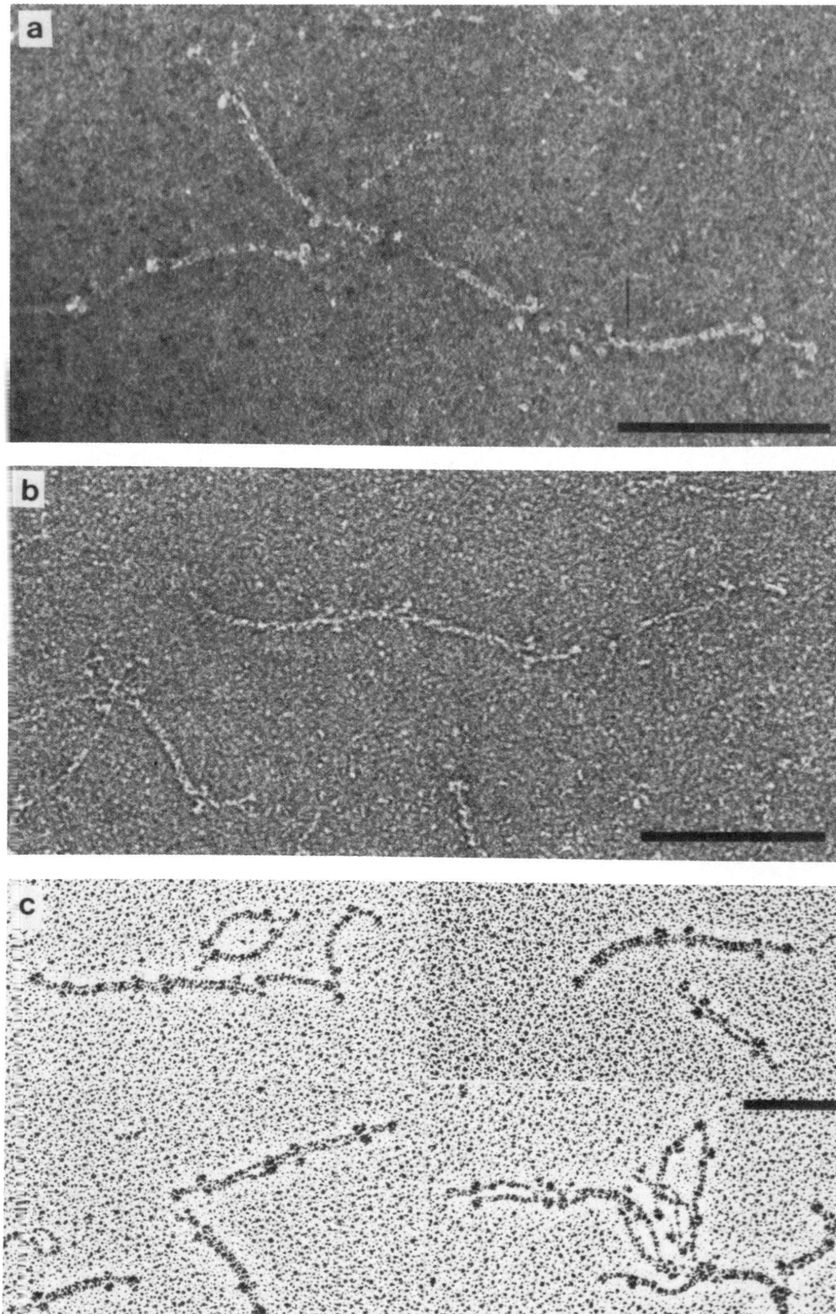
a
b
c

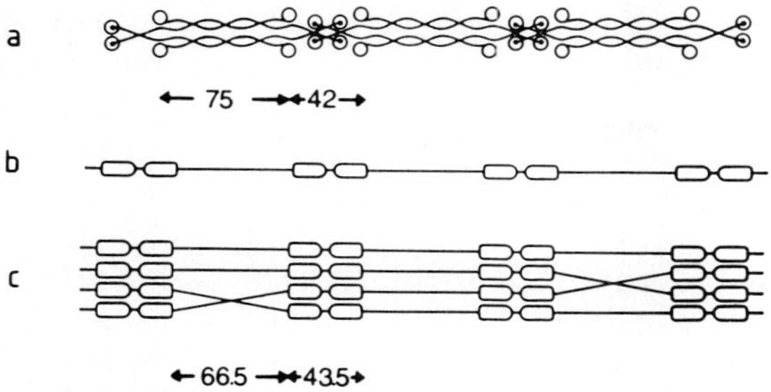

FIG. 6. Comparison of molecular models of linear aggregates of collagen VI tetramers (a), microfibrils (b), and aggregated microfibrils (c) observed in cell cultures. The lines and circles in the tetramer aggregates (a) correspond to the triple-helical and globular domains of collagen VI monomers, respectively, and the outer globular domains are marked by a central dot (Furthmayr *et al.*, 1983). Corresponding electron micrographs are shown in Fig. 5. The lines in the microfibril models (Bruns, 1984) represent the inner triple-helical domains of tetramers, while the two tapered beads represent, in an undefined fashion, eight globular domains originating from two overlapping tetramers. Electron micrographs of single and aggregated microfibrils are shown in Fig. 7. The precise orientation of the globular domains within the beaded regions is unknown, and several models have been discussed (Rauterberg *et al.*, 1986). Numbers between arrows refer to distances in nanometers.

(Marton and Arnason, 1982; von der Mark *et al.*, 1984) and cell matrices (Carter, 1982b) by denaturating solvents (6 *M* guanidine HCl, 8 *M* urea) in the absence of reducing agents. Larger yields were reported for nuchal ligament (Gibson and Cleary, 1982), indicating some tissue variations. Direct evidence for larger collagen VI structures were initially obtained for some material present in pepsin-solubilized collagen VI (Furthmayr *et al.*, 1983) which consisted of up to six tetramers linked to each other in end-to-end fashion (Fig. 5). The detailed electron microscopic analysis of these short filaments indicated an overlap of the outer triple-helical segments of two tetramers, shortening their distances between opposing inner globular domains from about 60 nm to either 42 nm (after negative staining) or 30 nm (after rotary shadowing). This arrangement suggested close contacts between outer globular domains of one tetramer with the inner globular domains of an adjacent tetramer (which may become disulfide-bonded) and a lateral repeat of this structure with a distance of 105–117 nm (Fig. 6). Such linear

---

FIG. 5. Electron microscopic visualization of linear aggregates of tetramers present in pepsin-solubilized collagen VI from human placenta by negative staining (a and b) or rotary shadowing (c). Bars, 100 nm. Reprinted by permission from *Biochem. J.* **211**, 303–311 copyright © 1983 The Biochemical Society, London.

aggregates of tetramers have not been observed in guanidine–HCl extracts of tissues which may contain, however, small amounts of dimers in linear association (Jander *et al.,* 1984). Such structures may have arisen from filaments in which the tetramers were imperfectly cross-linked in their scissors-like region.

Based on the model shown in Fig. 6, it was predicted that collagen VI could be a component of tissue microfibrils with a diameter of 5–10 nm and a periodicity of about 100 nm (Furthmayr *et al.,* 1983). This prediction gained some support from immunoelectron microscopy of tissues (von der Mark *et al.,* 1984; see Section V) and by electron microscopy of certain kinds of fibrils found in cornea, tendon, and skin fibroblast cultures (Bruns, 1984). The latter study identified negatively stained beaded microfibrils of 3 nm width with pairs of beads repeating at 100-nm intervals up to a length of 2 μm. A second structure consisted of bundles of laterally aligned microfibrils with all their beads in register (Fig. 7). The beads showed a tapered or rectangular shape and were often separated by a small central cleft. It was recognized by Bruns (1984) that the periodicity and dimensions of these fibrils closely resembled those of a linear array of collagen VI tetramers (Fig. 6). Their morphology was clearly different from those of collagen I precipitated in fibrous long spacing (FLS) form (periodicity 220–240 nm) and from microfibrils associated with elastin. That these fibroblast microfibrils contain, in fact, collagen VI was demonstrated by immunoelectron microscopy (Bruns *et al.,* 1986; see Fig. 8) and by isolating a similar fibril with monoclonal antibodies (Engvall *et al.,* 1986).

Microfibrils and cross-banded lateral aggregates essentially identical to those described by Bruns had been observed in previous studies with fibroblast cultures (Goldberg and Green, 1964; Yardley and Brown, 1965; Green and Meuth, 1974; Sear *et al.,* 1978). Morphologically similar cross-banded fibrils were also frequently observed in normal and pathological tissues (Table II) and first described to be present in Descemet's membrane (Jakus, 1956), in the trabecular meshwork of the eye (Garron *et al.,* 1958), and in several nerve tumors (Luse, 1960). These fibers show, after positive staining, a periodic arrangement of dark and light bands (Fig. 9) with a lateral repeat of 80–120 nm and vary in diameter from several hundred nanometers to a few micrometers. The electron-lucent zones of tissue fibers very often show the presence of fine filaments connecting the dark band zones, thus corresponding even in fine details to the collagen VI fibers in cell cultures (Fig. 7). These tissue structures were referred to as Luse bodies, fusiform bodies, and spiney or zebra collagen and were usually found close to collagen I fibrils, basement membranes, or cells. Their abun-

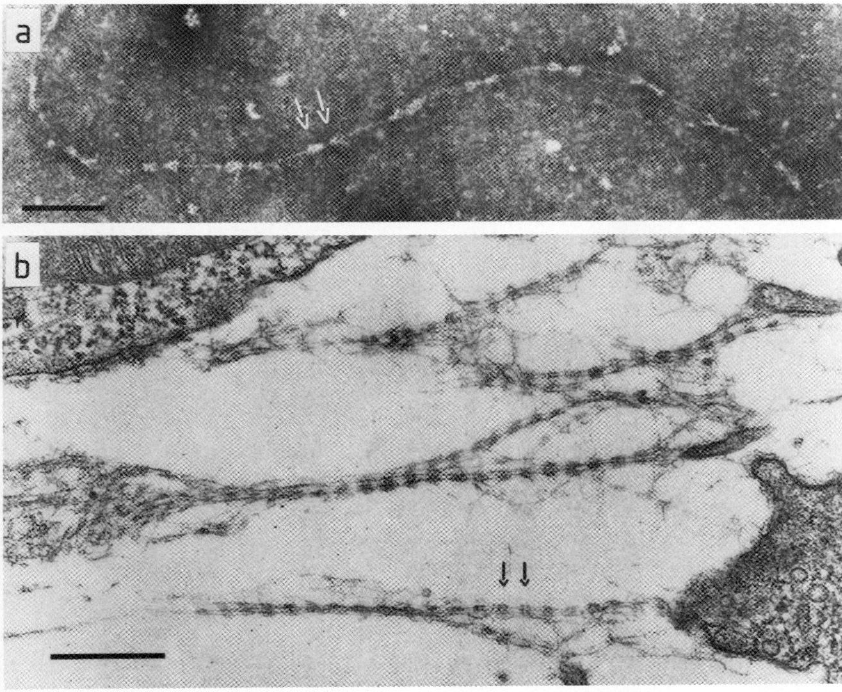

FIG. 7. Beaded filaments of collagen VI from chicken embryo tendon fibroblast culture medium. (a) Negatively stained beaded microfibril of 3-nm width with pairs of beads (arrows) repeating at 100-nm intervals. The elongated and tapered beads are about 18 nm in length, and their center-to-center distance is 25 nm. Bar, 100 nm. (b) Thin section of the fibroblast culture (embedded in Epon, stained with OsO_4, Mg acetate, and lead citrate) showing bundles of beaded filaments. Because of positive staining, the pair of beads appearing light in (a) are imaged as dark bands here. The 100-nm repeat is clearly apparent and gap regions (arrows) are also occasionally seen. Bar, 500 nm. The photographs were kindly supplied by Dr. R. Bruns. The original work was described by Bruns (1984). Reproduced with permission.

dance can be increased by treating cell matrices with ATP or by incubating tissues under culture conditions (Kajikawa *et al.*, 1980; Bruns *et al.*, 1986). Such fibers were occasionally considered to represent FLS collagen (Ramsey, 1965; Banfield *et al.*, 1973; Kajikawa *et al.*, 1980), which is presumably a misinterpretation since the periodicity of the tissue fibers is only half of that found for collagen I precipitated in FLS form (Hashimoto and Ohyama, 1974). The distinct staining of the cross-banded structures with ruthenium red (Buckwalter *et al.*, 1979; Kajikawa *et al.*, 1980) may indicate the presence of proteoglycan in the fiber complex.

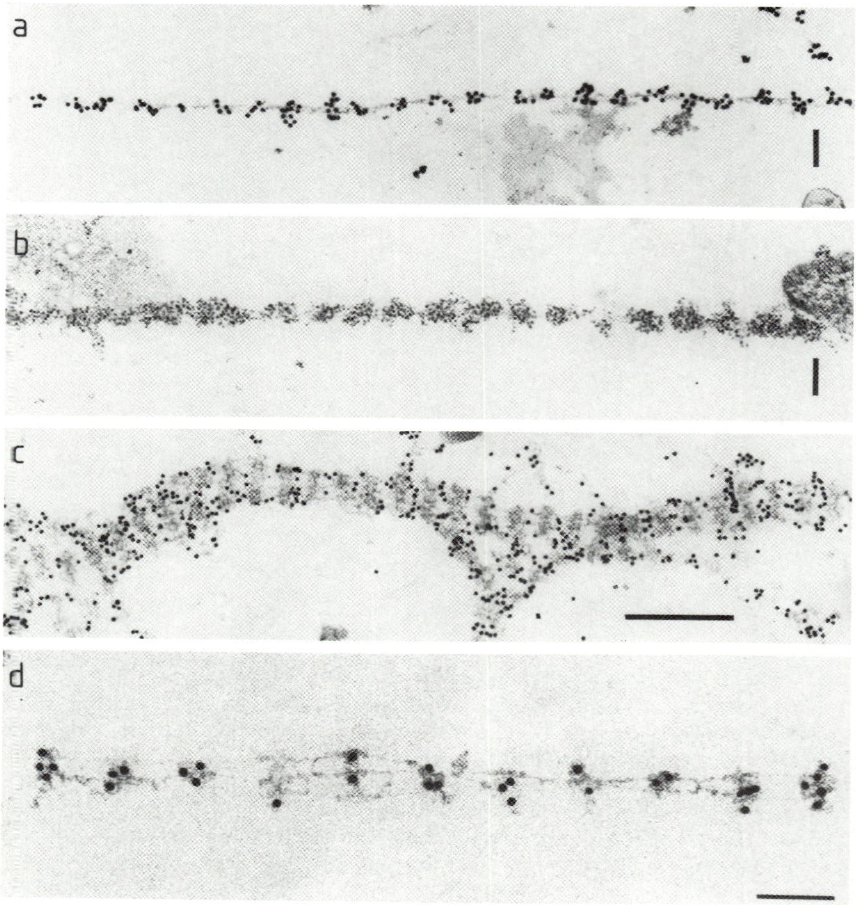

FIG. 8. Immunolabeling of collagen VI microfibrils present in cell culture matrix. (a) Thin section of a culture of unfixed human foreskin fibroblasts treated with monoclonal antibody to collagen VI and with secondary goat anti-mouse antibody conjugated with colloidal gold particles 10.7 nm in diameter. (b) Thin section of a culture fixed with glutaraldehyde and incubated with rabbit antiserum against collagen VI and with secondary goat anti-rabbit antibody coupled to ferritin. Note in a and b the clusters of immunolabel with a periodicity of 100 nm. (c) Thin section, as in a, showing gold particles labeling the filamentous interbands of the bundles of beaded filaments indicating that this monoclonal antibody is directed to the triple-helical region of collagen VI. (d) Thin section of a human skin fibroblast culture that was incubated with anti-GP-250 antibody and with secondary antibody conjugated with colloidal gold. Labeling of the dark bands instead of the filamentous interbands is observed in this case. The photographs were kindly supplied by Dr. R. Bruns. The original studies were described by Bruns *et al.* (1986). Reproduced from *The Journal of Cell Biology*, 1986, **103,** 393–404 by copyright permission of The Rockefeller University Press.

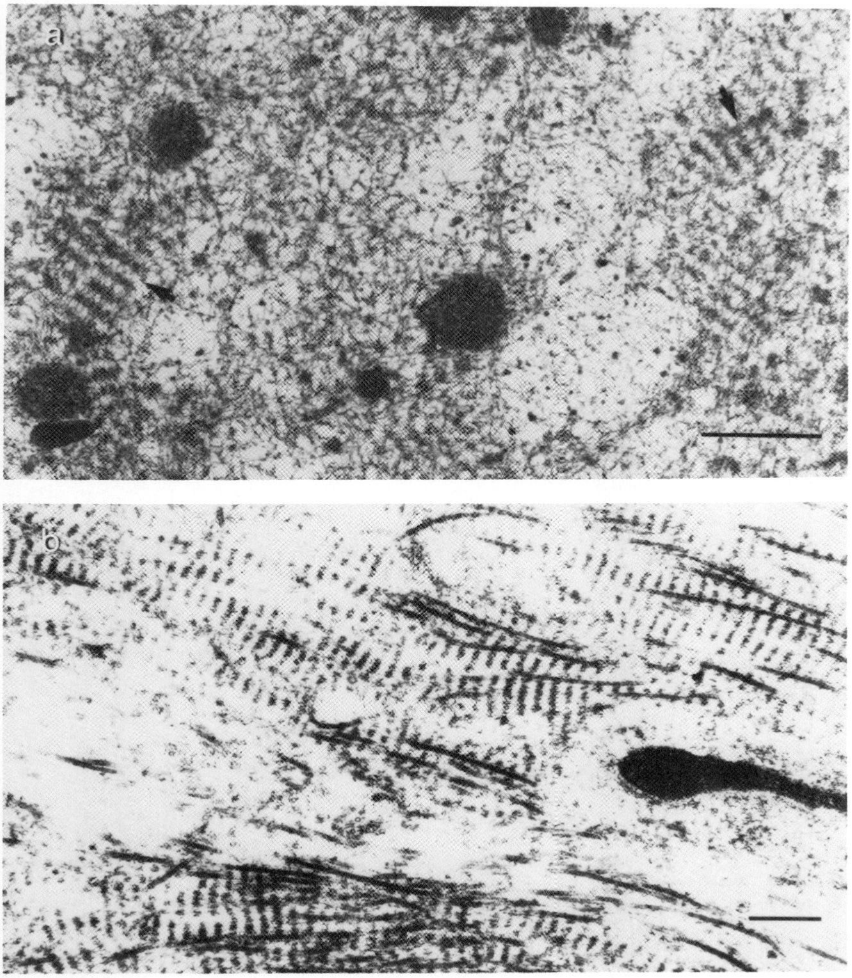

FIG. 9. Cross-banded structures which are possibly related to collagen VI aggregates in human nucleus pulposus (a) and neurofibroma skin (b). These structures show a periodicity of 100–110 nm and are less abundant in a (arrows) than in b. Bars, 500 nm. The electron micrographs were obtained after staining by lead citrate and uranyl acetate and kindly provided by Dr. Raul Fleischmajer and Mrs. Hanna Wiedemann.

Biochemical and immunohistological evidence for the presence of collagen VI in these cross-banded tissue fibers is still scarce. Ludowieg and Parker (1973) solubilized such fibers from the nucleus pulposus and found, after reconstitution by dialysis, that they contain significant amounts of hydroxyproline and glycosaminoglycans. Other stud-

TABLE II

CROSS-BANDED FIBERS OBSERVED IN TISSUES RESEMBLING IN MORPHOLOGY LATERAL AGGREGATES OF COLLAGEN VI MICROFIBRILS

Type	Periodicity (nm)	Localization, special features	References
Alternating dark and light bands of about equal size, fine filaments in the light bands	107–110	Descemet's membrane	Jakus (1956, 1962)
	120	Rare in corneal stroma, hexagonal patterns in tangential sections	Sawada (1982)
	100	Nucleus pulposus (human, rabbit)	Smith and Serafini-Fracassini (1968), Cornah *et al.* (1970)
	85		Buckwalter *et al.* (1979)
	80	Pig skin homogenates	Hayes and Allen (1967)
	90–120	Cultured rat skin	Kajikawa *et al.* (1980)
	120	Human leprosy skin	Edwards (1975)
	90	Human skin, tendon	Nemetschek-Gansler *et al.* (1977)
	70–100	Human myocardium	Banfield *et al.* (1973)
	124	Tibial nerve, rat	Pillai (1964)
	100	Trabecular meshwork of human eye	Garron *et al.* (1958), Rohen (1963)
	120	Schwannoma, acoustic nerve tumor	Luse (1960)
	107		Raimondi *et al.* (1962)
	100–117		Ramsey (1965)
	120		Craviotto and Lockwood (1968)
	100	Human lymph node tumors	Mollo and Monga (1971)
	85–125	Squamous cell carcinoma	Hashimoto and Ohyama (1974)
	85–100	Mouse osteosarcoma	Marquart *et al.* (1976)
Thin dark band (10 nm), light zones are bisected by additional dark bands	107	Rat brain	Naumann and Wolfe (1963)
	120	Mouse articular cartilage	Silberberg *et al.* (1963)
	80–112	Rat adenohypophysis	Chatterjee (1973)

ies (H. von der Mark, H. Wiedemann, K. Remberger, and R. Timpl, unpublished results, 1986) demonstrated the presence of some degraded collagen VI in such reconstituted material which was supported by immunogold staining with specific antibodies. Definitive proof that collagen VI really generates such cross-banded fibers is, however, still lacking. Other cross-banded structures may have no relationship to collagen VI at all. This appears likely for Descemet's membrane, which in tangential sections shows a characteristic hexagonal lattice of dumbbell-shaped elements (Jakus, 1956, 1962). These structures are sensitive against collagenase (Sawada, 1982), and their potential precursors observed in cell culture are larger than collagen VI (Sawada *et al.,* 1984). Still another form of fibril found in a few tissues (second section in Table II) shows a periodicity of 80–120 nm but differs distinctly in segmentation from the collagen VI fibrils present in cell cultures (Bruns, 1984).

The reconstitution of microfibrils from soluble, purified collagen VI monomers or oligomers has so far not been reported. Dialysis of pepsin-solubilized collagen VI against ATP at acidic pH produces fibrillar aggregates with a periodicity of 80 nm and a complex symmetric cross-striation pattern (Jander *et al.,* 1981). A different fibril pattern, but no SLS segments, was observed after precipitating the triple-helical domain of collagen VI with ATP. The existence of such structures in tissues is doubtful. The formation of cross-banded fibrils by dialyzing pepsin-solubilized collagen VI against neutral buffers has been reported (Rauterberg *et al.,* 1986), but their relationship to tissue structures remains to be established.

III. Biosynthesis

Biosynthesis of collagen VI has been amply documented in cultured fibroblasts obtained from adult and embryonic skin (Carter and Hakomori, 1981; Trüeb and Bornstein, 1984; von der Mark *et al.,* 1984; Aumailley *et al.,* 1985), lung (Carter, 1982a,b; Engvall *et al.,* 1986), nuchal ligament (Sear *et al.,* 1978, 1981a; Gibson and Cleary, 1985), tendon (Taylor *et al.,* 1982), and dental pulp (Shuttleworth *et al.,* 1982) as well as for aortic smooth muscle cells (von der Mark *et al.,* 1984; Gibson and Cleary, 1985) and a myoblast cell line (Engvall *et al.,* 1986). Endothelial and epithelial cells and various tumor cell lines were found to produce little or no collagen VI and to show a defect in matrix deposition (Carter, 1982b, 1984; Hessle and Engvall, 1984; von der Mark *et al.,* 1984). Similar observations were reported for SV40-

transformed fibroblasts (Carter, 1982b) and attributed to a more than 20-fold reduction in the amount of collagen VI mRNA (Trüeb *et al.,* 1985). However, a few tumor cell lines, such as osteosarcoma and leiomyosarcoma, may still produce distinct amounts of collagen VI (Engvall *et al.,* 1986). Production of collagen VI was monitored by labeling with radioactive amino acids or by using radioactive fucose (Sear *et al.,* 1978, 1981a), glucosamine, or surface labeling of galactose residues (Carter and Hakomori, 1981). Collagen VI chains were then identified either by their characteristic electrophoretic mobility or, more recently, by antibody reactions using immunoprecipitation (Sear *et al.,* 1981a,b; Trüeb *et al.,* 1985; Ayad *et al.,* 1985), binding to monoclonal antibodies (Hessle and Engvall, 1984; Engvall *et al.,* 1986), or immunoblotting (Carter, 1982b; Trüeb and Bornstein, 1984; Aumailley *et al.,* 1985).

All these data allowed a rather broad but still incomplete insight into the biosynthesis and matrix deposition of collagen VI. As expected for a collagenous protein, about 25–36% of proline and lysine in the α(VI) chains are subjected to hydroxylation as shown by radioactive labeling (Sear *et al.,* 1981a) and chemical analysis of purified material (Carter, 1982a; see Table I). Extensive posttranslational modifications were also demonstrated by a reduction in size of α(VI) chains from a M_r of 140,000 to 125,000 by culturing cells in the presence of tunicamycin and α,α-dipyridyl and in cell-free translation (Trüeb *et al.,* 1985). The effect of both drugs was additive, which indicated a blockade of incorporation of both N-linked and hydroxylysine-linked carbohydrates. Treatment with α,α-dipyridyl also interfered with triple-helix formation (due to under-hydroxylation), as judged by reaction with conformation-specific antibodies, but did not interfere with disulfide-bonding of α(VI) chains (Engvall *et al.,* 1986). Some (Sear *et al.,* 1981a; Engvall *et al.,* 1986) but not all studies (Carter, 1984; Leushner and Haust, 1986) found reduced secretion and matrix deposition of collagen VI in the absence of ascorbic acid.

Collagen VI shows a greater deposition in the cell matrix (50%) produced by cultured cells, in contrast to collagens I and III which, in large part, are lost to the medium (Carter, 1984). The remaining collagen VI is about equally distributed between intracellular compartments and the culture medium. After labeling, collagen VI appears as a major cell matrix component together with GP-250 (see below) and some fibronectin and procollagens I and III (Sear *et al.,* 1978; Carter and Hakomori, 1981). Pulse–chase studies demonstrated that the intracellular form, but not the medium form, of collagen VI is the precursor of the matrix form and that cell contact with the matrix is essential for deposition (Carter, 1984). Deposition is accompanied by increased disulfide bonding which results in collagen VI becoming insoluble in urea and detergents (Carter, 1982a). These structures appear fibrillar by immunofluorescence and electron microscopy while intracellular collagen VI shows a granular staining pattern around the nucleus (Carter, 1984; von der Mark *et al.*, 1984; Bruns, 1984; Bruns *et al.*, 1986; Engvall *et al.*, 1986). Synthesis and deposition of collagen VI usually starts soon after attachment of cultured cells but may be preceeded by the deposition of fibronectin (Carter, 1982a, 1984; Bruns *et al.*, 1986).

The electrophoretic comparison of α(VI) chains present within cells, in cell culture medium, and in the cell matrix showed a single or a doublet band with a M_r of 140,000 (globular standards), indicating only minor differences in size which may be entirely accounted for by differences in posttranslational modifications (Carter, 1984; Aumailley *et al.*, 1985; Engvall *et al.*, 1986). A similar size of α(VI) chains was also observed by comparing extracts obtained from cell cultures and tissues (von der Mark *et al.*, 1984; Heller-Harrison and Carter, 1984; Gibson and Cleary, 1985). The data, therefore, indicate the lack of a precursor form (procollagen) of collagen VI or only minor processing of a putative precursor involving the removal of a segment with a M_r less than 10,000, at least for the M_r 140,000 α(VI) chains. A single study (Trüeb and Bornstein, 1984) is at variance with the other data by claiming a M_r of 180,000–190,000 for tissue α(VI) chains and a 20% increase in mass for culture-derived α(VI) chains. Conceivably, these values were overestimated, since a different size (M_r = 140,000) was reported in subsequent studies (Trüeb *et al.*, 1985; Crawford *et al.*, 1985).

Disulfide bonding of newly synthesized α(VI) chains is presumably a fast process as shown for intracellular material isolated with monoclonal antibodies (Engvall *et al.*, 1986). Appropriate pulse–chase studies for determining kinetic parameters of chain synthesis and assembly still remain to be done. The electron microscopic analysis of antibody-isolated material indicated abundant amounts of globules together with some monomers, aligned and staggered dimers, and tetramers within intracellular compartments (Engvall *et al.*, 1986). The interpretation of these data is still difficult and will require the identification of the nature of the globules and better evidence for the presence of aligned dimers. Collagen VI in the culture medium is already highly disulfide linked to tetramers as shown by gradient centrifugation (Aumailley *et al.*, 1985) and electron microscopy (Engvall *et al.*, 1986). The latter study also demonstrated mainly tetramers and linear aggregates of tetramers in urea extracts of the cell matrix. This material may represent only a small portion of the deposited collagen VI as indicated by the studies discussed above. The data taken together, however, are compatible with an assembly process proceeding from monomers through dimers and tetramers which, in the extracellular space, are connected into microfibrils. Evidence exists for some free α(VI) chains in cell culture medium (Carter, 1984; Aumailley *et al.,* 1985), but these may arise as assembly artifacts.

The amounts of collagen VI produced in cell culture relative to those of other extracellular matrix proteins are unknown. Radioimmunoassays demonstrated concentrations of collagen VI in fibroblast medium in the range of 100–200 ng/ml (Schuppan *et al.,* 1985). These values are 10 to 100-fold lower than those obtained for procollagens I and III and agree with the data which estimate the concentration of collagen VI in tissues to be low (Section V). They indicate that collagen VI is, in most cases, a minor component of the extracellular matrix.

The most surprising observation from cell culture studies was the apparently intimate association of α(VI) chains with another polypeptide chain. This polypeptide, which we will refer to as GP-250 (Carter and Hakomori, 1981), has in reduced form an apparent M_r of 240,000–270,000 and was shown to be distinct from fibronectin. It was found to (1) be resistant to bacterial collagenase, (2) be disulfide bonded to larger complexes, and (3) remain associated with collagen VI during chromatography and immunoprecipitation (Sear *et al.,* 1981a; Carter, 1982b). The existence of complexes between GP-250 and collagen VI was then convincingly demonstrated by immunoprecipitation with affinity-purified antibodies against α(VI) chains and with monoclonal antibodies (Trüeb *et al.,* 1985; Engvall *et al.,* 1986). These complexes persisted during reduction under nondenaturing but not denaturing conditions and were also formed with under-hydroxylated collagen VI (Engvall *et al.,* 1986). GP-250 was also found to behave as a glycoprotein but not as a proteoglycan in CsCl gradient centrifugation (Shuttleworth *et al.,* 1982). A component with a M_r of 260,000 was recently reported to be the precursor form of the α3(VI) chain (Trüeb and Winterhalter, 1986). Even though it seems very likely that this precursor chain is identical with GP-250, it still remains to be proven experimentally.

The nature of the collagen VI chains was also recently studied by using various cDNA clones isolated from a human express library (Chu *et al.,* 1987). These probes reacted in Northern hybridization with three different mRNA species which, by matching peptide sequences, could be identified as encoding the α1(VI) (4.2 kb), α2(VI) (3.5 kb), and α3(VI) chain (8.5 kb). The sizes of the mRNAs are therefore, in good agreement with those determined at the peptide level (see above).

Among the M_r 140,000 components characterized from fibroblast cultures are also polypeptide chains now considered to constitute the fibronectin receptor (Pytela *et al.,* 1985; Horwitz *et al.,* 1985; Akiyama *et al.,* 1986; Patel and Lodish, 1986). These receptor chains differ from collagen VI by their solubility in detergents, lack of disulfide bonding, and lack of reaction with concanavalin A. Lack of relationship between the receptor and collagen VI was most convincingly demonstrated by sequence analysis of cDNA, which failed to detect nucleotide sequences coding for Gly-X-Y repeats in one of the subunits of the receptor (Tamkun *et al.,* 1986).

IV. Immunology

Polyclonal antisera specific for collagen VI have been useful tools in analyzing various forms of the protein and were instrumental in establishing the correlation between intact and degraded forms of collagen VI (Section I,A). They were raised by using pepsin-solubilized collagen VI (von der Mark *et al.,* 1984; Trüeb and Bornstein, 1984; Hessle and Engvall, 1984; Schuppan *et al.,* 1985), chain constituents of this material (Jander *et al.,* 1981, 1984; von der Mark *et al.,* 1984), and intact α(VI) chains (Marton and Arnason, 1982; Carter, 1982b; Gibson and Cleary, 1983) as immunogens. Most of the studies demonstrated lack of cross-reactions with collagens I to V, 7 S collagen, laminin, and fibronectin by enzyme immunoassays, immunoblotting, radioimmunoassays, and passive hemagglutination. If such cross-reactions were observed (collagen V, fibronectin), cross-adsorptions indicated that they arose due to contamination of the immunogen and not by shared epitopes (Risteli *et al.,* 1980; Marton and Arnason, 1982; Carter, 1982b; Gibson and Cleary, 1983; Jander *et al.,* 1984; von der Mark *et al.,* 1984; Trüeb and Bornstein, 1984; Schuppan *et al.,* 1985). Monospecific antibodies were also prepared from these polyclonal antisera by immunoadsorption and were particularly useful in immunohistological studies (Jander *et al.,* 1981; Carter, 1982b; von der Mark *et al.,* 1984).

Immunochemical studies demonstrated that collagen VI possesses potent sequential epitopes which are shared by the triple-helical molecule and its constituent chains as well as conformational epitopes which are dependent on triple-helical folding and/or disulfide bonding (von der Mark *et al.*, 1984). This made polyclonal rabbit antisera raised either against triple-helical material or unfolded chains both useful for immunoblotting of even small peptides (Heller-Harrison and Carter, 1984) and for the characterization of various forms of collagen VI antigens in biological materials (see below and Section V). Epitopes could also be located to the triple-helical and globular domains of pepsin-solubilized collagen VI, which are presumably of different nature (von der Mark *et al.*, 1984). Diversity of epitopes was also indicated in immunostaining of both beads and filaments of cell-derived collagen VI microfibrils by polyclonal antibodies (Bruns *et al.*, 1986). A more restricted reaction pattern was reported for some anti-chain antisera which stained only α1(VI)p and α3(VI)p chains in immunoblotting (Jander *et al.*, 1984). A similar restriction in immunoblot reaction was also observed in other (Trüeb and Bornstein, 1984; Crawford *et al.*, 1985; Gibson and Cleary, 1985) but not all studies. Chain-specific antibodies can, however, be generated by affinity chromatography on α(VI)p chains (Trüeb and Winterhalter, 1986). The identification of the reacting chains as originating from α1(VI), α2(VI), or α3(VI) has, in most cases, not been confirmed with purified material and should, therefore, be considered tentative.

The development of sensitive and quantitative radioimmunoassays for collagen VI appears to be difficult due to a high background binding observed for pepsin-solubilized collagen VI (von der Mark *et al.*, 1984). The choice of high-salt buffers and careful storage conditions may overcome these difficulties and allow the quantitation of as little as 3 ng/ml of collagen VI by inhibition assays (Schuppan *et al.*, 1985). This assay was useful for the detection of collagen VI antigen in human serum (22 ± 8 ng/ml) and bile but not in urine. Most of the circulating collagen VI has a $M_r > 400{,}000$, indicating the presence of oligomers in serum.

The production of monoclonal antibodies specific for collagen VI was achieved by using spleen cells from mice immunized either with whole human placenta (Hessle and Engvall, 1984) or pepsin-solubilized collagen V from chicken gizzard (Linsenmayer *et al.*, 1986). Since both materials used for immunization are presumed to contain only small amounts of collagen VI, the presence of antibodies indicates a high immunogenic potential for collagen VI compared to other collagenous proteins. The monoclonal antibodies specific for placenta collagen VI were selected for by their ability to bind a complex of α(VI) chains and GP-250 from cell cultures. Some, but not all, antibodies also reacted with pepsin-solubilized collagen VI, which indicates the recognition of several epitopes differing in stability against proteases. Some epitopes are characterized by high thermal stability but were also found to be nonreactive in under-hydroxylated collagen VI (Hessle and Engvall, 1984). Whether these data indicate the crucial involvement of hydroxylated amino acids or the conformational nature of epitopes is not quite clear. A distinct conformation dependence of epitopes was demonstrated in inhibition assays with monoclonal antibodies against chick collagen VI (Linsenmayer *et al.,* 1986).

The various polyclonal and monoclonal antibodies raised against collagen VI have been used in many applied studies, such as tissue localization (Section V) and establishing the relationship between degraded and intact collagen VI from tissues and cells using immunoblotting, immunoprecipitation, and enzyme immunoassays (Heller-Harrison and Carter, 1984; Jander *et al.,* 1984; Hessle and Engvall, 1984; Trüeb and Bornstein, 1984; von der Mark *et al.,* 1984; Aumailley *et al.,* 1985; Gibson and Cleary, 1985; Engvall *et al.,* 1986; Trüeb and Winterhalter, 1986). Due to their restricted specificity, monoclonal antibodies were also useful for mapping epitopes at the ultrastructural level. One of the antibodies showed a distinct binding to the filamentous elements of collagen VI microfibrils (Fig. 8) indicating its reaction with the triple-helical domain (Bruns *et al.,* 1986). Monoclonal antibody binding could also be demonstrated after rotary shadowing (Fig. 10), indicating the reaction with a triple-helical epitope about 25 nm away from the inner globular domain in collagen VI dimers (Linsenmayer *et al.,* 1986).

The collagen VI-associated GP-250 component has so far not been comprehensively studied by immunological methods. An antiserum raised against lung fibroblasts showed, after affinity purification on a cell matrix-adsorbent immunoblot, reactions with GP-250 and several other polypeptides but not with α(VI) chains (Carter, 1982b). Coprecipitation of collagen VI and possibly GP-250, which were named MFP I and MFP II in other studies (Sear *et al.,* 1978, 1981a), was observed with antiserum against a crude preparation of so-called microfibrillar protein. Whether the antiserum contains antibodies against both components is not known. A single monoclonal antibody with a undefined origin of the immunogen was also found to bind the collagen VI–GP-250 complex (Engvall *et al.,* 1986) and to react with the globular beads of collagen VI microfibrils (Bruns *et al.,* 1986; see Fig. 8). This antibody still reacted with GP-250 after complete reduction under denaturing conditions but not with α(Vl) chains, indicating the recogni-

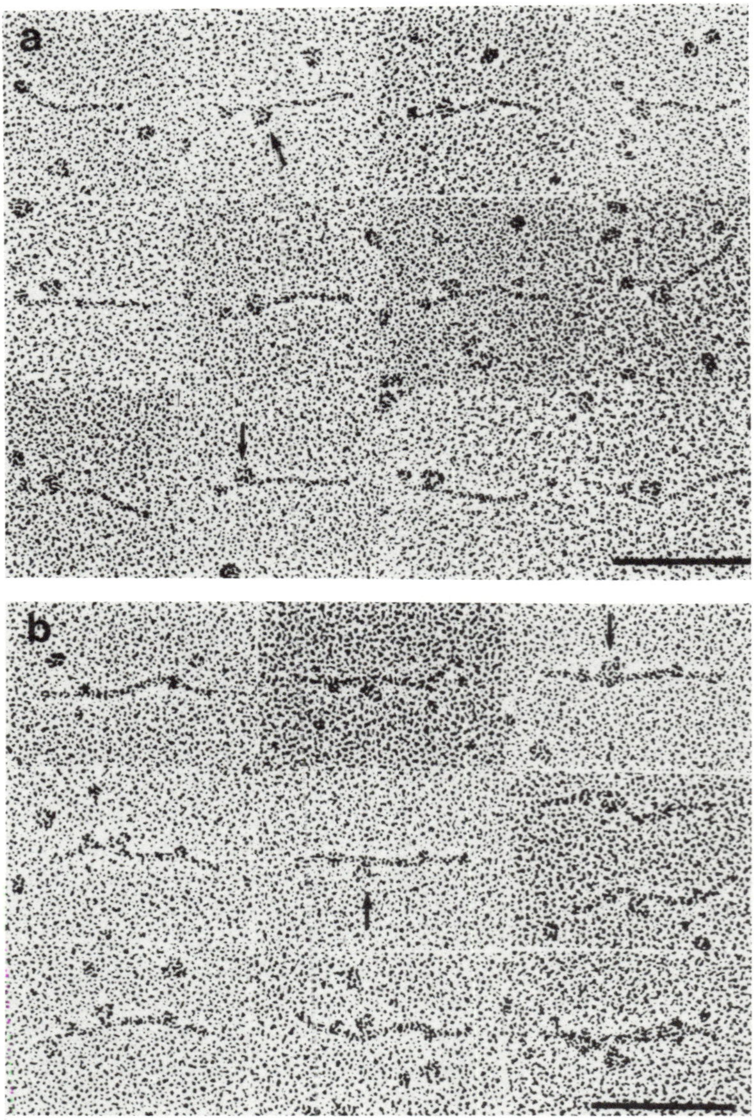

FIG. 10. Visualization by rotary shadowing of monoclonal antibodies bound to monomeric (a) and dimeric (b) chick collagen VI. In each case, the photographs show 12 selected particles, and the position of the antibody molecule is indicated in a few instances by an arrow. Bars, 100 nm. The photographs were kindly supplied by Dr. Richard Mayne. Reproduced with permission from Linsenmayer *et al.* (1986).

tion of sequential epitopes. The data would be compatible with the assumption that GP-250 and the precursor of the α3(VI) chain are identical.

V. Tissue Distribution

A broad tissue distribution of collagen VI was originally recognized in biochemical studies which isolated the protein from pepsin digests or denaturing extracts obtained from a variety of sources. The tissues used included placenta (Furuto and Miller, 1980; Risteli *et al.,* 1980; Jander *et al.,* 1981; Abedin *et al.,* 1982; Odermatt *et al.,* 1983; von der Mark *et al.,* 1984), large blood vessels (Chung *et al.,* 1976; Gibson and Cleary, 1982; Jander *et al.,* 1984; Trüeb and Bornstein, 1984), nuchal ligament (Gibson and Cleary, 1982, 1985; Knight *et al.,* 1984; Jander *et al.,* 1984), uterus (Abedin *et al.,* 1982; Trüeb and Bornstein, 1984; Ayad *et al.,* 1985), liver (Rojkind *et al.,* 1979), kidney (Risteli *et al.,* 1980), skin (Laurain *et al.,* 1980), muscle (Marton and Arnason, 1982), neurofibroma (von der Mark *et al.,* 1984; Fleischmajer *et al.,* 1985), and cornea (Linsenmayer *et al.,* 1986; Zimmermann *et al.,* 1986). The amounts of collagen VI relative to that of total collagen were determined only for a few tissues, and the values reported range from 1.5–3% for placenta and uterus, to less than 0.5% for the chorionic and amniotic membranes (Abedin *et al.,* 1982), to 0.1% for normal liver (Schuppan *et al.,* 1985). Cornea seems to be a tissue particularly rich in collagen VI, where it was estimated to represent one-quarter of total collagen (Zimmermann *et al.,* 1986).

The observations of an ubiquitous tissue occurrence of collagen VI were confirmed and extended by immunostaining at the light-microscopic level (indirect immunofluorescence or peroxidase technique) using affinity-purified polyclonal antibodies against pepsin-solubilized collagen VI (von der Mark *et al.,* 1984) or its constituent chains (Jander *et al.,* 1981), monoclonal antibodies (Hessle and Engvall, 1984; T. F. Linsenmayer, R. R. Bruns, A. Mentzer, and R. Mayne, unpublished results, 1986), or antisera specific for α(VI) chains (Marton and Arnason, 1982; Gibson and Cleary, 1983). These studies demonstrated mainly fibrillar staining patterns (Fig. 11) of interstitial septa and sinusoids of parenchymal organs (liver, lung, kidney, spleen), of the three layers of large arteries, of small blood vessels, and of fetal membranes. Intensive staining was also observed in tissues with abundant extracellular matrix, such as skin, tendon, and nuchal

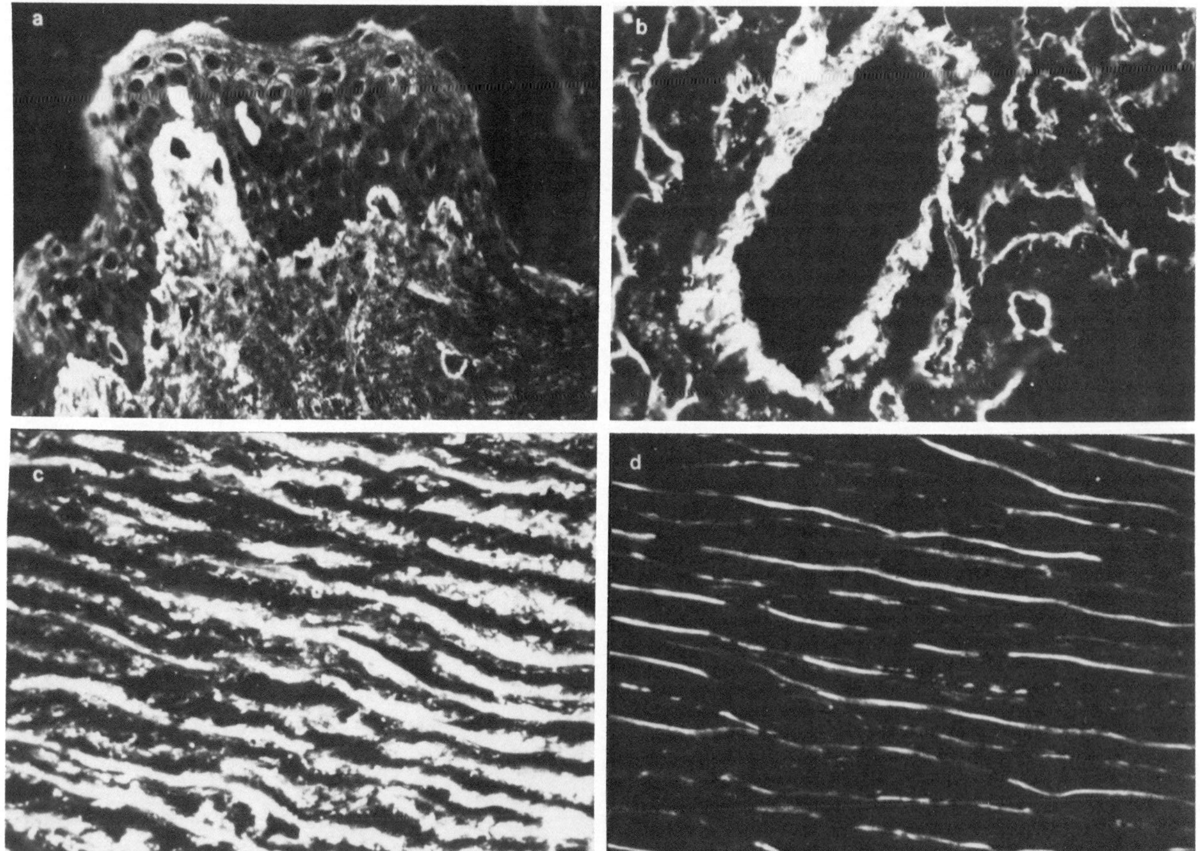

FIG. 11. Immunofluorescent localization of collagen VI in human skin (a), liver (b), and aortic media (c). The aorta section (d) was treated with nonimmune IgG as a control and shows distinct autofluorescence of elastic lamellae. ×250 (a and b) and ×500 (c and d). The photographs were kindly provided by Dr. Georg Wick and were reproduced with permission from von der Mark *et al.* (1984).

ligament, in the endoneurium of nerves, in the perichondrium of cartilage, and around muscle cells. Controversy still exists about positive (Hessle and Engvall, 1984; Ayad *et al.*, 1984; Linsenmayer *et al.*, 1986) or negative reactions (Marton and Arnason, 1982; von der Mark *et al.*, 1984) of muscle perimysium and hyaline cartilage. Nucleus pulposus, however, seems to contain collagen VI or a related protein with a staining pattern indistinguishable from that observed for collagen II (H. von der Mark, H. Wiedemann, K. Remberger, and R. Timpl, unpublished results, 1986). A brilliant staining was also observed with corneal stroma but apparently did not include Descemet's membrane (Schittny *et al.*, 1985; T. F. Linsenmayer, R. R. Bruns, A. Mentzer, and R. Mayne, unpublished results, 1986).

The staining patterns observed for collagen VI located the protein in most of the extracellular spaces which also contain collagens I and III and fibronectin. Differences in details of the staining patterns indicated, however, a lack of close codistribution of these proteins. Studies of aortic media and nuchal ligament (Gibson and Cleary, 1983; von der Mark *et al.*, 1984) demonstrated a distinct separation of regions stained for collagen VI from elastic lamellae (see Fig. 11). The data so far also indicate the absence of collagen VI from basement membranes except for renal mesangium. A possible presence in cartilage matrix cannot be excluded but will depend on biochemical data for a convincing demonstration.

Evidence for a clear anatomical separation of collagen VI structures from cross-striated collagen fibrils (types I and III) and elastic fibers was obtained by immunoelectron microscopy (von der Mark *et al.*, 1984). Studies with aorta and placenta frequently showed the ferritin stain along microfibrils of 5–10-nm diameter in a regular fashion (Fig. 12). Other regions which were stained included tissue sections which appeared unstructured as well as compartments close to smooth muscle cells. Staining patterns were often found to be oriented perpendicular to the direction of major collagen and elastin fibrils. Staining of microfibrils and amorphous regions, but not of basement membranes, was also reported for neurofibroma tissue (Fleischmajer *et al.*, 1985) and placenta (Amenta *et al.*, 1986). The data agree with those obtained from cell cultures which also demonstrated the assembly of collagen VI into microfibrillar structures (Fig. 7) independent from other fibrillar elements and fibronectin (Bruns *et al.*, 1986). A recent study with monoclonal antibodies demonstrated, by immunogold staining, regular patterns with 100-nm periodicity in the corneal stroma (T. F. Linsenmayer, R. R. Bruns, A. Mentzer, and R. Mayne, unpublished results, 1986), which is in agreement with the fibrillar models proposed for collagen VI (Section II,D).

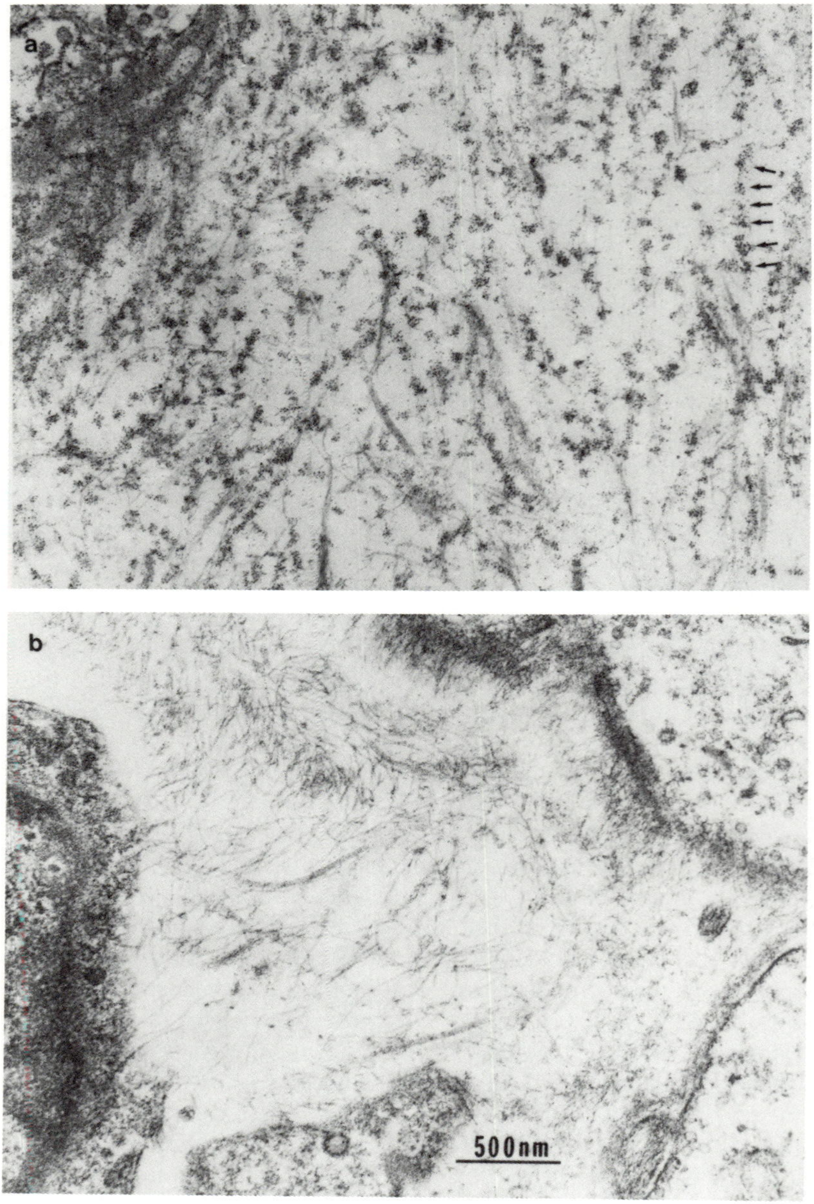

FIG. 12. Ultrastructural localization of collagen VI microfibrils in human placenta (a) by ferritin immunoelectron microscopy. The control section (b) was treated with nonimmune IgG and shows abundant amounts of microfibrils. Most of these microfibrils are stained by anti-collagen V antibodies in a periodic fashion (a) as indicated by arrows in a typical example. The photographs were kindly provided by Dr. Raul Fleischmajer and reproduced with permission from von der Mark *et al.* (1984).

A few earlier studies (Sear *et al.*, 1978, 1981a,b) based on antibody precipitation considered the polypeptides which are now known as α(VI) chains to be subunits of microfibrils usually associated with developing elastin fibers (Cleary and Gibson, 1983). This possibility appears rather unlikely considering the unique morphology observed for collagen VI fibrils (Bruns, 1984). This interpretation is also not supported by immunostaining (see above) and some previous data indicating the noncollagenous nature of elastin microfibrils. A more recent study (Ayad *et al.*, 1985) indicated that α(VI) chains and microfibrillar protein MFP I are different structures, but it is not clear how this interpretation can be related to previous data (Sear *et al.*, 1981a). Several other microfibrillar structures, such as oxytalan fibers or microfibrils stained by ruthenium red, have been identified in the extracellular matrix by morphological studies (Cleary and Gibson, 1983). Whether these fibrils correspond to collagen VI microfibrils (Fig. 12) or contain collagen VI in a different macromolecular organization remains, at present, a matter of speculation.

VI. Possible Functions and Involvement in Disease

The ubiquitous occurrence of collagen VI in extracellular matrices indicates an important biological role for this protein, but the understanding of its function is still based on guesses rather than solid facts. The localization of collagen VI in tissues by immunoelectron microscopy suggested a network of microfibrils which appears to be independent in orientation from the major fibrillar elements (von der Mark *et al.*, 1984). The possible contribution of such a network to the mechanical and biological properties of connective tissue is still ill defined. Trüeb and Bornstein (1984) suggested an adapter function, with collagen VI serving as a link for fibrous and globular proteins with cells. Strong interactions with collagen VI are so far indicated for GP-250 (Section III) but are weak or absent for fibronectin and procollagens (Carter, 1982b).

Another interesting possibility was indicated by the work of Carter (1982a), who demonstrated attachment and spreading of fibroblasts on substrates prepared from native collagen VI. This interaction was dependent on intact disulfide bonds and could not be blocked by antibodies against fibronectin. Similar interactions may exist *in vivo* or in cell cultures since collagen VI fibrils are often deposited in close vicinity to cells (Carter, 1984; von der Mark *et al.*, 1984; Bruns *et al.*, 1986). Some preliminary indications that collagen VI structures may also be involved in platelet aggregation exist (Morton *et al.*, 1985). A further surprising feature of collagen VI is the obvious absence of cross-links other than disulfide bonds or unknown reducible cross-links. This suggests that collagen VI networks may be readily modulated in a reversible manner based, for example, on particularly labile intermolecular bonds reducible with 1–10 m*M* cysteine (Odermatt *et al.*, 1983). Whether the formation of cross-banded structures by minor experimental treatments (Section II,D) reflects such a possibility remains to be demonstrated.

Further clues about the function of collagen VI may be expected from studies of changes found under pathological conditions. The available data are rather scarce. Immunofluorescence demonstrated increased collagen VI staining in cirrhotic liver (Biempica *et al.*, 1977), neurofibroma (Fleischmajer *et al.*, 1985), and elastotic areas of sun-damaged skin (Chen *et al.*, 1986). An increase of collagen VI in fibrotic livers was also demonstrated by radioimmunoassay while changes of serum levels were variable in these patients (Schuppan *et al.*, 1985). This suggests increased production of collagen VI in some, but perhaps not all, fibrotic diseases.

The best evidence for a pathological increase in collagen VI synthesis (4 to 6-fold) is available from a study of cultured fibroblasts of a patient with cutis laxa (Crawford *et al.*, 1985), a rare, acquired or inherited syndrome characterized by loss of skin elasticity. It is not unlikely that patients with other inherited connective tissue disorders (i.e., Marfan syndrome) show aberrant collagen VI synthesis or even possess mutant genes for collagen VI. These intriguing possibilities remain to be studied in the near future.

Acknowledgments

We wish to thank Drs. R. Bruns, R. Fleischmajer, R. Mayne, G. Wick, and Mrs. H. Wiedemann for providing photographs, Drs. G. R. Martin and M.-L. Chu for critical reading of the manuscript, and Mrs. M. Sasso for expert secretarial assistance. As far as the authors' own work is concerned, it was supported by grants of the Deutsche Forschungsgemeinschaft and the Swiss National Science Foundation.

References

Abedin, M. Z., Ayad, S., and Weiss, J. B. (1982). *Biosci. Rep.* **2,** 493–502.

Akiyama, S. K., Yamada, S. S., and Yamada, K. M. (1986). *J. Cell Biol.* **102,** 442–448.

Amenta, P. S., Gay, S., Vaheri, A., and Martinez-Hernandez, A. (1986). *Collagen Relat. Res.* **6,** 125–152.

Aumailley, M., von der Mark, H., and Timpl, R. (1985). *FEBS Lett.* **182,** 499–502.

Ayad, S., Evans, H., Weiss, J. B., and Holt, L. (1984). *Collagen Relat. Res.* **4,** 165–168.

Ayad, S., Chambers, C. A., Shuttleworth, A., and Grant, M. E. (1985). *Biochem. J.* **230,** 465–474.

Banfield, W. G., Lee, C. K., and Lee, C. W. (1973). *Arch. Pathol.* **95,** 262–266.
Biempica, L., Morecki, R., Wu, C. H., Giambrone, M. A., and Rojkind, M. (1977). *Gastroenterology* **77,** 1213 (abstr.).
Bruns, R. R. (1984). *J. Ultrastruct. Res.* **89,** 136–146.
Bruns, R. R., Press, W., Engvall, E., Timpl, R., and Gross, J. (1986). *J. Cell Biol.* **103,** 393–404.
Buckwalter, J. A., Maynard, J. A., and Cooper, R. R. (1979). *Clin. Orthop. Relat. Res.* **139,** 259–266.
Carter, W. G. (1982a). *J. Biol. Chem.* **257,** 3249–3257.
Carter, W. G. (1982b). *J. Biol. Chem.* **257,** 13805–13915.
Carter, W. G. (1984). *J. Cell Biol.* **99,** 105–114.
Carter, W. G., and Hakomori, S. (1981). *J. Biol. Chem.* **256,** 6953–6960.
Chatterjee, P. (1973). *J. Anat.* **115,** 415–420.
Chen, V. L., Fleischmajer, R., Schwartz, E., Palaia, M., and Timpl, R. (1986). *J. Invest. Dermatol.* **87,** 334–337.
Chŭ, M.-L., Mann, K., Deutzmann, R., Pribula-Conway, D., Hsu-Chen, C. C., Bernard, M. P., and Timpl, R. (1987). *Eur. J. Biochem.,* submitted.
Chung, E., Rhodes, R. K., and Miller, E. J. (1976). *Biochem. Biophys. Res. Commun.* **71,** 1167–1174.
Cleary, M. G., and Gibson, M. A. (1983). *Int. Rev. Connect. Tissue Res.* **10,** 97–209.
Cornah, M. S., Meachim, G., and Parry, E. W. (1970). *J. Anat.* **107,** 351–362.
Craviotto, H., and Lockwood, R. (1968). *J. Ultrastruct. Res.* **24,** 70–85.
Crawford, S. W., Featherstone, J. A., Holbrook, K., Yong, S. L., Bornstein, P., and Sage, H. (1985). *Biochem. J.* **227,** 491–502.
Edwards, R. P. (1975). *Br. J. Dermatol.* **93,** 175–182.
Engel, J., Furthmayr, H., Odermatt, E., von der Mark, H., Aumailley, M., Fleischmajer, R., and Timpl, R. (1985). *Ann. N.Y. Acad. Sci.* **460,** 25–37.
Engvall, E., Hessle, H., and Klier, G. (1986). *J. Cell Biol.* **102,** 703–710.
Fleischmajer, R., Timpl, R., Dziadek, M., and Lebwohl, M. (1985). *J. Invest. Dermatol.* **85,** 54–59.
Furthmayr, H., Wiedemann, H., Timpl, R., Odermatt, E., and Engel, J. (1983). *Biochem. J.* **211,** 303–311.
Furuto, D. K., and Miller, E. J. (1980). *J. Biol. Chem.* **255,** 290–295.
Furuto, D. K., and Miller, E. J. (1981). *Biochemistry* **20,** 1635–1640.
Garron, L. K., Feeney, M. L., Hogan, M. J., and McEwen, W. K. (1958). *Am. J. Ophthalmol.* **46,** 27–35.
Gibson, M. A., and Cleary, E. G. (1982). *Biochem. Biophys. Res. Commun.* **105,** 1288–1295.
Gibson, M. A., and Cleary, E. G. (1983). *Collagen Relat. Res.* **3,** 469–488.
Gibson, M. A., and Cleary, E. G. (1985). *J. Biol. Chem.* **260,** 11149–11159.
Goldberg, B., and Green, H. (1964). *J. Cell Biol.* **22,** 227–258.
Green, H., and Meuth, M. (1974). *Cell* **3,** 127–133.
Hashimoto, K., and Ohyama, H. (1974). *J. Invest. Dermatol.* **62,** 106–112.
Hayes, R. L., and Allen, E. R. (1967). *J. Cell Sci.* **2,** 419–434.
Heller-Harrison, R. A., and Carter, W. G. (1984). *J. Biol. Chem.* **259,** 6858–6864.
Hessle, H., and Engvall, E. (1984). *J. Biol. Chem.* **259,** 3955–3961.
Hofmann, H., Voss, T., Kühn, K., and Engel, J. (1984). *J. Mol. Biol.* **172,** 325–343.
Horwitz, A., Duggan, K., Greggs, R., Decker, C., and Buck, C. (1985). *J. Cell Biol.* **101,** 2134–2144.
Jakus, M. A. (1956). *J. Biophys. Biochem. Cytol.* **2** (Suppl.), 243–252.
Jakus, M. A. (1962). *Invest. Ophthalmol.* **1,** 202–225.

Jander, R., Rauterberg, J., Voss, B., and von Bassewitz, D. B. (1981). *Eur. J. Biochem.* **114,** 17–25.
Jander, R., Rauterberg, J., and Glanville, R. W. (1983). *Eur. J. Biochem.* **133,** 39–46.
Jander, R., Troyer, D., and Rauterberg, J. (1984). *Biochemistry* **23,** 3675–3681.
Kajikawa, K., Nakanishi, I., and Yamamura, T. (1980). *Lab. Invest.* **43,** 410–417.
Knight, K. R., Ayad, S., Shuttleworth, C. A., and Grant, M. E. (1984). *Biochem. J.* **220,** 395–401.
Lafuma, C., Moczar, M., and Robert, L. (1982). *Biochem. J.* **203,** 593–601.
Laurain, G., Delvincourt, T., and Szymanowicz, A. G. (1980). *FEBS Lett.* **120,** 44–48.
Lehto, V. P. (1983). *Exp. Cell Res.* **143,** 271–286.
Lehto, V. P., Vartio, T., and Virtanen, I. (1981). *FEBS Lett.* **124,** 289–292.
Leushner, J. R. A., and Haust, M. D. (1986). *Biochim. Biophys. Acta* **883,** 284–292.
Linsenmayer, T. F., Mentzer, A., Irwin, M. H., Waldrep, N. K., and Mayne, R. (1986). *Exp. Cell Res.* **165,** 518–529.
Ludowieg, J. J., and Parker, J. (1973). *J. Cell Sci.* **12,** 925–932.
Luse, S. A. (1960). *Neurology* **10,** 881–905.
Marquart, K. H., Luz, A., and Gössner, W. (1976). *Virchows Arch. A: Pathol. Anat. Histol.* **370,** 177–183.
Marton, L. S. G., and Arnason, B. G. W. (1982). *J. Cell. Biochem.* **19,** 363–381.
Mollo, F., and Monga, G. (1971). *Virchows Arch B* **7,** 356–366.
Morton, L. F., Young, R. D., and Barnes, M. J. (1985). *Collagen Relat. Res.* **5,** 493–503.
Naumann, R. A., and Wolfe, D. E. (1963). *Nature (London)* **198,** 701–703.
Nemetschek-Gansler, H., Meinel, A., and Nemetschek, T. (1977). *Virchows Arch. A: Pathol. Anat. Histol.* **375,** 185–196.
Odermatt, E., Risteli, J., van Delden, V., and Timpl, R. (1983). *Biochem. J.* **211,** 295–302.
Patel, V. P., and Lodish, H. F. (1986). *J. Cell Biol.* **102,** 449–456.
Pillai, P. A. (1964). *J. Ultrastruct. Res.* **11,** 455–468.
Pytela, R., Pierschbacher, M. D., and Ruoslahti, E. (1985). *Cell* **40,** 191–198.
Raimondi, A. J., Mullan, S., and Evans, J. P. (1962). *J. Neurosurg.* **19,** 731–753.
Ramsey, H. J. (1965). *J. Neuropathol. Exp. Neurol.* **24,** 40–48.
Rauterberg, J., Jander, R., and Troyer, D. (1986). *Front. Matrix Biol.* **11,** 90–109.
Risteli, J., Bächinger, H. P., Engel, J., Furthmayr, H., and Timpl, R. (1980). *Eur. J. Biochem.* **108,** 239–250.
Rohen, J. W. (1963). *Arch. Ophthalmol.* **69,** 335–349.
Rojkind, M., Giambrone, M. A., and Biempica, L. (1979). *Gastroenterology* **76,** 710–719.
Sawada, H. (1982). *Cell Tissue Res.* **226,** 241–252.
Sawada, H., Konomi, H., and Nagai, Y. (1984). *Eur. J. Cell Biol.* **35,** 226–234.
Schittny, J. C., Dziadek, M., Timpl, R., and Engel, J. (1985). *Biol. Chem. Hoppe-Seyler* **366,** 846–847.
Schuppan, D., Rühlmann, T., and Hahn, E. G. (1985). *Anal. Biochem.* **149,** 238–247.
Sear, C. H. J., Kewley, M. A., Jones, C. J. P., Grant, M. E., and Jackson, D. S. (1978). *Biochem. J.* **170,** 715–718.
Sear, C. H. J., Grant, M. E., and Jackson, D. S. (1981a). *Biochem. J.* **194,** 587–598.
Sear, C. H. J., Jones, C. J. P., Knight, K. R., and Grant, M. E. (1981b). *Connect. Tissue Res.* **8,** 167–170.
Shuttleworth, C. A., Berry, L., and Wilson, N. H. F. (1982). *Arch. Oral Biol.* **27,** 645–650.
Silberberg, R., Silberberg, M., and Feir, D. (1963). *Pathol. Microbiol.* **26,** 779–783.
Smith, J. W., and Serafini-Fracassini, A. (1968). *J. Cell Sci.* **3,** 33–40.
Tamkun, J. W., DeSimone, D. W., Fonda, D., Patel, R. S., Buck, C., Horowitz, A. F., and Hynes, R. D. *Cell* **46,** 271–282.

Taylor, C. M., Oelbaum, R. S., and Grant, M. E. (1982). *Connect. Tissue Res.* **10,** 319–331.
Trüeb, B., and Bornstein, P. (1984). *J. Biol. Chem.* **259,** 8597–8604.
Trüeb, B., Lewis, J. B., and Carter, W. G. (1985). *J. Cell Biol.* **100,** 638–641.
Trüeb, B., and Winterhalter, K. H. (1986). *EMBO J.* **5,** 2815–2819.
von der Mark, H., Aumailley, M., Wick, G., Fleischmajer, R., and Timpl, R. (1984). *Eur. J. Biochem.* **142,** 493–502.
Yardley, J. H., and Brown, G. D. (1965). *Lab. Invest.* **14,** 501–513.
Zimmermann, D. R., Trüeb, B., Winterhalter, K. H., Witmer, R., and Fischer, R. W. (1986). *FEBS Lett.* **197,** 55–58.

Type VII Collagen

Robert E. Burgeson

Shriners Hospital for Crippled Children and Departments of Biochemistry, Cell Biology, and Anatomy, Oregon Health Sciences University, Portland, Oregon 97201

I. Introduction

The ultrastructure of connective tissues varies dramatically with regard to the size, orientation, and morphology of structural elements. For example, a morphologically simple tissue such as tendon appears to be nearly exclusively composed of a single fiber type. These fibers have a uniform banded appearance, are of a similar diameter, and are all oriented parallel to the direction of applied stress. Disregarding a few fine filaments which occasionally can be seen in micrographs at high resolution, this ultrastructure could be relatively easily explained in terms of a single collagen molecule capable of a limited number of highly specific interactions. In contrast, the matrix of skin is highly complex, particularly at the dermal–epidermal junction. In skin, banded collagen fibers can be visualized which appear grossly similar to the fibers seen in tendon, although the variations of fiber diameters are far greater. Skin contains several fibrous elements in addition to the banded collagen fibers which are not obviously related to collagens based solely upon morphological criteria. However, over the past several years it has become increasingly clear that many of these fibers are unique forms of collagen.

Since the first report of collagen heterogeneity over a decade ago, it has been tacitly assumed that each additional collagen type could eventually be assigned a specific and unique function. In at least some cases this probably has been fulfilled. Type VII collagen is one of the newly identified members of the collagen family. In this article, a summary will be made of the existing data which indicate that this unique macromolecule is one of the largest of the known collagens. It is capable of forming antiparallel dimers stabilized by intermolecular disulfide bonds through a short overlap, which can then condense into an unstaggered lateral array just less than two molecules in length. The fibrils appear not to be capable of further linear growth, but

STRUCTURE AND FUNCTION OF
COLLAGEN TYPES

rather they interact with other molecules (including type IV collagen) to form more extended structures. On the basis of molecular dimensions, banding patterns, tissue distribution, and ultrastructural immunolocalization, type VII collagen has been recognized as a major and perhaps sole component of anchoring fibrils. The anchoring fibrils are believed to function by securing the lamina densa, which underlies certain epithelia, to its subjacent stroma.

II. The Structure of Type VII Collagen

A. *The Triple-Helical Domain*

1. Isolation of the Type VII Collagen Triple Helix

The type VII collagen triple helix has been solubilized from human skin or amniotic membrane by limited proteolysis using pepsin. The digestion conditions need to be optimized for each lot of pepsin, but in general the following procedure has been successfully utilized (Fig. 1). While type VII collagen can be isolated from skin, the use of a simplier tissue such as amnion facilitates the procedure. The amniotic membrane is separated from the chorion within 24 hr following parturition. The wet weight of each preparation should be between 180–200 g, which is the amniotic yield from approximately 25 placentas. Pooled membranes are washed extensively in phosphate-buffered saline to remove blood products. The membranes are then washed in ice-cold water and homogenized on ice using a polytron (Brinkmann Instruments). The final volume should be approximately 3.5 liters.

The homogenate is harvested by centrifugation, and resuspended in 3 liters of ice-cold 1 *M* NaCl, 50 m*M* Tris-HCl, pH 7.5. The suspension is stirred at 4°C for 2–3 hr. Particulate materials from the suspension are harvested by centrifugation, washed with 3 liters of ice-cold water, and reharvested by centrifugation. This material is resuspended in 1 liter of ice-cold 0.5 *M* acetic acid, and crystalline pepsin (Boehringer Mannheim) is added to 0.08 g per 100 g wet wight. Digestion is carried out at 4°C for 16 hr. Undigested materials are removed by centrifugation, and the clarified supernatant solution is gently stirred during the addition of solid NaCl to 10%. Protein precipitation is allowed to proceed for 20 min, at which time the precipitate is harvested by centrifugation. The precipitate is redissolved in 1 liter of 1 *M* NaCl, 0.1 *M* Tris-HCl pH 8.1. This solution is stirred for 72 hr at 4°C to inactivate pepsin.

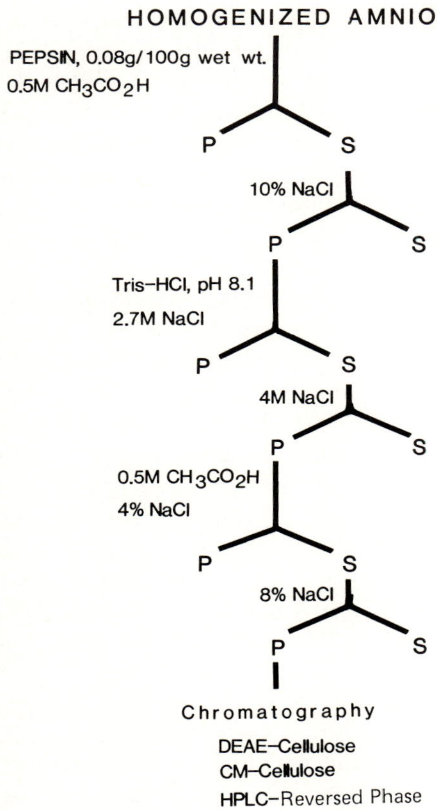

FIG. 1. Abbreviated flow diagram of the purification of type VII collagen from human amnion.

The solubilized collagens are then partially fractionated by differential salt precipitation. Solid NaCl is slowly added to 2.7 *M*. After stirring for several hours, the solution is clarified by centrifugation and by filtering through glass wool. Solid NaCl is then slowly added to 4 *M*, stirred for several hours or overnight, and the precipitate is harvested by centrifugation, redissolved in 250 ml of 0.5 *M* acetic acid, and stirred at 4°C until redissolved. Solid NaCl is slowly added to 4%, and the solution is clarified by centrifugation. Solid NaCl is added to the supernatant to a final concentration of 8%, and the resulting precipitate is harvested by centrifugation and redissolved in 50 ml of ice-cold 2× DEAE buffer (4 *M* urea, 0.5 *M* NaCl, 0.2 *M* Tris-HCl, pH 8.3). Strongly anionic species are removed from the solution by passing the mixture over DEAE-cellulose equilibrated with 2 *M* urea, 0.25 *M* NaCl, 0.1 *M* Tris-HCl, pH 8.3. The flow-through is acidified by the addition of acetic acid to 0.5 *M,* solid NaCl is added to a final concentration of 14%, and the precipitate is harvested by centrifugation. The precipitate is redissolved in 50 ml of cold 0.5 *M* acetic acid and reprecipitated by the addition of 200 ml of absolute ethanol. The harvested precipitate is redissolved in 2× CMC (carboxymethylcellulose) buffer (2 *M* urea, 0.08 *M* lithium acetate, 0.06 *M* LiCl, pH 4.8), and diluted with an equal volume of water.

Type VII collagen is separated from type V collagen by chromatography on carboxymethylcellulose (CM-cellulose). The diluted sample is applied to a 2.5 × 10 cm column equilibrated with 1× CMC buffer at ambiant temperature. Elution is achieved by a superimposed linear gradient from 0.03 to 0.28 *M* LiCl using a total gradient volume of 1 liter. Under these conditions, native type VII collagen elutes prior to type V collagen, at approximately 0.12 *M* LiCl. Column fractions are analyzed by polyacrylamide gel electrophoresis, and identified fractions are pooled, desalted by dialysis against 0.1 *M* acetic acid, and lyophilized. The lyophilizate has a dry weight of approximately 20 mg, about half of which is contributed by type VII collagen and the remainder by an unknown component which is not stained with Coomassie Blue.

Final purification of the native type VII collagen triple-helical domain can be obtained by HPLC-reverse phase chromatography on C18 or C4 columns equilibrated with 0.1% trifluoroacetic acid maintained at 4°C. Elution is performed using a superimposed linear gradient from 0 to 52% acetonitrile. Collected fractions are maintained on ice, pooled, and lyophilized.

2. Characterization of the Triple-Helical Domain

The product of the above isolation procedures is a high-molecular-weight, disulfide-bonded aggregate of identical collagen α chains. Following disulfide bond reduction and denaturation, only a single electrophoretic species is observed with an apparent relative molecular weight of 150,000 to 200,000 (Fig. 2, lane 2), depending upon which collagen chain is used for calibration. A somewhat more accurate estimation of 170,000 was obtained following HPLC-molecular sieve chromatography calibrated against known collagen cyanogen bromide peptides (Bentz *et al.*, 1983). This triple-helical domain peptide was demonstrated to be the product of a unique chain by peptide mapping

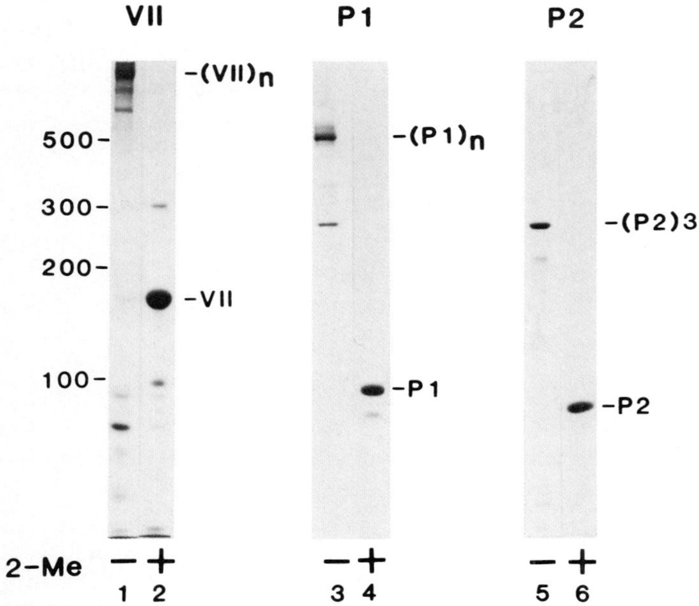

FIG. 2. The effects of disulfide bond reduction on the electrophoretic mobilities of type VII, P1, and P2. Type VII collagen was substantially converted to peptides P1 and P2 by pepsin proteolysis. The reaction products, VII, $(P1)_n$, and $(P2)_n$ were fractionated by HPLC–reversed phase chromatography under nondenaturing conditions. Portions of type VII (lanes 1 and 2), of $(P1)_n$ (lanes 3 and 4), and of $(P2)_n$ (lanes 5 and 6) were analyzed by electrophoresis without reduction (lanes 1, 3, and 5) and following complete reduction (lanes 2, 4, and 6) with 2-mercaptoethanol (2-Me). Relative molecular weights (scale, $M_r \times 10^{-3}$) were determined from the mobilities of disulfide-bonded and reduced type III collagen (not shown). The results show that type VII, P1, and P2 are all independently disulfide bonded. P2 alone is covalently associated into trimeric aggregates by interchain disulfide bonds, whereas both type VII and P1 are covalently associated into higher order aggregates. From Morris *et al.* (1986) with permission.

using cyanogen bromide or V8-protease fragmentation. The peptide also has a unique amino acid composition (Table I). Based on the content of glycine, the composition suggests that the peptide is approximately 90% triple helical and is characterized by a relatively high content of hydroxylysine and a distinctive ratio of leucine to isoleucine. It contains cysteine, but is otherwise unremarkable.

The type VII collagen triple helix has always been present in small amounts in preparations of type V collagen. The amount present varies widely depending on the extent of proteolysis occurring during the solubilization. This is due to the presence of a discontinuity in the

TABLE I
AMINO ACID COMPOSITION OF TYPE VII COLLAGEN[a]

Amino acid	Type VII	P1	P2
Asx	60	63	68
Glx	94	94	102
Hyp	83	105	92
Ser	41	38	35
Gly	303	327	338
His	6.2	Tr[b]	Tr[b]
Arg	65	67	59
Thr	23	18	11
Ala	61	59	60
Pro	82	91	81
Tyr	6.9	3.6	2.4
Val	22	18	21
Met	8.9	3.9	6.9
Cys[c]	5.0	6.8	4.7
Ile	16	7.5	11
Leu	48	42	49
Hyl	41	41	41
Phe	11	6.5	9.3
Lys	13	8.7	8.7

[a]Residues per 1000 residues. From Morris *et al.* (1986) with permission.
[b]Tr, Trace.
[c]Determined as carboxymethylcysteine.

triple helix nearly midway along its length. This discontinuity can be easily demonstrated by reexposure of native isolated type VII collagen triple-helical domain to pepsin (100 μg per ml, 4 hr at 4°C). The type VII collagen triple helix can be quantitatively converted to two pepsin products, noted as P1 and P2 (Fig. 2, lanes 4 and 6). P1 and P2 are limit digestion products at this temperature (Bentz *et al.*, 1983). Densitometric quantitation of the conversion indicates that it occurs with a loss of no more than 5% of the initial mass. The relative molecular weights of P1 and P2 are 89,000 and 73,000 respectively.

Native P1 and P2 peptides can be separated from each other and from type VII collagen by HPLC–reverse phase chromatography at 4°C as described above. However, this separation is extremely sensitive to the extent of pepsin digestion, and the digestion conditions must be optimized before routinely utilizing this methodology. Denatured P1 and P2 peptides are readily separated by HPLC–molecular sieve chromatography using TSK-3000. The uniqueness of P1 and P2 was demonstrated by the observation of different banding patterns of segment long spacing (SLS) crystallites prepared from native P1 and P2 (Burgeson *et al.*, 1985) and by the observation of nonidentical fragmentation patterns observed for P1 and P2 following digestion with cyanogen bromide or V8 protease (unpublished results).

3. Type VII Collagen Solubilized from Tissue Using Limited Pepsin Digestion Is a Disulfide-Bonded Antiparallel Dimer

As indicated previously, the disulfide bond reduction and denaturation product of isolated type VII collagen is a single electrophoretic species. However, prior to disulfide bond reduction, it migrates more slowly than expected for three covalently associated 170 kDa peptides (Fig. 2), suggesting that the chains are associated by intermolecular as well as interchain disulfide bonds. Similarly, both P1 and P2 peptides are independently disulfide bonded. Prior to reduction, P1 peptides migrate more slowly than expected for three covalently associated 89 kDa peptides, but P2 peptides are associated as trimers with no indication of higher molecular weight aggregates. These data suggested that intermolecular disulfide bonds occur between the P1 regions of the triple-helical domain but not between regions containing P2. Using a combination of velocity sedimentation and rotary shadowing, we were able to demonstrate that both whole type VII collagen triple helix and P1 peptides are associated as hexamers, while P2 peptides are associated as trimers (Morris *et al.,* 1986). The sedimentation profiles of unreduced P1, P2, and the uncleaved triple helix are sufficiently monodispersed to suggest that no covalently stabilized aggregate larger than a dimer of two molecules commonly occurs.

The type VII collagen triple helix, P1, and P2 were observed by transmission electron microscopy as SLS crystallites (Burgeson *et al.,* 1985) and as individual molecules by rotary shadowing. Whole triple-helical molecules (Fig. 3) and P1 peptides are seen as dimeric structures interacting through an overlap region. The dimensions of these molecules and of the overlap were calculated and are shown in Table II. The banding patterns of SLS crystallites made from whole type VII triple helix and from P1 are centrosymmetric, again indicating an antiparallel arrangement of two molecules stabilized by intermolecular disulfide bonds within an overlap of 60 nm of the P1 region.

The amino acid sequences of the amino-terminal residues of the whole triple helix, and of P1 and P2, were determined. The sequences obtained from the whole triple helix and from P1 were identical and

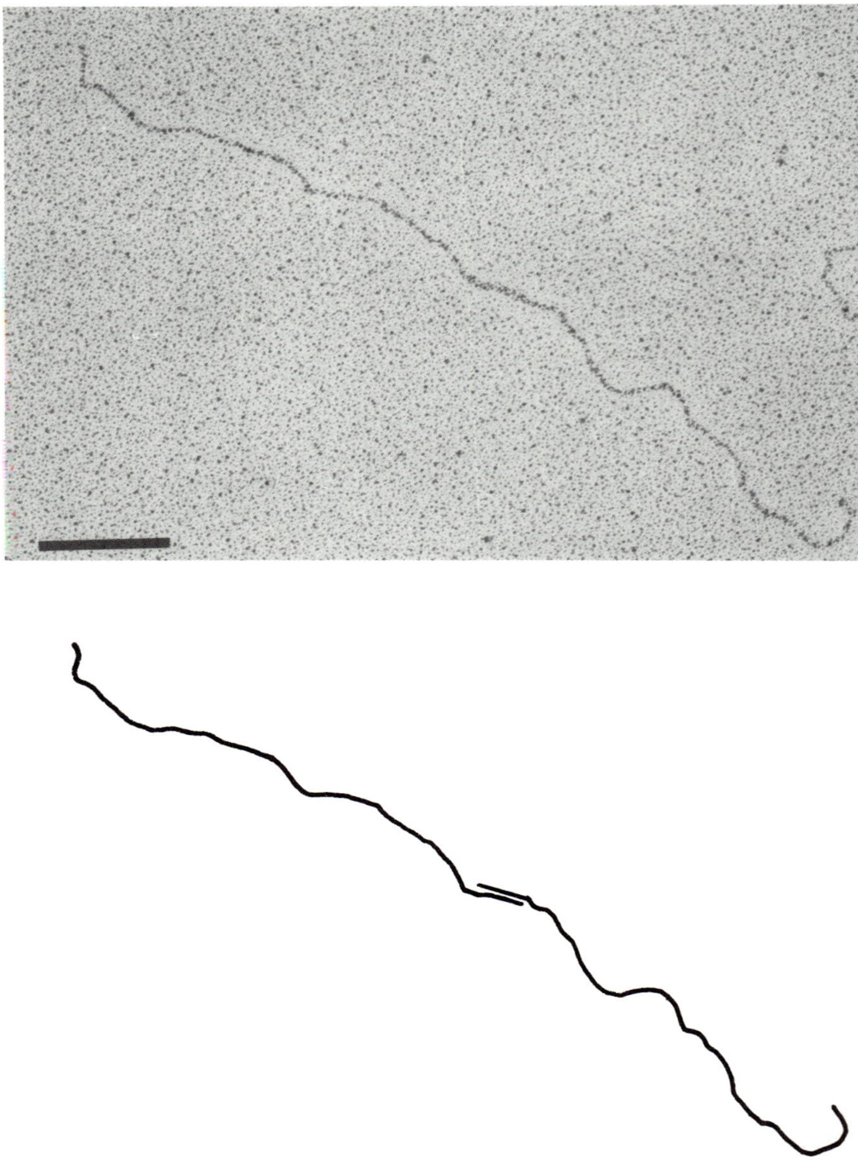

FIG. 3. Dimers of type VII collagen as visualized by transmission electron microscopy following rotary shadowing. Purified native type VII collagen was examined by electron microscopy. The image corresponds to a 780-nm dimer of two type VII collagen triple-helical domains, each approximately 425 nm in length, overlapping by a distance of 60 nm. Bar, 100 nm.

TABLE II
MOLECULAR DIMENSIONS AND OVERLAP[a]

	SLS measurement (nm)	Rotary shadowing measurement (nm)
Type I	295 (70%)	297
Type V	302	
Type VII (individual molecule)	424 (100%)	467
P1	231	
P2	175	
Total (P1 + P2)	406 (96%)	
Type VII (dimer)		
Calculated (end to end)	848[b]	
Measured	785	796
Calculated overlap[c]	63	
P1 dimer		
Calculated (end to end)	462[d]	
Measured	405	
Calculated overlap[c]	57	
Measured overlap		59–75

[a]From Burgeson *et al* (1985), with permission.
[b]Represents 2 × 424.
[c]Overlap is obtained by subtracting measured from calculated values.
[d]Represents 2 × 231.

distinct from the sequences obtained from P2, indicating that the overlap between type VII collagen molecules occurs at the amino-terminus (Morris *et al.,* 1986).

The helix-to-coil transition of the type VII collagen triple-helical domain when measured by circular dichroism is biphasic, with a midpoint of 41°C for the major transition accounting for approximately 85% of the amplitude of the total transition and a midpoint of approximately 55°C for the minor transition (N. P. Morris and H. P. Bächinger, personal communication). The biphasic nature of the transition profile is entirely dependent upon intact disulfide bonds. These data suggest that the intermolecular disulfide bonds contained within the overlap region confer additional stability to that portion of the triple helix. By analogy to studies of the relative insensitivity of the "7 S" domain of type IV collagen to bacterial collagenase digestion, the overlap region of native type VII collagen was isolated following clostridial collagenase digestion. These studies allowed the isolation of a disulfide-bonded hexamer with a relative molecular weight for each peptide of approximately 20,000 (Morris *et al.,* 1986). This peptide accounts for approximately 90% of the overlap region and is termed C-1

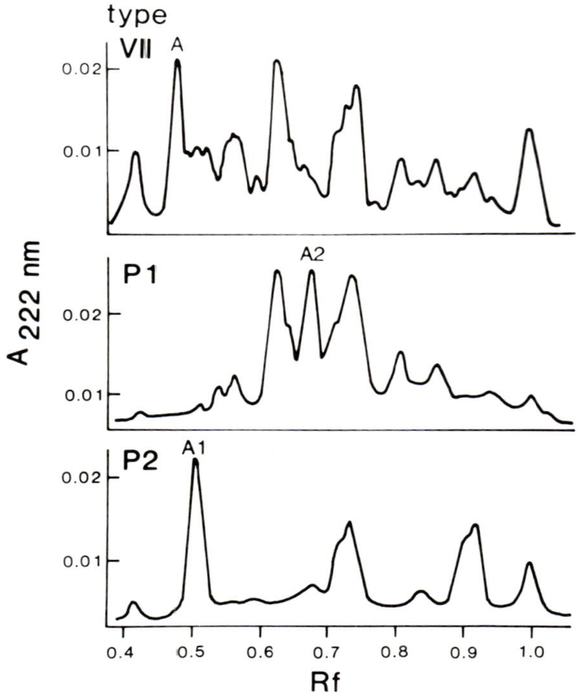

FIG. 4. Cyanogen bromide peptide profiles of type VII, P1, and P2 as resolved by HPLC–molecular sieve chromatography. Cyanogen bromide (CNBr) peptides of type VII, P1, and P2 were individually prepared and analyzed under identical conditions on TSK-3000. The patterns indicate that P1 and P2 are distinct peptides which sum to give the pattern of whole type VII. In addition, peptide A in type VII contains the pepsin cleavage site, as this peptide disappears following reexposure of type VII to pepsin, and the peptides A1 in P2 and A2 in P1 appear following cleavage. From Morris *et al.* (1986) with permission.

(see Fig. 9). Thus, the intermolecular disulfide bonds between type VII collagen triple-helical domains are confined to a 60-nm-long segment at the amino-terminus of each molecule. Interchain disulfide bonds may also occur in this region but thus far have not been directly demonstrated.

As described previously, P2 peptides are covalently associated by interchain disulfide bonds. In order to investigate the location of the interchain disulfide bonds, cyanogen bromide peptides of whole type VII collagen triple helix were analyzed before and following disulfide bond reduction. Reduction-induced changes in electrophoretic mobility were confined to peptides smaller than 20 kDa (Morris *et al.*, 1986). A large fragmentation-resistant peptide (M_r 82,000, termed CB-A), which was unchanged by reduction (Fig. 4), has consistently been observed. This peptide lacks a stoichiometric relationship to the other peptides between digests and is therefore likely to be an uncleaved peptide with a partially digestion-resistant methionine residue. This large peptide is absent from digests of P1 or P2, indicating that the site of pepsin-mediated protease activity occurs within this large fragment. Following pepsin digestion, CB-A partitions between P1 and P2 as CB-A2 and CB-A1, respectively. CB-A1 accounts for approximately 79% of the mass of P2, beginning at the amino-terminus. CB-A1 and CB-A2 are not disulfide bonded; therefore, the interchain disulfides occurring in P2 must be located in the carboxy-terminal 18% of the peptide. Interchain disulfide bonds do not occur within the pepsin-sensitive triple-helical discontinuity between P1 and P2. These data are summarized in Fig. 9.

B. *The Structure of the Type VII Collagen Biosynthetic Product*

Cell Line Selection

Identifiable type VII collagen is not easily solubilized from either skin or amnion by extraction using acetic acid or nondenaturing neutral solutions. Extractions using 6 *M* guanidine hydrochloride allowed the observation by rotary shadowing of dimeric molecules with a helix length identical to type VII collagen, which contained large disperse globules at the distal ends (R. Glanville, personal communication). The visualization of molecules similar to type VII collagen, with large terminal globular domains and the substantial insolubility of type VII collagen, strongly suggested that pepsin-mediated solubilization of type VII collagen occurs by removal of a noncollagenous domain present within the tissue. In order to fully assess this possibility, we sought to obtain a cell culture system which would allow us to isolate and characterize the type VII collagen biosynthetic product. To aid this search, we prepared a monoclonal antibody using whole type VII collagen triple helix as antigen. We demonstrated the antibody to be specific for type VII collagen and, by rotary shadowing of the antigen–antibody complex, demonstrated that the epitope occurs at the carboxyl-terminus of type VII collagen dimers (Sakai *et al.*, 1986). Using this antibody, we screened a variety of primary cell cultures and established cell lines by indirect immunofluorescence for their ability to secrete the antigen in culture. Primary cultures of human amniotic epithelial cells and skin basal cells demonstrated reactivity with the antibody as did the cell lines KB (Eagle, 1955) and WISH (Hayflick, 1961). Because of the ease of propagating the established cell lines, these were chosen for further study. The KB cell line is derived from a human oral basal cell carcinoma, while the WISH line arises from a spontaneous alteration of cultured amniotic epithelial cells. By a variety of criteria, the biosynthetic products of these two cell lines are extremely similar, and the cell lines were used indiscriminately for the following studies. Either cell line was grown to confluency in 50 T-100 flasks using standard tissue culture medium supplemented with ascorbic acid. On alternate days, the cell medium was harvested. The biosynthetic profile over 10 days appeared unchanged. Media proteins were precipitated using ammonium sulfate, and type VII collagen was partially purified by extraction of the precipitate, followed by a combination of DEAE-chromatography and velocity sedimentation (Lunstrum *et al.,* 1986). Separations were monitored by ELISA assay utilizing the monoclonal antibody. The partially purified type VII collagen and total unfractionated media proteins were analyzed by Western blotting using the monoclonal antibody (Lunstrum *et al.,* 1986). In both cases, the antibody recognizes a single peptide that has an apparent relative molecular weight of approximately 350,000 following disulfide bond reduction (Fig. 5). Since this peptide contains both triple-helical and globular domains, the true molecular weight could not be assessed by this method. To identify the globular domain, the electrophoretic gel band recognized by the antibody was excised from polyacrylamide gel patterns obtained from the purest preparations of biosynthetic type VII and used to produce polyclonal antibodies in rabbits. The resulting antibodies produced identical immunofluorescence and Western blotting patterns to the monoclonal antibody and did not recognize any species in unfractionated total cell protein other than type VII collagen biosynthetic product (Lunstrum *et al.,* 1986). Partially purified biosynthetic type VII collagen was subjected to limited pepsin digestion under native conditions. The product was resolved by electrophoresis following disulfide bond reduction and analyzed by Western blotting using the polyclonal antibody. As expected, the antibody recognized a pepsin-resistant region identical in size to the tissue solubilized type VII collagen triple helix (Fig. 6). In addition, several semistable species were observed with relative molecular weights between 50,000 and 150,000 as calibrated using globular standards.

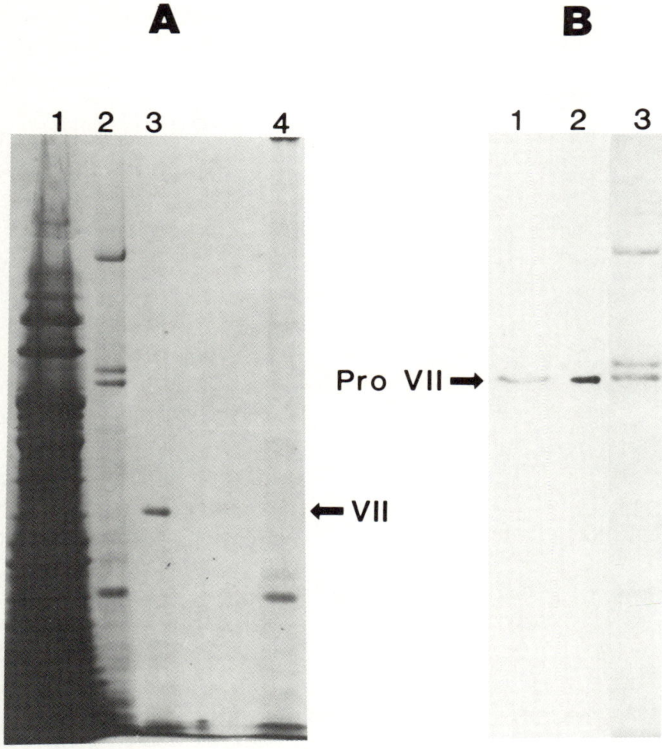

FIG. 5. Immunoidentification of type VII procollagen. (A) KB cell extract (lane 1), a preparation of secreted media proteins (lane 2), and type VII collagen (lane 3), pepsin extracted from human amniotic membrane, were separated by polyacrylamide gel electrophoresis, and proteins were visualized by Coomassie Blue staining. The media fraction was also analyzed under nonreducing conditions (lane 4). (B) Identical cell (lane 1) and media (lane 2) samples were transferred to nitrocellulose and immunoblotted with the type VII collagen-specific monoclonal antibody. A direct comparison is made to media proteins transferred and stained with amido black (lane 3). The data demonstrate that a type VII collagen biosynthetic product is substantially larger than that obtained from whole tissues by pepsin extraction. From Lunstrum *et al.* (1986) with permission.

These were subsequently demonstrated to be semistable fragments of the carboxy-terminal globular domain (Lunstrum *et al.*, 1986). A similar preparation of biosynthetic type VII collagen was subjected to limit bacterial collagenase digestion. The reduction and denaturation product was then analyzed by Western blotting using the same polyclonal antibody. The antibody recognizes a single species with an M_r of

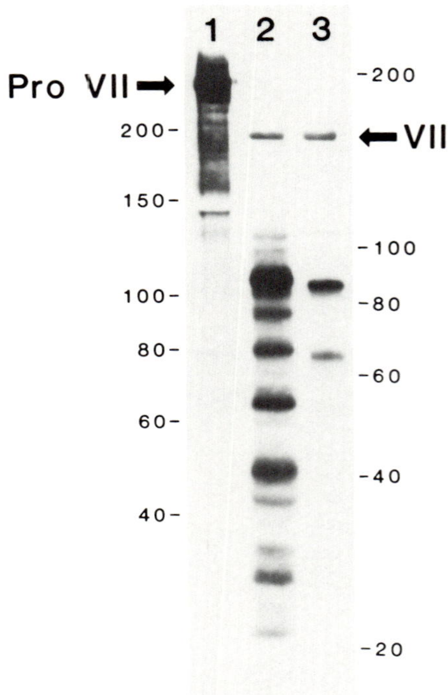

FIG. 6. Immunoblot analysis of pepsin-digested type VII procollagen. KB cell media fractions were incubated without (lane 1) and with (lane 2) pepsin. These samples and type VII collagen (lane 3) were fractionated by gel electrophoresis and immunoblotted with a type VII procollagen-specific antiserum. The relative-molecular-weight scales ($M_r \times 10^{-3}$) for globular (left) and collagen helical (right) peptides are indicated and were determined using globular protein standards and type I collagen CNBr peptides. The antibody identifies a pepsin product identical in size to type VII collagen obtained from whole tissue using pepsin solubilization. In addition, the antisera identifies a series of pepsin digestion intermediates with consistent and reproducible molecular weights. These results suggest that the non-triple-helical domains of type VII collagen are complex and may have subdomain structure. From Lunstrum *et al.* (1986) with permission.

approximately 150,000 following calibration with globular standards (Fig. 7). The results indicated that biosynthetic type VII collagen contains a triple-helical domain identical to that isolated from whole tissue by pepsin digestion and at least one globular domain containing three identical chains with an M_r of approximately 150,000 each.

To assess the location of this large globular region, partially purified biosynthetic type VII collagen was visualized by transmission electron microscopy following rotary shadowing (Fig. 8). The images obtained

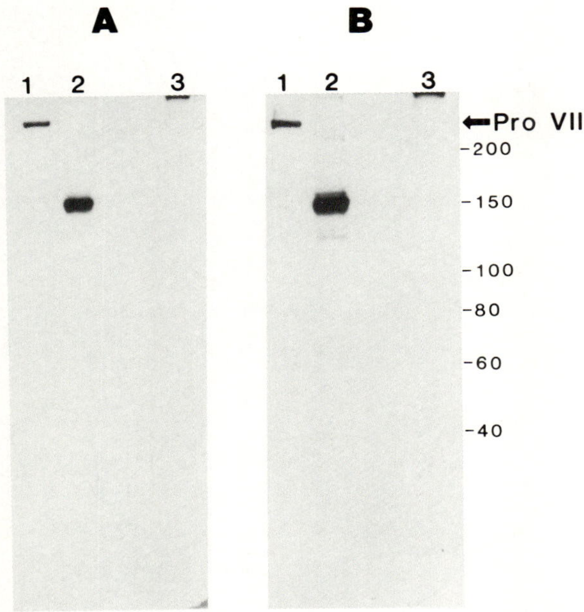

FIG. 7. Identification of an unusually large carboxy-terminal non-triple-helical domain in type VII procollagen. Media fractions were incubated without (lane 1) and with (lane 2, +2-Me; lane 3, −2-Me) clostridial collagenase. Duplicate samples were fractionated by gel electrophoresis. Immunoblot analysis was performed with both the type VII collagen-specific monoclonal antibody (A) and the type VII procollagen-specific polyclonal antiserum (B). The relative-molecular-weight scales ($M_r \times 10^{-3}$) were determined using globular standards. The results indicate that both antibodies identify a single peptide after disulfide bond reduction with an apparent relative molecular weight of approximately 150,000. From Lunstrum *et al.* (1986) with permission.

indicate a mixture of monomeric and dimeric molecules containing triple-helical regions equivalent in length and overlap distance to those obtained from purified, pepsin-solubilized molecules. Both the biosynthetic monomers and dimers demonstrate large and complicated arrays of globules at the carboxyl-terminus of the molecules. The complexity of these regions strongly suggests a substantial subdomain structure, with each carboxyl-terminal region demonstrating three identifiable arms, each of which appears to be composed of one or more globules. This observation of multiple subdomains at the carboxyl-terminus is consistent with the immunological identification of several semistable pepsin fragments generated from the digestion of biosynthetic type VII collagen. In addition to the large globules at the carboxyl-terminus (NC1, Fig. 9), a small globule is obvious in both the

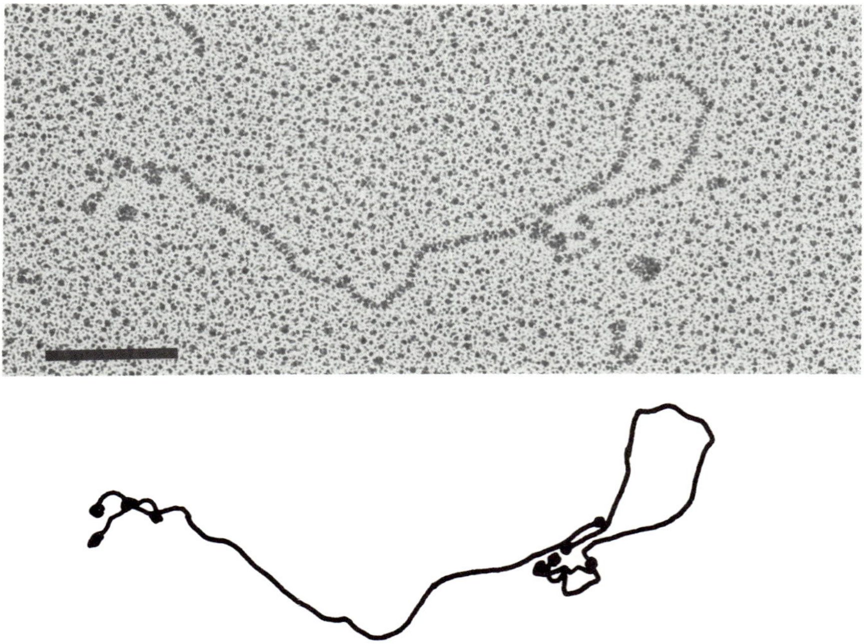

FIG. 8. Transmission electron microscopical visualization of type VII procollagen following rotary shadowing. Partially purified type VII procollagen obtained from the media of cultured KB cells was examined by electron microscopy. Molecules were identified that had triple-helix and overlap lengths identical to pepsin-solubilized type VII collagen. In addition, these procollagen molecules contain large and complex carboxy-terminal globular domains, as well as small amino-terminal globules. Bar, 100 nm.

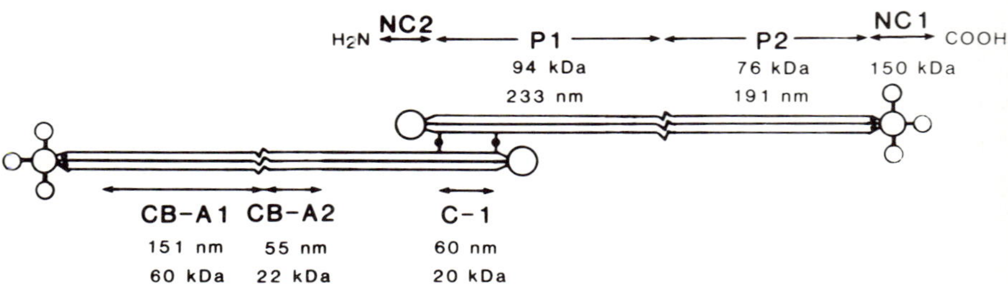

FIG. 9. Diagrammatic representation of our current understanding of type VII procollagen. Disulfide bonds are indicated by small solid circles. The model depicts the relationships in size and relative molecular weight of the various peptide regions which have been characterized. The structures and relative sizes of the non-triple-helical domains are not to scale, and the actual substructures of these regions are far more complex than depicted.

monomers and dimers at the amino ends flanking the overlap region (NC2, Fig. 9). These amino-terminal globular domains are apparently not recognized by the polyclonal antibodies, and therefore have not yet been further characterized.

The observation of an amino-terminal globular domain (NC2) was not entirely unexpected. Velocity sedimentation analyses of pepsin-solubilized type VII collagen triple helix have suggested that these domains are required for dimerization and subsequent covalent stabilization (Morris *et al.,* 1986). Velocity sedimentation of the triple helix following disulfide bond reduction without denaturation of the triple helix demonstrates that the dimers dissociate to monomers even under nearly physiological conditions. These data suggest that the interactions between the triple helices within the overlap region are not sufficiently stable to resist dissociation under these conditions. It is likely, then, that the initial association of the overlap region is mediated and stabilized by the amino-terminal globules prior to disulfide bond formation.

The observation of both biosynthetic monomers and dimers by rotary shadowing suggests that dimerization can occur in the cell culture media of both KB and WISH cells. At this time, it is not clear whether these dimers are stabilized by disulfide bonding. Velocity sedimentation analyses occasionally suggest the presence of covalently stabilized dimers, but at this time these data are equivocal. The biosynthetic studies do suggest that the synthesis and secretion of biosynthetic type VII collagen is a slow process. Although the kinetics of secretion, dimer formation, and disulfide-bond stabilization have not yet been rigorously studied, the present data suggest that substantial amounts of biosynthetic type VII collagen do not accumulate in cell culture media during the first half of the 24-hr collection period.

Thus far, biosynthetic studies have shown no evidence of proteolytic processing of biosynthetic type VII collagen in cell culture, although these studies have not been rigorously pursued. As has been the case with other collagen types, the use of cell culture systems is usually not optimal for detection of processing, and these studies are more suitable to organ culture systems utilizing long term pulse–chase protocols. However, the electrophoretic mobility of biosynthetic type VII collagen derived from cell culture has been compared by Western blot analysis to that of molecules extracted from whole tissue without pepsin digestion (G. Lunstrum, personal communication). These preliminary studies suggest that the extracted molecule is slightly smaller than that accumulated within the cell culture media. Rotary shadowing of extracted molecules demonstrates images lacking part or all of the amino-terminal globular domain, suggesting that if processing occurs physiologically, it may be restricted to the amino-terminal end of the molecules. Since we believe these globular domains to be responsible for type VII collagen dimerization, we would predict that processing would occur only following intermolecular disulfide bond formation within the overlap region.

III. The Function of Type VII Collagen

A. Tissue Distribution

As with all collagens, the fiber, not the individual molecule, is the physiological subunit of connective tissue function. To ascertain the function of type VII collagen, the fiber form of this macromolecule was investigated, initially by determining the tissue distribution of type VII collagen (Sakai *et al.,* 1986) and subsequently by attempting to identify the ultrastructural correlate of the type VII collagen fiber (Sakai *et al.,* 1986; Keene *et al.,* 1987). Finally, an attempt was made to identify inherited disease conditions in which accumulation of the macromolecule is deficient.

The tissue distribution of type VII collagen was determined by indirect immunofluorescence using the monoclonal antibody to the carboxyl-terminus of the triple-helical domain of type VII collagen as previously described (Sakai *et al.,* 1986). A variety of tissues were surveyed, including skin, chorioamnion, placenta, lung, oral mucosa, tongue, bladder, vaginal mucosa, cartilage, and cornea (Table III). In general, the antibody produced crisp, brilliant fluorescence in a linear pattern underlying the epithelial cells of many external tissues. In skin, the basement membrane zone at the dermal-epidermal junction was brightly fluorescent, but no staining of the dermal vasculature or adnexal structures was observed. In chorioamnion, linear staining was found immediately under the amniotic epithelial cells corresponding to the basement membrane zone, but was not seen anywhere within the stroma or the chorionic pseudo basement membranes. No staining of placenta, muscle, normal lung, bladder, or cartilage was observed. The data indicate that type VII collagen is found within the basement membrane zone of certain epithelial cell layers. Staining could not be directly correlated with the complexity of the epithelial cell layer nor with the embryologic origin of the cells presumed to be responsible for its synthesis.

As discussed earlier, the banding pattern of SLS crystallites made

TABLE III
CORRELATION OF THE PRESENCE OF TYPE VII COLLAGEN AS DETECTED BY IMMUNOFLUORESCENCE WITH ANCHORING FIBRILS DETECTED ULTRASTRUCTURALLY[a]

	Immunofluorescence	Anchoring fibrils
Skin	+	+ Brody (1960)
Chorioamnion	+	+ Burgeson *et al.* (1986)
Placenta	−	−
Muscle	−	−
Eye	+	+ McTigue and Fine (1966)
Oral mucosa	+	+ Susi *et al.* (1967)
Cervix	+	+ Laguens (1972)
Kidney	−	−
Lung	−	−
Liver	−	−
Stomach	−	−
Intestine	−	−
Elastic cartilage	−	−

[a]From Sakai *et al.* (1986) with permission.

from pepsin-solubilized type VII collagen shows a distinctive centrosymmetric pattern. A striking similarity between this pattern and that observed ultrastructurally along anchoring fibrils was noted by Dr. Romain Bruns. Since the ultrastructural identification of anchoring fibrils correlated well with the tissue distribution of type VII collagen as demonstrated by indirect immunofluorescence, the possible relationship between the anchoring fibril and type VII collagen was further investigated.

B. Anchoring Fibrils

Anchoring fibrils are broad, centrosymmetrically banded structures approximately 800 nm in length (Palade and Farquhar, 1965; Bruns, 1969). They occur within the subbasal lamina of numerous external tissues and appear to have two distinct morphologies. They are often seen as "horseshoe-shaped" structure with both distal ends inserted into the lamina densa and the banded portion of the structure looping into the subbasal lamina often surrounding large banded collagen fibrils. More frequently, they appear to have one end inserted into the lamina densa and extend perpendicularly to the basement membrane with the opposite end inserted into an electron-dense amorphous patch (Keene *et al.*, 1986). These amorphous patches have been thought to be tangential sections of basement membrane undulations (Palade and Farquhar, 1965). Anchoring fibrils are believed to function by providing additional structural stability to the subbasal lamina of selected tissues by physical entrapment of dermal matrix components. The proposed attachment function must either be in addition to or substitute for other attachment mechanisms, as not all epithelia overlying a connective tissue stroma contain anchoring fibrils. A comparison of the banding patterns of anchoring fibrils with those observed for type VII collagen SLS crystallites suggests that the main body of the anchoring fibril is composed of the type VII collagen triple-helical domain. The carboxy-terminal globular domain (NC1) would appear to insert and be encapsulated within the basement membrane or the amorphous electron-dense patches. If this model were correct, monoclonal antibodies directed to the carboxy-terminal end of the triple-helical domain should recognize an epitope at the distal ends of banded anchoring fibrils and localize the site of the insertion of the anchoring fibril into the amorphous electron-dense patches within the subbasal lamina. This hypothesis was tested by ultrastructural immunolocalization using *en bloc* impregnation of the above-described monoclonal antibody into unfixed tissue followed by reaction with a second antibody conjugated with colloidal gold (Sakai *et al.*, 1986). As shown in Fig. 10A, gold deposits are found in the predicted sites. The intensity of gold labeling at the basal lamina is far less than expected. However, this would appear to be a result of the *en bloc* antibody penetration procedure, as antibodies to basal lamina components such as type IV collagen or laminin also have difficulty penetrating the lamina densa of the dermal–epidermal junction under identical conditions (Keene *et al.*, 1986). The diffusional barrier to antibody penetration is different at the dermal–epidermal junction than at other skin basement membranes, for uniform labeling with anti-type IV collagen was easily obtained for vascular basement membranes present in skin even using the *en bloc* penetration method. Surface labeling of either Lowacryl-embedded skin directed by antibodies to either type IV collagen or to type VII collagen (figure 10B) show intense labeling of the lamina densa, but banded anchoring fibrils are somewhat more difficult to identify (Keene *et al.*, 1987). These data demonstrate that anchoring fibrils are specifically labeled as predicted by the proposed structure of type VII collagen. This specific labeling, together with the strong similarity in banding patterns between anchoring fibrils and type VII collagen SLS crystallites in addition to the correlation of tissue distribution shared by anchoring fibrils of type VII, strongly support the identification of type VII collagen as a major component of anchoring fibrils.

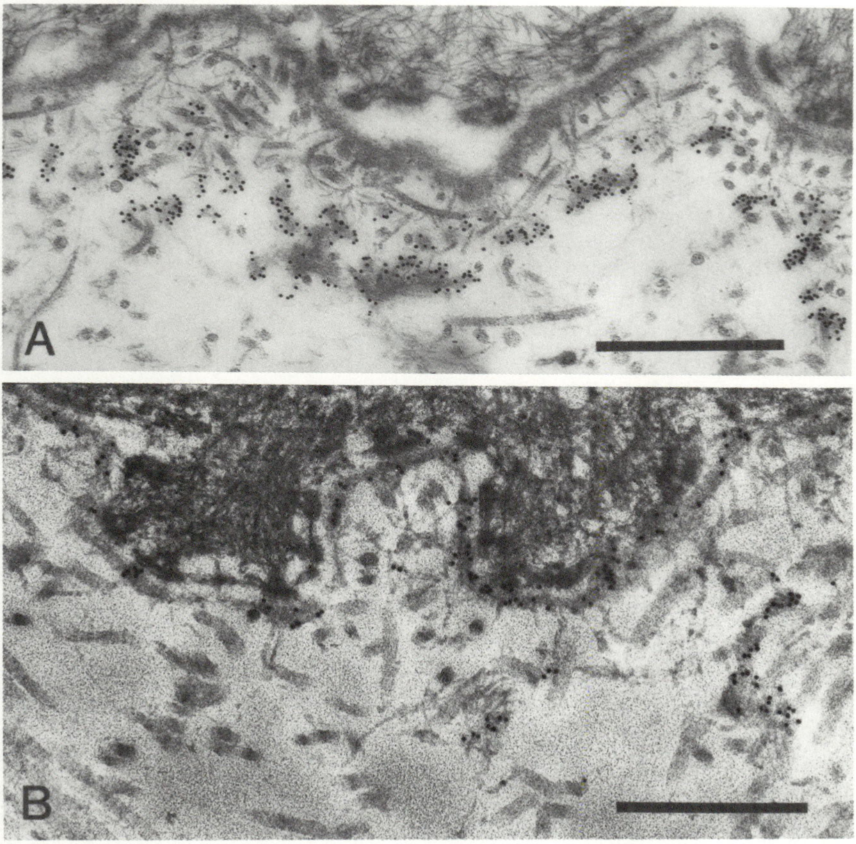

FIG. 10. Immunoelectron microscopic localization of type VII procollagen at the dermal–epidermal junction of human skin. (A) Monospecific anti-type VII procollagen antisera was infused into unfixed tissue. Following washing, colloidal gold-containing second antibodies were similarly applied, and unbound antibody was removed by extensive washing. The tissue was then fixed, stained, and sectioned for examination. By this technique, gold is consistently localized to electron-dense patches at the dermal termini of anchoring fibrils. (B) The same antisera was applied to the surface of sections cut from Lowacryl-embedded foreskin. By this technique, gold is deposited upon the basal lamina in addition to the dermal plaques. Bar, 500 nm.

C. *The Anchoring Fibril Network*

In the course of these experiments, it was consistently observed that intense antibody-directed gold deposition was seen over the amorphous electron-dense patches at the dermal termini of anchoring fibrils. Furthermore, these patches occur frequently along the basement membrane within the subbasal lamina and were most commonly found within 0.5–1.5 μm of the basal lamina but occasionally were seen as far as 2 μm distant. The regularity of their appearance appeared to be inconsistent with their identification as tangential sections of basement membrane undulations as previously suggested. In order to test this concept, human skin which had been labeled *en bloc* with antibodies to type VII collagen, and therefore showed intense labeling of these regions, were serial sectioned, and the appearance and disappearance of these regions were observed (Keene *et al.*, 1987). The electron-dense amorphous regions appear and disappear within these serial sections and demonstrate no continuity with the basement membrane other than that provided by interconnecting anchoring fibrils. These observations led to the conclusion that these patches were independent of the lamina densa but retained an ultrastructural resemblance to that structure. We have termed these electron-dense regions as anchoring plaques. These plaques appear throughout the subbasal lamina and are interconnected by anchoring fibrils. They are frequently seen at a distance of more than one anchoring fibril length from the lamina densa and nearly always appear to be the sites of anchoring fibril insertion. These observations led to the hypothesis that the most common form of anchoring fibril is that which extends perpendicularly from the lamina densa, inserting into the lamina densa at one end and into the anchoring plaque at the other. Anchoring plaques are interconnected by additional anchoring fibrils which both insert and terminate in plaques. Thus, we hypothesize an extended network of anchoring fibrils forming a large scaffold throughout the subbasal lamina (illustrated in Fig. 11). This extended architectural superstructure is capable of entrapping large numbers of stromal connective tissue elements including large, banded collagen fibrils and noncollagen microfibrils. This concept is particularly appealing in that the attachment function of the anchoring fibril has been questioned because only a few of the anchoring fibrils could be demonstrated to physically entrap dermal connective tissue components (Wasano and Yamamoto, 1985). The concept of an extended superstructure circumvents these objections.

Since the anchoring plaques grossly resemble the basal lamina, their composition was investigated using several polyclonal and monoclonal antibodies (Keene *et al.*, 1987). Several monoclonal antibodies were previously produced to the carboxy-terminal globular domain of type VII collagen. A polyclonal antisera, which recognizes the carboxy-terminal globular domain as well as the carboxy-terminal end of the triple-helical domain of type VII collagen, was also produced in rabbits

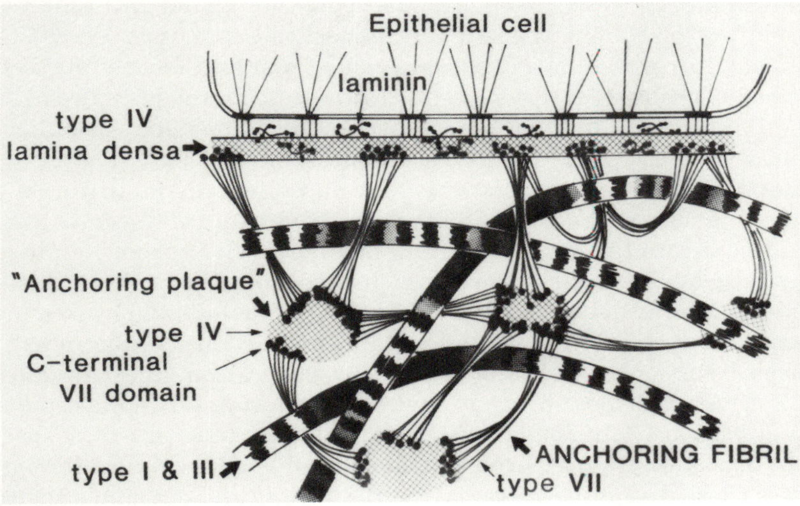

FIG. 11. Diagrammatic representation of the anchoring fibril network. Our current hypothesis for the structure of the anchoring fibril network predicts that anchoring fibrils are composed of lateral aggregates of type VII procollagen aligned in a nonstaggered array. Anchoring fibrils most frequently originate within the lamina densa and extend perpendicularly to the basement membrane, inserting into anchoring plaques within the subjacent matrix. Additional anchoring fibrils originate from these plaques and insert into other anchoring plaques farther into the matrix. The anchoring plaques contain both the carboxy-terminal of type VII procollagen and type IV collagen. The resulting structure is an extended network forming a complex scaffolding within the subbasal lamina. This extended network is capable of physical entrapment of large numbers of matrix components within this region forming a firm connection between the lamina densa and the stromal matrix. From Keene *et al.* (1987) with permission.

(Lunstrum *et al.*, 1986). Immunoelectron microscopic localization studies using both the polyclonal antibody and the monoclonal antibodies to the carboxy-terminal globular domain produce identical results to those obtained using the monoclonal antibody which recognizes the carboxyl-terminus of the triple-helical domain. A monoclonal antibody to type IV collagen was previously described (Sakai *et al.*, 1982) which recognizes an epitope on each arm of type IV collagen "spiders" about 60 nm from the 7 S region (Dieringer *et al.*, 1985). This anti-type IV collagen monoclonal antibody produced gold deposition on the anchoring plaques, indicating that they contain type IV collagen. To be certain that the plaques identified by the type IV and type VII antibodies were identical, double-labeling experiments were performed using the polyclonal rabbit antisera to the carboxy-terminal domain of type VII collagen identified by 20-nm colloidal gold-conjugated anti-rabbit IgG second antibody and the murine monoclonal anti-type IV collagen antibody recognized by a 5-nm gold-conjugated anti-mouse IgG second antibody. The results (Keene *et al.*, 1987) indicate the presence of both type IV collagen and type VII collagen carboxy-terminal globular domain within the same anchoring plaques.

Because of the ultrastructural and compositional similarities of the anchoring plaques and the lamina densa, the origin of these structures could be accounted for by a variety of events. The anchoring plaques may arise by tensional stress in which anchoring fibrils may pull away pieces of lamina densa into the subbasal lamina, followed by basement membrane repair. The plaques could also arise from tissue injury, or during basement membrane repair or regeneration occurring during growth. Another possibility is that the anchoring plaque is a physiologically significant component of the subbasal lamina and is specifically deposited by regulated developmental events. In an attempt to assess these possibilities, we have ultrastructurally examined the subbasal lamina regions from a variety of tissues, in skin at a variety of locations, in skin as a function of age, and in skin of varying species. These surveys, while not exhaustive, indicate that the anchoring fibril network as detected by the presence of anchoring plaques exists continuously in all tissues examined. It appears to be most extensive in the cornea (Fig. 12), but is clearly obvious in all other tissues containing anchoring fibrils. In skin, the presence or extent of the network does not vary with the site of biopsy, as skin samples taken from the lower thorax, leg, thigh, footpad, or lip are ultrastructurally equivalent with regard to the anchoring plaques. Dorsal skin from neonates and from 16-, 31-, and 70-year-old individuals are extremely similar in this respect, although some basal lamina reduplication was present in the specimen from the 70-year-old individual. No difference in the basic occurrence of the anchoring fibril network was observed between human, bovine, chick, or mouse skins. These results strongly suggest that the anchoring plaques arise from an organized developmental program and serve a specific physiological function. Their continuous and consistent observation suggest that they do not arise as a function of injury and subsequent repair processes.

There has been considerable speculation that the anchoring fibril passes from the dermis, through the lamina densa, and into the lamina rara where it inserts into the hemidesmosomes of the basal cell surface of skin. Our structural analyses of type VII collagen as related to anchoring fibril structure are inconsistent with this concept. To the extent that structure can actually be predicted from images produced

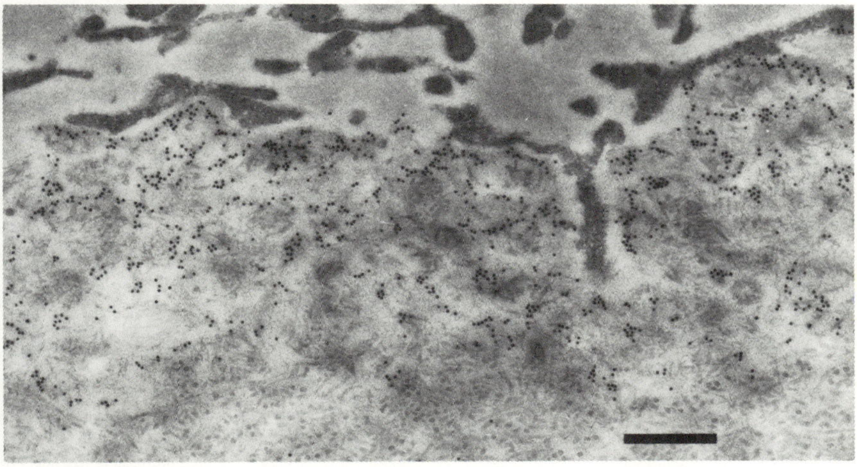

FIG. 12. Electron microscopic immunolocalization of type VII collagen in adult human cornea. Adult human cornea was lightly fixed and embedded in Lowacryl resin at low temperature. Sections were surface stained by a monospecific anti-type VII procollagen antisera followed by colloidal gold-conjugated second antibody. Gold deposition is seen upon the lamina densa as well as over a broad region at the epithelial edge of Bowman's layer. Cornea demonstrates the most extensive anchoring fibril network thus var visualized. Bar, 500 nm.

by rotary shadowing, our observations suggest that the complex carboxyl globular domain is entirely contained within the lamina densa and is not directly attached to the epithelial cell surface. We cannot rule out the possibility that the very fine filaments called anchoring filaments (Briggaman and Wheeler, 1975a), which span the lamina rara from the hemidesmosome to the lamina densa, may be contained within the type VII collagen carboxy terminal globular domain but cannot be distinguished by the rotary shadowing technique. At present, we consider this possibility unlikely, as no fine filaments have been observed nor would be required within the dermal plaques. We believe that it is more likely that the anchoring filament is a separate gene product with the specific function of linking the hemidesmosome with the lamina densa. This structure may have an affinity for the portions of anchoring fibrils contained within the lamina densa as anchoring fibrils are frequently seen directly opposite hemidesmosomes. However, anchoring fibrils in the absence of anchoring filaments are also frequently seen along the basal lamina.

The presence of the type VII collagen carboxy-terminal domain within the lamina densa does appear to influence the formation of hemidesmosomes. The studies of Gipson *et al.* (1983) indicate that epithelial cell sheets denuded of hemidesmosomes will regenerate those hemidesmosomes when applied *in vitro* to an intact basal lamina containing anchoring fibrils. The sites of newly formed hemidesmosomes correspond well to the locations of anchoring fibrils along the existing basal lamina. One of the first steps in this process appears to be the formation of anchoring filaments between the cell surface and the sites along the basal lamina expected to be occupied by the type VII collagen carboxy-terminal globular domain. The final structure of the hemidesmosomes is then elaborated upon this anchoring filament network. These studies suggest that the anchoring filaments originate along the basal surface of epithelial cells and accumulate opposite anchoring fibrils, possibly by recognition of the type VII collagen carboxyl-terminus. The apparent multidomain structure of the terminal ends of type VII collagen molecules is consistent with the possibility of multiple specific recognition sites.

The Function of Type VII Collagen

Our present knowledge of the structure of type VII collagen as it relates to the anchoring fibril network suggests that this collagen type is one of several molecules which form a specific architectural gridwork directly linking the basal surface of epithelial cells with the underlying dermis. The most obvious consequence of this series of associated structural elements is to provide a firm attachment of the epithelial cell to its underlying stromal matrix. It is also possible that this network provides a special trans-basement-membrane route for the transmission of information from the dermis to the epithelial cell or vice versa. The mechanism of this information transfer or its possible content is completely unknown. The requirement for this anchoring complex is also unclear. Other epithelia appear firmly affixed to their subjacent stroma without the presence of either hemidesmosomes, anchoring filaments, or the anchoring fibril network (Kawanani *et al.,* 1978). It is possible that this additional structural complex is required for tissues subjected to relatively increased frictional or tortional forces.

There are additional, less well-characterized components which appear to be specific to these basement membrane regions as well. For example, our use of the *en bloc* antibody penetration procedure has consistently indicated a size-dependent diffusion barrier immediately underlying the dermal surface of the dermal-epidermal junction of the basement membrane. Other investigators (Wasano and Yamamoto, 1985) have observed the deposition of ruthenium red along this region, suggesting the presence of strongly anionic dye binding sites. The identity of this diffusion barrier with its apparently highly polar sites is unknown. It is possible that this unknown basement membrane zone element may be a component of this anchoring complex.

IV. Pathological Findings Related to Disruptions of the Anchoring Fibril Network

If the anchoring fibril network functions to secure the basal lamina to the underlying matrix, then any perturbation of this network would be expected to weaken the adhesion of the lamina densa to the stroma. There are a number of inherited conditions which are characterized by separation of skin basal cells from the dermal stroma. These diseases comprise the general classification of epidermolysis bullosa. The dystrophic varieties of this disorder are characterized by separations between the lamina densa and the dermis consistent with a potential defect in the anchoring fibril network (Goldsmith and Briggaman, 1983). The recessively inherited dystrophic form in its most severe occurrence is characterized by a marked decrease or absence of anchoring fibrils (Briggaman and Wheeler, 1975b) in addition to increased collagenolytic activity (Valle and Bauer, 1980). The dominantly inherited forms also show differences in anchoring fibril morphology and may be reduced in number as well (Anton-Lamprecht and Hashimoto, 1976).

Monoclonal antibodies that recognize anchoring fibrils in human skin have been used to detect these antigens in a number of patients with various types of epidermolysis bullosa. The results (Goldsmith and Briggaman, 1983) indicate the absence of immunofluorescence in severe generalized gravis type recessive dystrophic epidermolysis bullosa. Other types of the disease showed the presence of the antigenic determinants. The data suggest a role for anchoring fibrils in this disease process. However, the primary cause of the dermal–epidermal separation is unclear. It is quite possible that structural defects in type VII collagen could be a primary cause of this disease in some pedigrees. It is also possible that loss of the anchoring fibril network could be due to a more generalized loss of all junctional components, and the involvement of type VII collagen in this destruction is entirely secondary. The substantial complexity of the anchoring fibril network predicted by the studies of the structure of type VII collagen strongly suggest that perturbations in this network could arise from multiple causes. Further study will be required to understand this involvement of type VII collagen with this disease condition at the molecular level.

References

Anton-Lamprecht, I., and Hashimoto, I. (1976). *Hum. Genet.* **32,** 69–76.

Bentz, H., Morris, N. P., Murray, L., Sakai, L. Y., Hollister, D. W., and Burgeson, R. E. (1983). *Proc. Natl. Acad. Sci. U.S.A.* **80,** 3168–3172.

Briggaman, R. A., and Wheeler, C. E., Jr. (1975a). *J. Invest. Dermatol.* **65,** 71–84.

Briggaman, R. A., and Wheeler, C. E., Jr. (1975b). *J. Invest. Dermatol.* **65,** 203–211.

Brody, I. (1960). *J. Ultrastruct. Res.* **4,** 264–297.

Bruns, R. R. (1969). *J. Cell Biol.* **42,** 418–430.

Burgeson, R. E., Morris, N. P., Murray, L. W., Duncan, K. G., Keene, D. R., and Sakai, L. Y. (1985). *Ann. N.Y. Acad. Sci.* **460,** 47–57.

Dieringer, H., Hollister, D. W., Glanville, R. W., Sakai, L. Y., and Kühn, K. (1985). *Biochem. J.* **227,** 217–222.

Eagle, H. (1955). *Proc. Soc. Exp. Biol. Med.* **89,** 362–364.

Gipson, I. K., Grill, S. M., Spurr, S. J., and Brennan, S. J. (1983). *J. Cell Biol.* **97,** 849–857.

Goldsmith, L. A., and Briggaman, R. A. (1983). *J. Invest. Dermatol.* **81,** 464–466.

Hayflick, L. (1961). *Exp. Cell Res.* **23,** 14.

Kawanani, O., Ferans, V. J., Roberts, W. C., Crystal, R. G., and Fulmer, J. D. (1978). *Am. J. Pathol.* **92,** 389–410.

Keene, D. R., Sakai, L. Y., Lunstrum, G. P., Morris, N. P., and Burgeson, R. E. (1987). *J. Cell. Biol.* **104,** 611–621.

Laguens, R. (1972). *J. Ultrastruct. Res.* **41,** 202–208.

Lunstrum, G. P., Sakai, L. Y., Keene, D. R., Morris, N. P., and Burgeson, R. E. (1986). *J. Biol. Chem.* **261,** 9042–9048.

McTigue, J. W., and Fine, B. S. (1966). *Electron Microsc., Proc. Int. Congr., 6th* **2,** 775.

Morris, N. P., Keene, D. R., Glanville, R. W., Bentz, H., and Burgeson, R. E. (1986). *J. Biol. Chem.* **261,** 5638–5644.

Palade, G. E., and Farquhar, M. G. (1965). *J. Cell Biol.* **27,** 215–224.

Sakai, L. Y., Engvall, E., Hollister, D. W., and Burgeson, R. E. (1982). *Am. J. Pathol.* **108,** 310–318.

Sakai, L. Y., Keene, D. R., Morris, N. P., and Burgeson, R. E. (1986). *J. Cell Biol.* **103,** 1577–1586.

Susi, F. R., Belt, W. D., and Kelly, J. W. (1967). *J. Cell Biol.* **34,** 686–690.

Valle, K.-J., and Bauer, E. A. (1980). *J. Clin. Invest.* **66,** 176–187.

Wasano, K., and Yamamoto, T. (1985). *Cell Tissue Res.* **239,** 485–495.

Type VIII Collagen

Helene Sage

Department of Biological Structure, University of Washington, Seattle, Washington 98195

and

Paul Bornstein

Department of Biochemistry, University of Washington, Seattle, Washington 98195

I. Introduction

Type VIII collagen was first discovered during the course of biosynthetic studies on bovine aortic endothelial (BAE)[1] cells *in vitro* (Sage *et al.*, 1979). Although these cells produced primarily the interstitial, fibrillar type III collagen, approximately 20% of the secreted protein that was sensitive to bacterial collagenase displayed unusual properties for a collagen. Three different molecular-weight forms were present, ranging from approximately 180,000 to 100,000, and the collagen itself was very labile to pepsin. A survey of its biosynthetic distribution showed that this collagen was secreted by most, but not all, endothelial cells that had been established in culture and, for this reason, the protein was originally termed EC (endothelial collagen) (Sage *et al.*, 1984). EC is now used to denote the individual chains of type VIII collagen, a structurally distinct member of the collagen gene family (Sage *et al.*, 1983).

In addition to its unusual structural properties, there were several aspects related to the biosynthetic regulation of type VIII collagen that appeared intriguing: (1) rapidly proliferating/migrating cells plated at low density secreted the highest levels of type VIII collagen into the

[1]Abbreviations: β-APN, β-aminopropionitrile fumarate; DTT, dithiothreitol; BAE, bovine aortic endothelial; CM, carboxymethyl; BCE, bovine corneal endothelial; PRF, pepsin-resistant fragment; DM, Descemet's membrane; ELISA, enzyme-linked immunosorbent assay.

culture medium, while virtually no type VIII was recovered from the medium of quiescent, confluent cells (Alitalo *et al.*, 1983; Sage *et al.*, 1986); (2) the collagen was also a biosynthetic product of both normal and malignant cells of nonendothelial origin (Sage *et al.*, 1984); and (3) secretion of type VIII collagen was modulated with respect to the growth state and morphology of endothelial cells *in vitro* (Sage, 1985a). Some of these phenotypic switches are listed in Table I. Bovine endothelial cells from aorta and brain capillary synthesized primarily the interstitial type I collagen and ceased production of type VIII collagen when they adopted a "sprouting" or elongate morphology in the presence of serum. Interstitial collagen production was favored by capillary cells grown in serum and tumor cell-conditioned medium, while cells cultured in plasma-derived serum (lacking platelet components, including the platelet-derived growth factor) secreted exclusively basement membrane (type IV) and type VIII collagens.

In this review, we focus primarily on the structural aspects of type VIII collagen. Most of these studies have been performed on the metabolically radiolabeled protein produced by bovine endothelial and human astrocytoma cells *in vitro*. Recent studies on bovine corneal Descemet's membrane have provided evidence that type VIII collagen is a major component of this tissue as well (Kapoor *et al.*, 1985, 1986; Labermeier and Kenney, 1983).

TABLE I
MODULATION OF TYPE VIII COLLAGEN SYNTHESIS BY BOVINE ENDOTHELIAL CELLS *in Vitro*[a,b]

Source of endothelium	Culture conditions	Morphology	Collagen type
Aorta	Confluence, FCS[c]	Sprouting	VIII → I
Brain capillary	Prolonged exposure to FCS	Elongate	VIII → I
Adrenal capillary	FCS, TCM[d], gelatin substrate	Elongate	I, III
Retinal capillary	PDS[e], fibronectin substrate	Cobblestone	IV, VIII

[a]Sage (1985a).

[b]Collagen types were identified in the culture medium of metabolically labeled endothelial cells by DEAE-cellulose chromatography and SDS–PAGE as described by Sage and Bornstein (1982).

[c]FCS, Fetal calf serum.

[d]TCM, Tumor cell-conditioned medium.

[e]PDS, Plasma-derived serum.

II. Type VIII Collagen from Cells in Culture

A. *Identification and Isolation of the Biosynthetic Product*

Type VIII collagen has several distinguishing characteristics that collectively provide a set of criteria by which it can be identified. In Fig. 1A is shown the secretory profile of a human astrocytoma cell line. Type VIII collagen was secreted rather rapidly (within 1 hr after a 1-hr pulse with [^{3}H]proline) primarily as EC2 (M_r 125,000). Both EC2 and a lower molecular-weight form, EC3 (M_r 100,000), were degraded by bacterial collagenase (Fig. 1A, "+" lanes), and their mobility on SDS–polyacrylamide gels (SDS–PAGE) was not altered by reduction (not shown; Alitalo *et al.,* 1983). Secretion of type VIII collagen by BAE cells is shown in Fig. 1B. In this experiment, an initial chromatographic procedure on DEAE-cellulose provided an enrichment in type VIII collagen, which was present as EC1 (M_r 177,000) and EC2 (I and II, left lanes). These components were secreted in both the presence and absence of ascorbate, a cofactor necessary for the complete hydroxylation of procollagens by the enzymes prolyl and lysyl hydroxylase (Fig. 1B, II). The properties of bovine type VIII collagen that are readily distinguishable in metabolic radiolabeling studies are, therefore, (1) secretion as EC1, EC2, and EC3, (2) lack of interchain disulfide bonds, (3) secretion in the absence of ascorbate , (4) lability to bacterial collagenase, and (5) failure to bind to DEAE-cellulose.

An initial fractionation step in the purification of type VIII collagen includes precipitation of radiolabeled culture medium proteins with ammonium sulfate. A concentration of 20% (weight-to-initial volume ratio) favored the recovery of EC2 and EC3, while greater amounts of ammonium sulfate (20–50%) were generally more effective for precipitation of EC1 (Sage *et al.,* 1980). Partial purification of type VIII collagen was achieved by chromatography on DEAE-cellulose at 4°C in 6 *M* urea, as shown in Fig. 2. The initial peak, which eluted before the inception of the gradient, contained EC1, EC2, and EC3 (inset, lane I as identified by SDS–PAGE). Type VIII collagen was effectively separated from several other proteins secreted in greater abundance by bovine endothelial cells, namely fibronectin and type III procollagen (inset, lanes II and III).

Table II lists the cell types that have been shown to synthesize type VIII collagen *in vitro.* This collagen was isolated and characterized from both normal and malignant cells representing four mammalian genera (mouse, cow, rabbit, and human). Type VIII collagen was the major collagenous protein secreted by murine brain capillary endo-

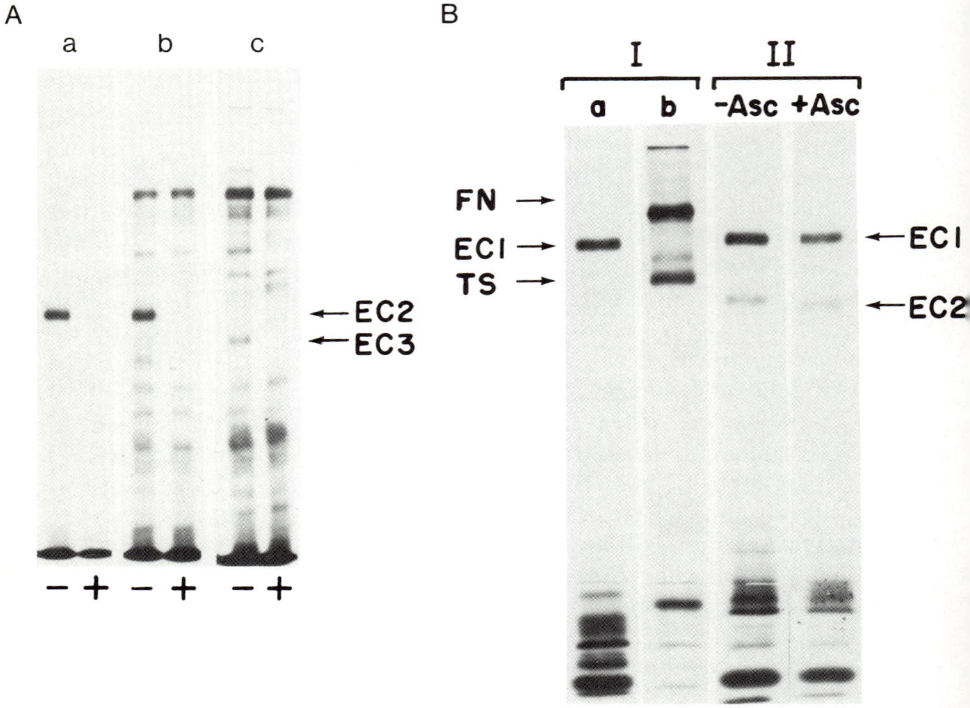

FIG. 1. Secretion of type VIII collagen by normal and malignant cells *in vitro*. (A) A human astrocytoma cell line was pulsed with [^{3}H]glycine in serum-free medium containing ascorbate and β-APN for 1 hr. This solution was subsequently replaced with medium containing ascorbate, β-APN, and glycine, and radiolabeled proteins were chased into the culture medium. At 1 hr, 6 hr, and 3 days (a, b, and c, respectively), equal aliquots were removed and incubated with bacterial collagenase (+) or enzyme buffer alone (−). The digests and controls were precipitated with 10% trichloroacetic acid and were analyzed by SDS–PAGE (5% gel) in the presence of DTT. Type VIII collagen, present as EC2 (M_r 125,000) and EC3 (M_r 100,000), is identified. Within 1 hr after a 1-hr pulse, a collagenase-sensitive component, EC2, was the major product secreted by the astrocytoma cells. Between 24 hr (not shown) and 3 days, EC2 was not present in the culture medium, although small amounts of EC3 were apparent. Reproduced in part from Alitalo *et al.* (1983) with permission. (B) Bovine aortic endothelial (BAE) cells were incubated in serum-free medium containing ascorbate, β-APN, and [^{3}H]proline or in medium with only the isotope. Radioactive proteins were chromatographed on DEAE-cellulose as described by Sage *et al.* (1980) and were analyzed by SDS–PAGE after reduction. I, (a) denotes the fraction that was not bound by the DEAE-cellulose; (b) shows proteins that were bound and subsequently eluted with 0.5 *M* NaCl. II, Fractions that did not bind to DEAE-cellulose containing proteins metabolically radiolabeled in the presence (+Asc) or absence (−Asc) of ascorbate. Type VIII collagen (EC1 and EC2) is identified, as are two major secretory products of BAE cells: fibronectin (FN) and thrombospondin (TS). Type VIII collagen elutes from DEAE-cellulose under conditions which promote binding of

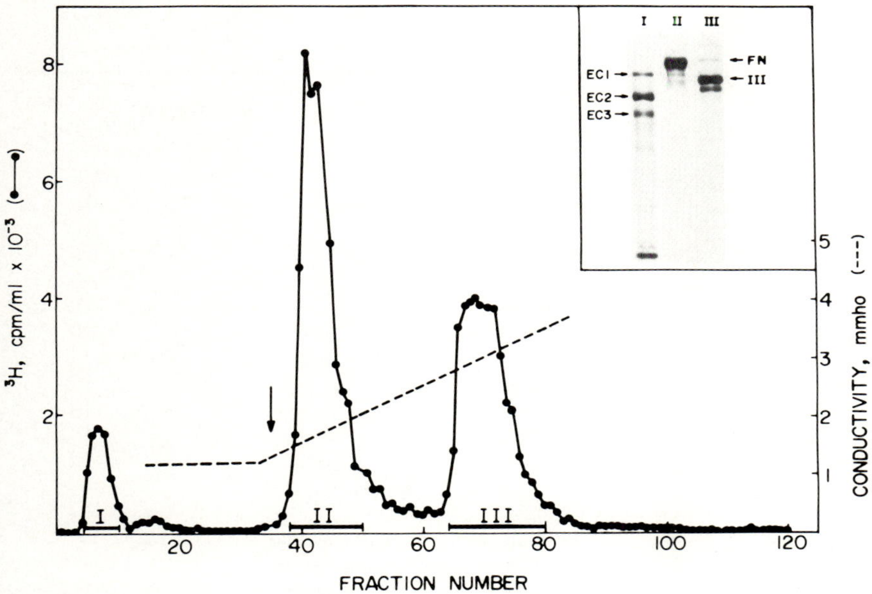

Fig. 2. Isolation of type VIII collagen from the culture medium of bovine endothelial cells by chromatography on DEAE-cellulose. Radiolabeled culture medium proteins, secreted by bovine vena cava endothelial cells, were precipitated with 20% ammonium sulfate and fractionated on DEAE-cellulose at 4°C (Sage *et al.*, 1980). Arrow indicates inception of 0–200 m*M* NaCl gradient. Peak-containing fractions were analyzed by SDS–PAGE (6%/10% gel) under reducing conditions. Inset shows fluorescence autoradiogram of unbound fraction (I) (EC1, EC2, and EC3) and of proteins eluted with increasing concentrations of NaCl [peak II contained fibronectin (FN) and peak III contained type III procollagen (III)]. From these cells, all three forms of type VIII collagen were recovered in the 20% ammonium sulfate fraction that did not bind to the DEAE-cellulose. Reproduced from Sage *et al.* (1981) by permission of the American Heart Association, Inc.

thelial cells and by three cell lines derived from human tumors (Ewing's sarcoma, astrocytoma, and hepatocellular carcinoma) (Sage *et al.*, 1984). From these studies it was also apparent that not all endothelial cells synthesized type VIII collagen (for example, human aortic, vena cava, and umbilical vein endothelial cells, and porcine

most native procollagens. This property, which enables type VIII collagen to be separated from other secreted proteins of similar M_r, was used to show that the secretion of VIII collagen was not dependent on inclusion of ascorbate in the culture medium. Reproduced from Sage *et al.* (1982) with permission.

TABLE II
CELLS WHICH SYNTHESIZE TYPE VIII COLLAGEN *in Vitro*[a]

Cell type or line	Origin	Major secreted procollagen type
Endothelial, vascular	Bovine aorta	III
	Bovine pulmonary artery	III
	Bovine vena cava	III
	Bovine mesenteric vein	III
	Bovine retinal capillary	IV
	Murine brain capillary	VIII
Endothelial, corneal	Bovine Descemet's membrane	III
	Rabbit Descemet's membrane	IV
EW-A2	Human Ewing's sarcoma	VIII
U-251MG	Human astrocytoma	VIII
HS 0578T	Human carcinosarcoma (breast)	I
Hep 3B21-7	Human hepatocellular carcinoma	VIII
A549	Human alveolar cell carcinoma	IV
HS 0696T	Human adenocarcinoma (bone)	IV
HS 0700T	Human adenocarcinoma (pelvis)	IV
HS 0746T	Human adenocarcinoma (muscle)	I
Fibroblast	Human fetal skin	I

[a]Sage (1984); Sage *et al.* (1984).

aortic endothelial cells), and that type VIII collagen was not restricted solely to the vascular endothelium (Sage *et al.*, 1984). In addition to its presence as a minor collagen in human fetal fibroblasts (Table II), both Benya (1980) and Sage and co-workers (1981) described the secretion of significant amounts of type VIII (EC) collagen by corneal endothelial cells. These results have prompted several investigators to examine the Descemet's membrane as a source of this unusual collagen (see Section III). Another conclusion that can be drawn from the data presented in Table II is that the synthesis of type VIII collagen is not contingent upon expression of particular collagen types, since it was present in the culture medium concomitantly with interstitial (types I and III) as well as with basement membrane-associated collagens (types IV and V) (Sage *et al.*, 1984).

B. *Secretory Properties*

The secretion of type VIII collagen by aortic and corneal endothelial cells and by astrocytoma cells was found to be independent of a requirement for prolyl hydroxylation (Sage *et al.*, 1982; Alitalo *et al.*, 1983) (see also Fig. 1B). Inhibition of hydroxylation by 98% with α,α'-dipyridyl, a chelator of the ferrous iron cofactor required for prolyl and lysyl hydroxylase activity, caused a marked reduction in the secretion of type III procollagen by BAE cells with no apparent effect on the secretion of type VIII collagen (Sage *et al.*, 1982). This property is similar to that observed for type IV procollagen but differs markedly from the secretory behavior of types I, II, and III procollagen (Crouch and Bornstein, 1979).

From pulse–chase experiments, a synthesis and secretion time of 75 min was estimated for the release of EC2 into the culture medium by BAE and astrocytoma cells (Alitalo *et al.*, 1983; Sage *et al.*, 1983). This interval is similar to that reported for type IV procollagen and is considerably longer than that required for the secretion of type I procollagen (Crouch and Bornstein, 1979). As shown in Fig. 3, EC2 (M_r 125,000) was the initial form detected in the culture medium (1.5 hr, lane A). At later time points, EC1 (M_r 177,000) and EC3 (M_r 100,000) were apparent. The molecular weights of these chains were unaltered in the presence of reducing agent [compare Fig. 3a (−DTT) with Fig. 3b (+DTT)], and quantitative, stepwise processing to stable, lower molecular-weight forms was not observed. This apparent lack of processing also distinguishes type VIII from the interstitial collagens. Recent experiments have shown that the release of type VIII collagen into the culture medium is favored significantly in rapidly proliferating BAE cells (Sage *et al.*, 1986). This result is in accordance with the data of Alitalo *et al.* (1983), who demonstrated that the amount of type VIII collagen present in the culture medium of astrocytoma cells was a function of the degree of confluence of the cells (see also Fig. 1A, compare 1 hr with 3 days). Type VIII collagen, therefore, appears to be sequestered preferentially in the cell layer in confluent, nonproliferating cultures. This characteristic could reflect its potential role as an extracellular matrix macromolecule that is closely associated with the cell surface.

C. *Biochemical Characteristics*

Type VIII collagen was shown to be a unique collagen by 1- and 2-dimensional peptide mapping (Sage *et al.*, 1980; Benya, 1980). As shown in Fig. 4, the peptide patterns, based on cleavage with cyanogen bromide (CNBr), are distinct for each collagen type and provide convincing evidence that type VIII collagen (EC2) is a novel gene product.

One interesting property of type VIII collagen is its sensitivity to

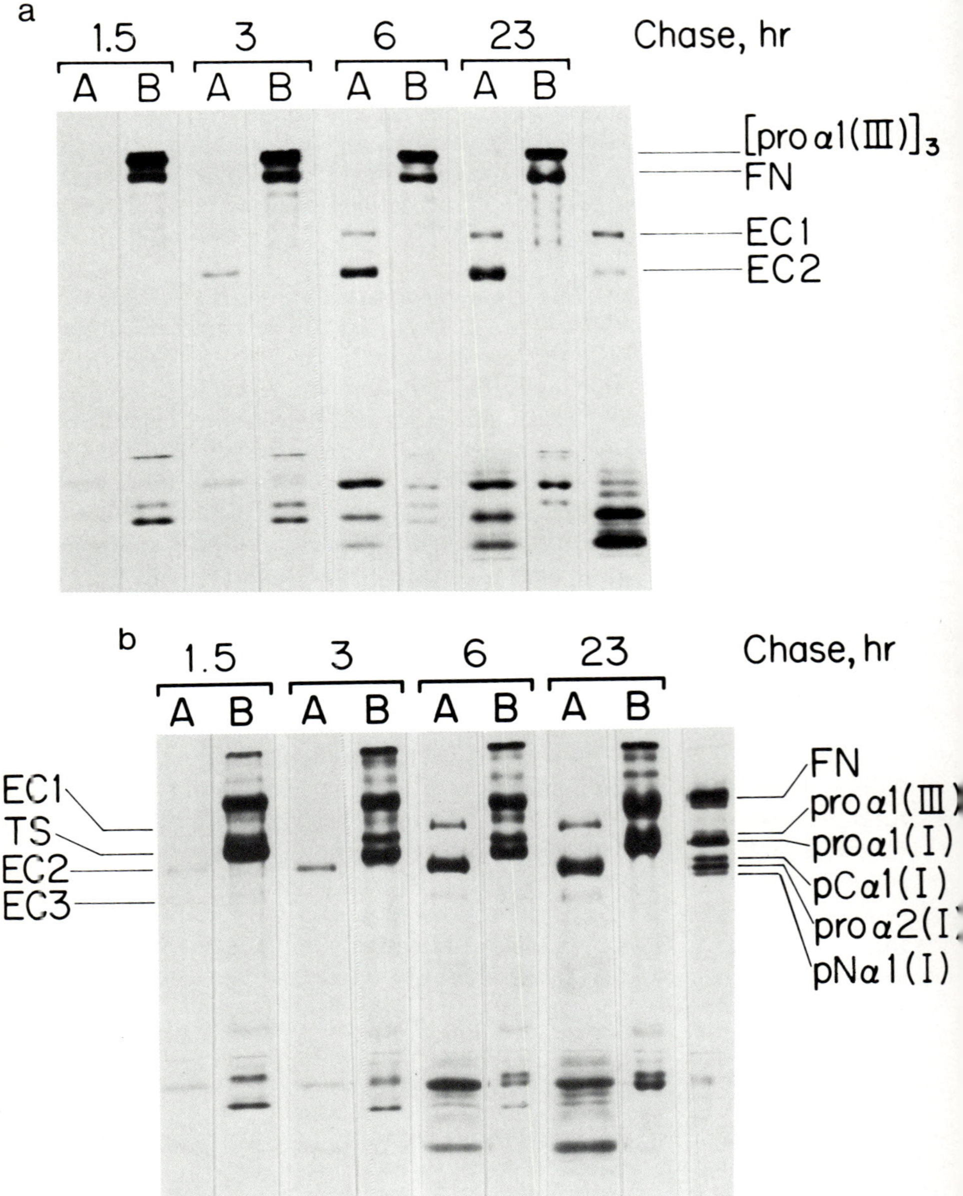

FIG. 3. Secretory characteristics of type VIII collagen: pulse–chase studies *in vitro*. BAE cells were pulsed with [^{3}H]proline and [^{3}H]glycine for 1 hr in serum-free medium supplemented with ascorbate and β-APN. After removal of pulse medium, cells were

pepsin (Sage *et al.*, 1980). Within 5 min following incubation of EC2 with the enzyme, there was complete degradation of the molecule to fragments smaller than M_r 50,000–60,000 (Fig. 5A) (Sage *et al.*, 1980; Benya, 1980). From the pepsin digest, a peptide of M_r 50,000 was purified to homogeneity by molecular sieve chromatography. This pepsin-resistant fragment contained nearly equal amounts of hydroxyproline and proline, with 6.6% of hydroxyproline present as the 3-isomer (Sage *et al.*, 1980).

Further studies on the triple-helical structure of type VIII collagen were performed on the native collagen by exposure of the culture medium and cell layer proteins directly to pepsin without prior precipitation or chromatographic procedures. The results have been summarized in Fig. 5B, which depicts elution profiles, from CM-cellulose chromatography, of BAE culture medium (top) and cell layer proteins (bottom) after pepsin digestion. Analysis of the peak-containing fractions by SDS–PAGE revealed type III collagen as well as the M_r 50,000 pepsin-resistant fragment (50K fragment) of type VIII collagen (inset, 50K). The results indicate that type VIII collagen was present both in the culture medium and in the cell layer (cells and extracellular matrix) of BAE cells (Sage *et al.*, 1983). The 50K fragment was also produced from the native, secreted molecule and, therefore, did not appear to be an artifact resulting from partial denaturation.

Additional studies were performed to assess the extent to which type VIII collagen was triple helical. BAE cells produced the collagen (EC1 and EC2) over a temperature range of 24–37°C. A gelatin-binding domain of fibronectin, which specifically interacts with denatured collagen chains but will not bind those in native conformation, did not associate with type VIII collagen as isolated from DEAE-cellulose

---

chased in the same medium containing 10 m*M* proline in place of the isotope. Each time point represents one dish of cells. Culture medium proteins were precipitated by ammonium sulfate and were chromatographed on DEAE-cellulose. The unbound (A) and bound (B) fractions were analyzed by SDS–PAGE (6%/10% gel) in both the presence (b, +DTT) and absence (a, −DTT) of DTT. Type VIII collagen (EC1, EC2, and EC3) has been identified on the fluorescence autoradiogram, as has fibronectin (FN), thrombospondin (TS), and type III procollagen. Type I procollagen and its processed intermediate forms (pC-, pN-) are shown on the right side of the reduced gel, and a standard of EC1 and EC2 from BAE cell culture medium after 18 hr continuous labeling is shown on the right side of the unreduced gel. Within 1.5 hr after a 1 hr pulse, EC2 was secreted into the culture medium (lane A). EC1, EC2, and EC3 were visible by 6 hr (lane A) and accumulated in the medium for 23 hr. None of these forms of type VIII collagen was disulfide bonded (compare +DTT with −DTT, lanes A). EC2 migrated between pCα1(I) (M_r 135,000) and pNα1(I) (M_r 120,000). Reproduced from Sage *et al.* (1983) with permission.

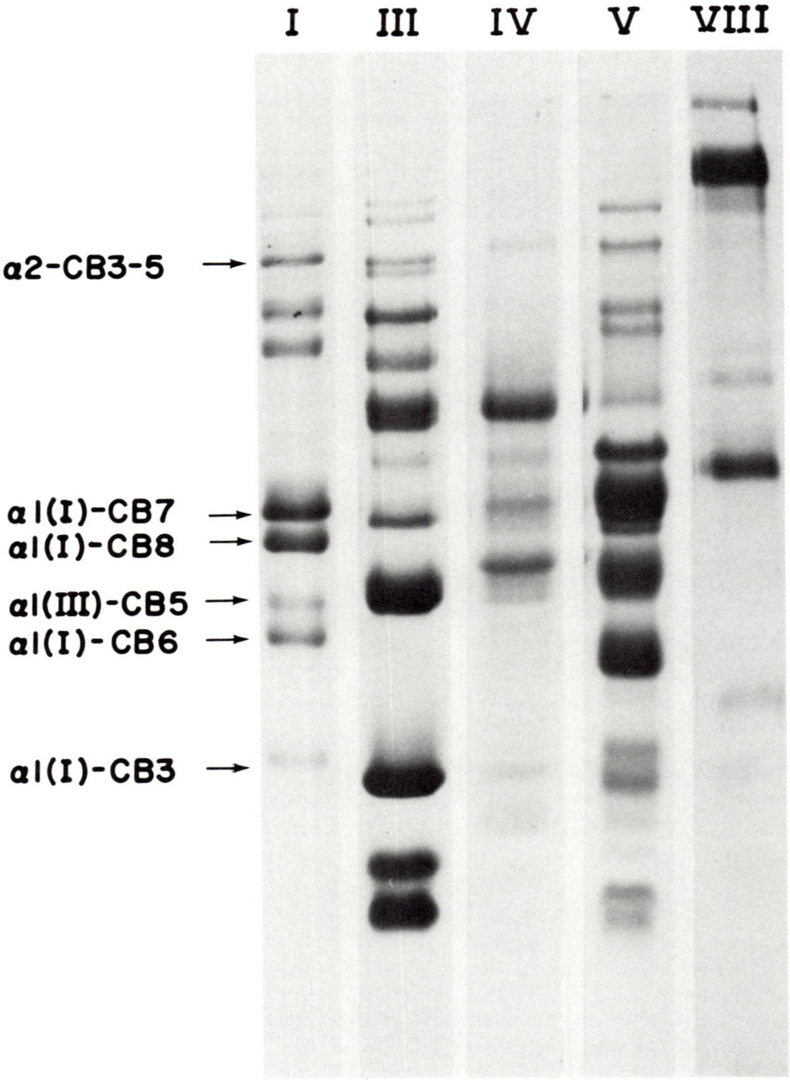

FIG. 4. Comparison of collagen types by CNBr-derived peptide maps. Bovine types I, III, IV, V, and VIII (EC2) collagen were cleaved with CNBr and the products resolved by SDS–PAGE (12.5% gel) in the presence of DTT. Some of the major collagen CNBr (CB) peptides have been identified. EC2 (VIII) has a unique distribution of CB peptides and appeared structurally distinct from the other collagen types. Reproduced with permission from Sage © (1982), The Williams & Wilkins Co., Baltimore, Maryland.

chromatography. Prior denaturation at 37°C for 10 min, followed by affinity chromatography at 4°C, resulted in binding of EC1 and EC2 to the fibronectin fragment. It therefore appeared that this association occurred only when the binding sites of type VIII collagen were exposed after a conformational change was induced by heat. Deliberate denaturation also altered the susceptibility of type VIII collagen to mast cell protease and to vertebrate collagenase (Sage *et al.,* 1983).

The presence of non-triple-helical globular domains, similar to the N- and C-terminal propeptides characterized from the interstitial collagens, was not observed in type VIII collagen. As shown in Fig. 6, incubation of EC1 and EC2 with bacterial collagenase for only 5 min did not release any protease-resistant peptides of M_r <10,000 (a maximum approximation, based on the limit of detection on a 12.5% polyacrylamide gel), although the EC bands were completely digested (lane 2). Incubation of the native protein with mast cell protease, an enzyme that, under appropriate conditions, specifically releases propeptides from native interstitial procollagens, produced no perceptible change in the molecular weight of type VIII collagen (lane 3). However, three major peptides were produced after denaturation of type VIII collagen (lane 4). Since mast cell protease cleaves within the triple-helical sequences of denatured procollagens, it is likely that the peptides (M_rs 70,000, 48,000, and 20,000) shown in lane 4 also contain Gly-X-Y sequence (Sage *et al.,* 1983). A similar spectrum of peptides was produced by cleavage of type VIII collagen with vertebrate interstitial collagenase (lanes 7–9). Based on the specificity of this enzyme, the cleavage points within the native type VIII molecule are most probably within Gly-X-Y domains that exhibit a somewhat relaxed triple-helical structure. Unlike types IV and V collagen, type VIII collagen is a substrate for interstitial collagenase (Sage, 1982; Sage *et al.,* 1983).

A summary of some biochemical properties of type VIII collagen is presented in Table III. Experimental data have indicated the following structural features for this unusual protein:

1. it is secreted as three molecular-weight forms (EC1, EC2, and EC3) which are not interchain disulfide bonded;
2. it is a glycoprotein containing significant levels of hydroxyproline;
3. it does not bind to heparin, fibronectin, or several lectins, although after denaturation it displays affinity for the gelatin-binding domain of fibronectin;
4. it is labile to several neutral proteinases;
5. its secretion is characterized by a minimum secretion time of 75

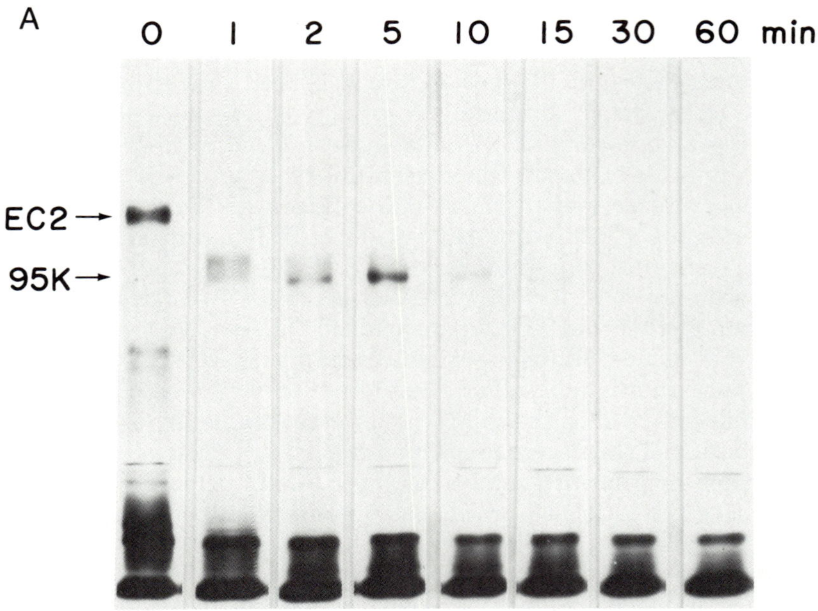

FIG. 5. Degradation of type VIII collagen by pepsin. (A) EC2 was isolated by DEAE-cellulose chromatography under conditions that maintain native conformation (Sage *et al.*, 1983). A time course of incubation with pepsin is shown (1:100 enzyme-to-substrate weight ratio, 4°C, pH 2.8). EC2 and degradation products of M_r 95,000 (95K) are indicated. After 5 min, EC2 was degraded to fragments of $M_r \leq 50{,}000$ under conditions in which stable α-chains were produced from interstitial procollagens. Reprinted in part with permission from Sage, H. Pritzl, P., and Bornstein, P. (1980). *Biochemistry* **19,** 5747–5755. Copyright (1980) American Chemical Society. (B) Isolation of type VIII collagen M_r 50,000 pepsin-resistant fragment from endothelial cell layers and culture medium. Radiolabeled BAE cell culture medium proteins were dialyzed directly against 0.5 *M* acetic acid, digested with pepsin for 6 hr at 4°C, and dialyzed against a denaturing buffer in preparation for chromatography on CM-cellulose (Sage *et al.*, 1983). Inset shows SDS–PAGE analysis (6%/10% gel) of peaks as numbered in the elution diagram. Peak I contained type III collagen [α1(III)] and peak II contained an M_r 50,000 fragment (50K) of type VIII collagen. Radiolabeled BAE cell layers (cells and extracellular matrix) were treated with pepsin and chromatographed as described above. Inset is a fluorescence autoradiogram of the major peak, which contained both type III collagen [α1(III)] and the 50K fragment of type VIII collagen. The discrepancy in gradient elution position between the medium and cell layer 50K fragment has not been explained. Reproduced from Kapoor *et al.* (1985) with permission.

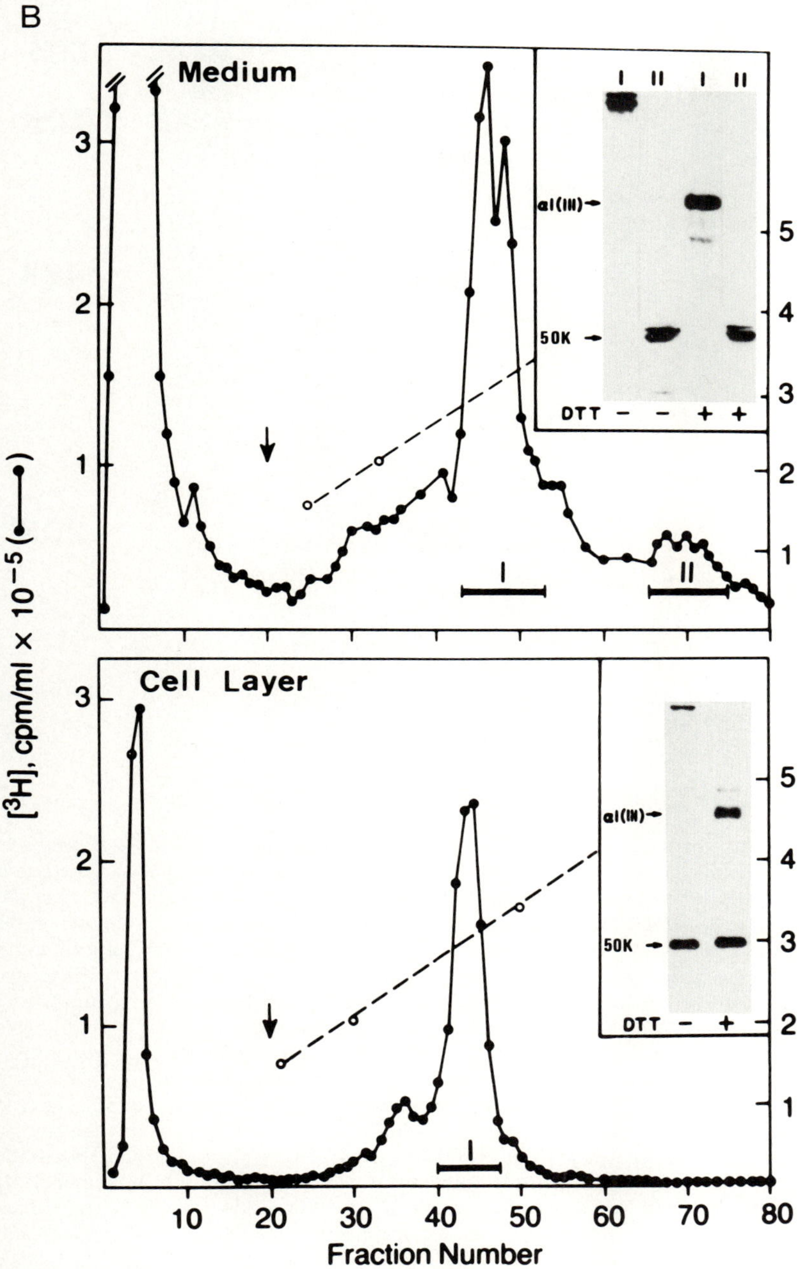
B
Medium
I
II
I
II
I
II
αI(III)
50K
DTT
Cell Layer
I
αI(III)
50K
DTT
10
20
30
40
50
60
70
80
Fraction Number
[3H], cpm/ml × 10−5
Conductivity, mmho

FIG. 5B.

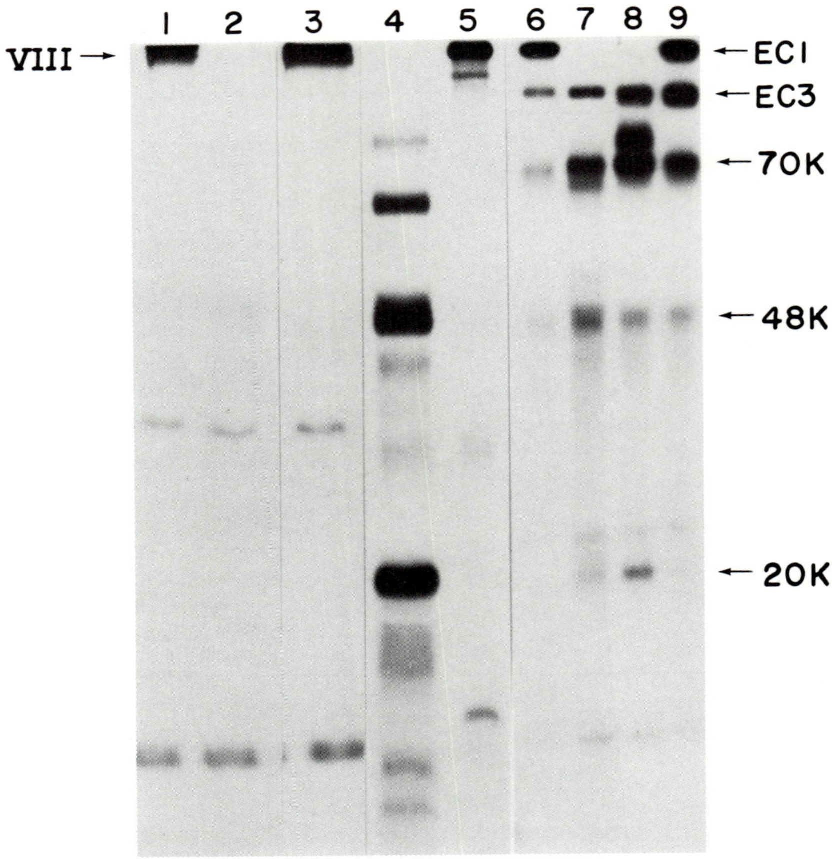

FIG. 6. Susceptibility of type VIII collagen to collagenases and mast cell protease. Native, radiolabeled type VIII collagen (lane 1, control) was incubated at 37°C for 5 min with bacterial collagenase (lane 2) or mast cell protease (lane 3). Type VIII, after denaturation at 45°C for 10 min (lane 5, control), was incubated with mast cell protease for 5 min at 37°C (lane 4). Another sample of native type VIII collagen containing EC1 and EC3 (lane 6, control) was incubated at 24°C with human skin (interstitial) collagenase for 12 hr (lane 7), 30 min (lane 8), or 5 min (lane 9). Reactions were terminated by addition of SDS–PAGE buffer containing DTT; products were resolved on 12.5% gels and visualized by fluorescence autoradiography (Sage *et al.*, 1983, 1984). Bacterial collagenase digested type VIII collagen completely; globular, noncollagenous propeptides of $M_r > 10,000$ were not apparent (lane 2). Similarly, mast cell protease did not release putative propeptides from native type VIII collagen (lane 3). After denaturation, however, the protein was cleaved into three major collagenous peptides of approximate M_r 65,000, 48,000, and 20,000 (lane 4). Interstitial collagenase initially produced a M_r 100,000 peptide (similar in size to EC3) and a M_r 70,000 peptide (lane 9). With increasing incubation times, other fragments were apparent (approximate M_r 50,000 and 20,000) (lanes 7 and 8). Reproduced in part with permission from Sage *et al.* © (1984), The Williams & Wilkins Co., Baltimore, Maryland.

TABLE III
STRUCTURAL PROPERTIES OF TYPE VIII COLLAGEN[a]

Molecular size[b]	EC1, 180,000
	EC2, 125,000
	EC3, 100,000
	EC-PRF[c], 50,000
Disulfide bonding	Intrachain
Carbohydrate	Mannose, glucosamine
Hydroxyproline:proline[d]	EC1, 0.5:1
	EC2, 1:1
	EC-PRF, 1.1:1
Degradation by proteases	
Pepsin	+ (M_r 50,000 PRF)
Vertebrate interstitial collagenase	+ (5–7 fragments, M_r 90,000–5,000)
Bacterial collagenase	+ (no peptides of $M_r > 10,000$)
Mast cell protease	− (24°C; no "propeptides" $M_r >$ 10,000)
	+ (37°C; 3–5 major fragments)
Neutral proteinases (elastase, chymotrypsin, trypsin)	+ (28°C, extensive)
Thrombin	+ (28°C, 5–7 major fragments)
Schistosomal protease[e]	+
Affinity properties	
Lectins, heparin, fibronectin	−
Fibronectin-GBD[f]	+ (after denaturation)
Secretion characteristics	
Ascorbate requirement	−
Minimum synthesis/secretion time	75 min[g]
Temperature range	24–37°C[g]
Location	Medium (subconfluent, growing cells)
	Cell layer (confluent, quiescent cells)

[a]Sage *et al.* (1983, 1984).
[b]Approximate M_r by SDS–PAGE, based on procollagen molecular weight standards.
[c]Pepsin-resistant fragment.
[d]3-Hydroxyproline comprised between 7 and 10% of the total hydroxyproline content.
[e]McKerrow *et al.* (1985).
[f]Gelatin-binding domain.
[g]BAE cells.

min, without the necessity for prolyl hydroxylation, over a temperature range of 24–37°C;

6. it is digested by pepsin to fragment(s) of M_r 50,000 that have a hydroxyproline to proline ratio of approximately 1:1;

7. limited proteolysis occurs in the presence of vertebrate interstitial collagenase; and

8. globular propeptide domains of M_r 10,000 are not evident after

digestion of native type VIII collagen by bacterial collagenase or mast cell protease.

A structural model that incorporates these experimental findings is presented in Section IV.

III. Type VIII Collagen from Tissue: Descemet's Membrane

Descemet's membrane (DM) is the basement membrane which separates corneal endothelial cells from the corneal stroma. Ultrastructurally, it appears as a hexagonal, netlike matrix that is considerably thicker than most basement membranes. This density is due to the presence of many lamellar units that are joined by short, cross-linking 'bridges' that are 110 nm apart and perpendicular to the lamellae (Murphy *et al.,* 1984). During human postnatal development, a nonlamellar, nonstriated component is added posteriorly to the stacks of striated lamellar units (Murphy *et al.,* 1984). It is generally regarded that the corneal endothelial cells secrete a major portion of the DM. Sawada and co-workers (1984) performed an ultrastructural analysis of the extracellular matrix produced by BCE cells *in vitro*. In the presence of the cross-link inhibitor β-APN, the cells deposited a hexagonal lattice of 150-nm periodicity that was composed of collagenase-sensitive, dumbbell-shaped structures 160 nm in length. Since both bovine and rabbit corneal endothelial cells produced type VIII collagen in culture (Sage *et al.,* 1981; Benya, 1980), several laboratories have examined the DM, an unusually dense basement membrane, as a source of this novel collagen. Although Sawada *et al.* (1984) claimed the dumbbell-shaped structures were composed of type IV collagen, recent studies have shown that type VIII, and not type IV, collagen is the major component of bovine DM (Labermeier and Kenney, 1983; Kapoor *et al.,* 1985).

A. *Isolation*

Labermeier *et al.* (1983) first described the release of collagenous polypeptides of M_r 50,000 from bovine DM after limited pepsin treatment. Although its possible relationship to type IV collagen was not clarified in this study, the authors proposed that DM was composed of a unique basement membrane collagen with a basic subunit of M_r 50,000. Further studies indicated that the major pepsin-resistant collagenous moiety in the bovine DM and in the thickened DM of corneas from patients with Fuchs' endothelial dystrophy was a fragment of a collagen, termed EC, of M_r 50,000 (Labermeier and Kenney, 1983; Kenney *et al.*, 1984). In recent studies, Kapoor *et al.* (1985) have purified pepsin-resistant fragments of M_r 50,000 from calf and adult bovine DM. In many respects, these polypeptides were structurally similar to the 50K fragment of type VIII collagen (see Section III,B).

B. Biochemical Characteristics

The polypeptides of M_r 50,000 that were solubilized by brief pepsin treatment of DM did not contain interchain disulfide bonds and were degraded by bacterial collagenase (Kapoor *et al.*, 1985). Peptide mapping (1- and 2-dimensional) of the 50K polypeptides showed that they were distinct from several other collagen types, including pepsin-resistant fragments of type IV collagen (M_rs 140,000, 70,000, and 50,000). In addition, antibodies raised against the 50K polypeptides from DM did not react with collagen types I, III, IV, V, and VI by ELISA and Western immunoblotting; conversely, antibodies specific for these other collagens did not cross react with the 50 K polypeptides. The unique nature of these pepsin-resistant polypeptides from DM was confirmed by amino acid analysis of 50K-A and 50K-B, two chains that were purified by high-performance liquid chromatography. In addition to differences between each other, 50K-A and 50K-B were distinct from the other collagen types that have been shown to be present in cornea (Kapoor *et al.*, 1985). The identity of these polypeptides as derivatives of type VIII collagen was confirmed by immunoblotting (1) BCE cell culture medium type VIII collagen, (2) the 50K pepsin-resistant fragment from BCE culture medium, and (3) the 50K fragment from the BCE cell layer, with anti-50K DM collagen antiserum. Positive reactions obtained from all three samples thus confirmed that the M_r 50,000 polypeptides purified from bovine DM and BCE cell type VIII collagen were structurally related (Kapoor *et al.*, 1986).

[^{125}I]-Peptide maps (2-dimensional) of bovine DM collagenous polypeptides (M_r 50,000) were also published by Labermeier and Kenney (1983) and provided the basis for their preliminary identification as EC, or type VIII collagen (Sage *et al.*, 1980). However, the amino acid composition of the M_r 50,000 material isolated by Labermeier and Kenney (1983) was significantly different from that published by Kapoor *et al.* (1985). Although there are obvious structural similarities among all of the M_r 50,000 collagenous components isolated from bovine DM, it is possible that several distinct chains are present in this tissue, all of which are related to type VIII collagen.

IV. Structure of Type VIII Collagen

A. *Cassette Model*

Based largely on data obtained from protease digestion studies, Sage *et al.* (1983) proposed a model for the structure of type VIII collagen. Three triple-helical domains of M_r 50,000, which are resistant to pepsin, nondisulfide bonded, and composed of approximately equal amounts of hydroxyproline and proline, are joined by noncollagenous segments of M_r <10,000 that are resistant to bacterial collagenase. Large, globular N- and/or C-terminal propeptides are absent. The "parent" chain, EC1 of M_r 175,000–180,000, would therefore contain three cassettes of M_r 50,000, two interconnecting noncollagenous sequences, and one or two N- or C-terminal short segments (maximum M_r of 10,000). EC2 and EC3 would have two triple-helical domains (M_r 50,000, with EC3 lacking any terminal extensions. Homology among EC1, EC2, and EC3 was indicated by CNBr and 2-dimensional peptide mapping studies (Sage *et al.,* 1980); however, processing of EC1 to smaller forms has not been demonstrated *in vitro.* That the 50K cassettes themselves might be different is suggested by peptide mapping, amino acid composition, and N-terminal sequence analyses of 50K-A and 50K-B from DM (Kapoor *et al.,* 1985, 1986). Conservation of the cleavage site for vertebrate collagenase in each cassette would result in multiples of fragments, as shown in Fig. 6 and by Sage (1982) and Sage *et al.* (1983).

A similar model was proposed by Benya (1980) for the largest triple-helical form of EC (type VIII collagen) of approximate M_r 540,000. Cleavage of native $(EC1)_3$ by pepsin would produce an M_r 180,000 fragment $[(60K)_3]$, while endogenous proteolytic activity could explain the presence of other fragments, including $(EC2)_3$ of M_r 360,000. After denaturation, three major peptides of M_r 180,000, 120,000, and 60,000 were recovered. The essential feature of both models is the existence of interruptions within the collagen triple helix and the presence of discrete collagenous domains of M_r 50,000–60,000. Since the designations EC1, EC2, and EC3 refer only to size, it is not known whether there is more than one unique chain that participates in the assembly of type VIII collagen. The existence of oligomeric forms of type VIII collagen has not been described.

Recently Benya and Padilla (1986) have suggested that bovine corneal endothelial (BCE) cells synthesize type VIII collagen in the form of a relatively small triple helix with a chain M_r of 61,000. These authors have found that a large fraction of the higher molecular-weight type VIII collagen chains in the culture medium of postconfluent BCE cells (similar in size to EC1 and EC2) can be dissociated with a combination of acid pH and SDS to chains of M_r 61,000. These higher molecular-weight forms, therefore, appear to be stabilized by strong, noncovalent, acid-labile interactions. A fraction of the higher molecular-weight type VIII collagen produced by BCE cells is, however, resistant to such dissociation. It is possible that the lower molecular-weight form of type VIII collagen studied by Benya and Padilla (1986) represents an interesting processed form of the protein, but a complete resolution of these new findings with the models summarized above is not possible at this time.

B. *Triple-Helical Domains of Approximate M_r 50,000*

Studies on the nonfibrillar collagen types have revealed triple-helical domains of approximately 50,000 in molecular weight. Type X collagen, which has a chain M_r of 59,000, was converted by pepsin to a triple-helical molecule of three M_r 45,000 chains (Schmid and Conrad, 1982; Gibson *et al.*, 1982). In contrast, type VI collagen has very large globular domains which account for approximately 50% of the total molecular mass of the protein (Trüeb and Bornstein, 1984; von der Mark *et al.*, 1984). Cleavage of the precursor form of type VI collagen with pepsin, followed by reduction and alkylation, produced three α chains of M_r 40,000–55,000 (Jander *et al.*, 1981), although Furthmayr and co-workers (Odermatt *et al.*, 1983) have estimated from electron microscopic measurements that the maximum size of the triple-helical region was 35,000–40,000 Da per chain.

Early studies on the pepsin-derived fragments of type IV collagen demonstrated that at least one polypeptide of M_r 50,000 was derived from each of the proα1 and proα2 chains of this protein (Kresina and Miller, 1979; Dixit, 1978). This basement membrane collagen contains several interruptions within the triple helix, although periodicity with respect to an M_r 50,000 Gly-X-Y sequence has not been found. Recently, Dixit and co-workers (Dixit *et al.*, 1985) have isolated a polypeptide of M_r 55,000 from guanidinium chloride extracts of bovine lens capsule. This chain might correspond to α3(IV) or could represent yet another collagen type.

It therefore appears that the existence of domains of M_r 50,000 that are capable of triple-helix formation is not an exclusive property of type VIII collagen. Such domains can occur singly (as in types VI and X collagen), or perhaps in multiples (types IV and VIII collagen?), and are typically flanked by noncollagenous "linkers." Neither immunologic nor protein-sequencing studies has revealed homology among these triple-helical domains, other than the conservation of the triplet repeat that characterizes all members of the collagen gene family.

C. *Inclusion in the Nonfibrillar Collagen Gene Family?*

The structure of type IX collagen has recently been elucidated (van der Rest *et al.*, 1985). It consists of three collagenous domains (of 137, 339, and 115 residues per chain, respectively) and four noncollagenous domains, in each of three unique α chains, that associate by disulfide bonds to form the type IX collagen molecule. Based on the exon structure of the α1(IX) and α2(IX) collagen genes, Olsen and coworkers have proposed a distinct class of collagen genes (B) that is separate from that of the fibrillar collagens [types I, II, and III (class A)] (Lozano *et al.*, 1985). The class A genes are characterized by exon sizes related to a basic 54-base pair coding unit, while the size of exons comprising the class B genes is more irregular. Type IV collagen, from which three exons of 64, 123, and 182 base pairs have recently been characterized, may also belong to the class B gene family (Kurkinen *et al.*, 1985).

Sage *et al.* (1984) proposed a novel class of collagenous proteins which were characterized by imperfect triple helices and a mode of secretion and assembly that was highly different from that of the fibrillar collagens. Several new collagens (or their genes) containing very short triple-helical domains have been described in *Drosophila, C. elegans,* and fetal calf ligament fibroblasts [see Sage (1985b) and references therein]. It is possible that these proteins with shortened or interrupted triple-helical sequences would interact preferentially with other macromolecules of the extracellular matrix. Selective pressure for the rapid folding, secretion, and assembly of perfect triple-helical molecules might then have occurred in only one class of collagen genes (class A), specifically for the fibrillar collagens that contribute in large part to the interstitium.

V. Summary and Conclusions

Based on its distribution *in vitro,* we predict that type VIII collagen (1) is a minor component of most mesenchymal tissues including the vascular tree, (2) is a major component of corneal Descemet's membrane, and (3) forms part of the extracellular matrix and may be closely associated with the cell surface. Its synthesis by differentiated endothelial cells and sequestration to confluent, quiescent cell layers suggest that this collagen could facilitate stabilization of the characteristic endothelial phenotype. Cessation of type VIII collagen synthesis might then be associated with cellular injury and/or loss of the differentiated phenotype (as occurs during angiogenesis).

Type VIII collagen belongs to a unique subclass of the collagen gene family that is distinguished, in part, by an interrupted triple helix and a mode of secretion that differs from those of the fibrillar interstitial collagen types. The presence of several triple-helical domains (of M_r 50,000) and apparent lack of large, globular propeptide sequences account for several of the unusual features of type VIII collagen. The selective lability shared by types IV, V, and VIII collagen to proteases associated with injury, such as thrombin and a schistosomal serine proteinase, suggests a role for cell surface and/or basement membrane collagens in the maintenance of cellular integrity and a differentiated phenotype.

Acknowledgments

Contributions from the authors' laboratories were supported in part by National Institutes of Health Grants HL 18645, GM 29853, AM 11248, and DE 02600, and by R. J. Reynolds, Inc. H.S. is an Established Investigator of the American Heart Association. We thank Donna Stewart for artwork and secretarial assistance.

References

Alitalo, K., Bornstein, P., Vaheri, A., and Sage, H. (1983). *J. Biol. Chem.* **258,** 2653–2661.

Benya, P. D. (1980). *Renal Physiol.* **3,** 30–35.

Benya, P. D., and Padilla, S. R. (1986). *J. Biol. Chem.* **261,** 4160–4169.

Crouch, E., and Bornstein, P. (1979). *J. Biol. Chem.* **254,** 4197–4204.

Dixit, S. (1978). *FEBS Lett.* **85,** 153–157.

Dixit, R., Harrison, M. W., and Dixit, S. N. (1985). *Biochem. Biophys. Res. Commun.* **130,** 1–8.

Gibson, G. J., Schor, S. L., and Grant, M. E. (1982). *J. Cell Biol.* **93,** 767–774.

Jander, R., Rauterberg, J., Voss, B., and van Bassewitz, D. (1981). *Eur. J. Biochem.* **114,** 17–25.

Kapoor, R., Bornstein, P., and Sage, H. (1985). *Biol. Chem. Basement Membr. [Proc. Int. Symp.], 1st* 105–118.

Kapoor, R., Bornstein, P., and Sage, H. (1986). *Biochemistry*, **25,** 3930–3937.

Kenney, M. C., Labermeier, U., Hinds, D., and Waring, G. O. (1984). *Exp. Eye Res.* **39,** 267–277.

Kresina, T. F., and Miller, E. J. (1979). *Biochemistry* **18,** 3089–3097.

Kurkinen, M., Bernard, M. P., Barlow, D. P., and Chow, L. T. (1985). *Nature (London)* **317,** 177–179.

Labermeier, U., and Kenney, M. C. (1983). *Biochem. Biophys. Res. Commun.* **116,** 619–625.

Labermeier, U., Demlow, T. A., and Kenney, M. C. (1983). *Exp. Eye Res.* **37,** 225–237.

Lozano, G., Ninomiya, Y., Thompson, H., and Olsen, B. R. (1985). *Proc. Natl. Acad. Sci. U.S.A.* **82,** 4050–4054.

McKerrow, J. H., Jones, P., Sage, H., and Pino-Heiss, S. (1985). *Biochem. J.* **231,** 47–51.

Murphy, C., Alvarado, J., and Juster, R. (1984). *Invest. Ophthalmol. Visual Sci.* **25,** 1402–1415.

Odermatt, E., Risteli, J., Van Delden, V., and Timpl, R. (1983). *Biochem. J.* **211,** 295–302.

Sage, H. (1982). *J. Invest. Dermatol.* **79,** 51s–59s.

Sage, H. (1984). *In* "Biology of Endothelial Cells" (E. A. Jaffe, ed.), pp. 161–177. Nijhoff, Boston.

Sage, H. (1985a). *In* "Vascular Endothelium in Thrombosis and Hemostasis" (M. A. Gimbrone, ed.), pp. 187–208. Churchill Livingstone, Edinburgh.

Sage, H. (1985b). *Biochemistry* **24,** 7430–7440.

Sage, H., and Bornstein, P. (1982). *Methods Enzymol.* **82,** 97–127.

Sage, H., Crouch, E., and Bornstein, P. (1979). *Biochemistry* **18,** 5433–5442.

Sage, H., Pritzl, P., and Bornstein, P. (1980). *Biochemistry* **19,** 5747–5755.

Sage, H., Pritzl, P., and Bornstein, P. (1981). *Arteriosclerosis* **1,** 427–442.

Sage, H., Pritzl, P., and Bornstein, P. (1982). *Collagen Relat. Res.* **2,** 465–479.

Sage, H., Trüeb, B., and Bornstein, P. (1983). *J. Biol. Chem.* **258,** 13391–13401.

Sage, H., Balian, G., Vogel, A. M., and Bornstein, P. (1984). *Lab. Invest.* **50,** 219–231.

Sage, H., Tupper, J., and Bramson, R. (1986). *J. Cell. Physiol.* **127,** 373–387.

Sawada, H., Konomi, H., and Nagai, Y. (1984). *Eur. J. Cell Biol.* **35,** 226–234.

Schmid, T. M., and Conrad, H. E. (1982). *J. Biol. Chem.* **257,** 12451–12457.

Trüeb, B., and Bornstein, P. (1984). *J. Biol. Chem.* **259,** 8597–8604.

van der Rest, M., Mayne, R., Ninomiya, Y., Seidah, N. G., Chrétien, M., and Olsen, B. R. (1985). *J. Biol. Chem.* **260,** 220–225.

von der Mark, H., Aumailley, M., Wick, G., Fleischmajer, R., and Timpl, R. (1984). *Eur. J. Biochem.* **142,** 493–502.

Type IX Collagen

Michel van der Rest

Genetics Unit, Shriners Hospital for Crippled Children, Montréal, Canada H3G 1A6

and

Richard Mayne

Department of Cell Biology and Anatomy, The University of Alabama at Birmingham, Birmingham, Alabama 35294

I. Introduction

In addition to type II collagen, several other collagenous molecules are present in cartilages and can be solubilized after extensive digestion with pepsin [reviewed by Miller (1976) and Mayne and von der Mark (1983)]. Burgeson and Hollister (1979) first discovered that the pepsin digest contains small amounts of a native collagen which is assembled from three additional α chains of sizes similar to the α1(II) chain. This molecule (or molecules) is now known as type XI or 1α2α3α collagen (see Eyre and Wu, this volume). In the early 1980s, several reports were published describing disulfide-bonded and highly soluble collagenous fragments that were also solubilized by pepsin digestion and whose helical domains were much smaller than the interstitial collagens (Shimokomaki *et al.,* 1980, 1981; Reese and Mayne, 1981; Ayad *et al.,* 1981, 1982; von der Mark *et al.,* 1982; Ricard-Blum *et al.,* 1982). Except for type X collagen, which is synthesized only by hypertrophic chondrocytes (see Schmid and Linsenmayer, this volume), it is now clear that these fragments originate from a single parent molecule, type IX collagen.

In this chapter, we will initially present a model for the structure of the intact type IX collagen molecule. This model was derived from both biochemical (Ninomiya and Olsen, 1984; van der Rest *et al.,* 1985; Mayne *et al.,* 1985) and rotary shadowing (Irwin *et al.,* 1985) observations. Subsequently, we will present the detailed evidence which supports this model together with some speculation as to the possible function of this collagen.

STRUCTURE AND FUNCTION OF
COLLAGEN TYPES

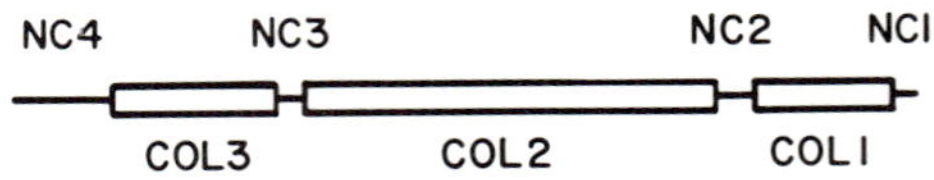

FIG. 1. Extended model for the structure of type IX collagen. The molecule consists of three collagenous, triple-helical domains (COL1–COL3; open bars) and four noncollagenous domains (NC1–NC4; solid lines). The model was initially derived from nucleotide sequencing of a cDNA clone that did not extend completely to the N-terminus (see Ninomiya and Olsen, 1984). Subsequent studies show that the NC4 domain of the α1(IX) chain is much larger than the NC4 domains of the α2(IX) or α3(IX) chains (Konomi *et al.*, 1987; Huber *et al.*, 1986). The N-terminus is to the left, the C-terminus to the right.

In Fig. 1, the structure of type IX collagen is drawn in an extended form and shown to consist of three collagenous domains (COL1–COL3) and four noncollagenous domains (NC1–NC4). The numbering of these domains is from the carboxyl-terminus. This occurs as the model was initially derived from the sequencing of cDNA clones and the conversion of the nucleotide sequence of these clones to an amino acid sequence (Ninomiya and Olsen, 1984; van der Rest *et al.,* 1985; Ninomiya *et al.,* 1985). Such clones always begin at the 3′ end of the nucleotide sequence or, at the protein level, from the carboxyl-terminus. Further results, based on biochemical analyses of the pepsin-resistant fragments, show that the molecule consists of three genetically distinct chains, designated α1(IX), α2(IX) and α3(IX), and that these three chains are present in equimolar amounts in a single molecule (Mayne *et al.,* 1985; van der Rest *et al.,* 1985; Ninomiya *et al.,* 1985). Evidence in support of this model also comes from rotary shadowing observations of the intact type IX collagen molecule as secreted by a suspension culture of embryonic chick chondrocytes (Fig. 2; also see Irwin *et al.*, 1985). The NC4 domain is visualized as a small knob at one end of the molecule. In addition, the NC3 domain is highly flexible and, in many of the molecules, is viewed as a kink of no fixed angle. The NC1 domain cannot be visualized by rotary shadowing, and a kink is not easily recognized at the NC2 domain. However, the overall length of the molecule corresponds closely to the predicted lengths of the three collagenous domains (Mayne *et al.,* 1986).

II. The Pepsin-Resistant Fragments

A. Mammalian Species

The first fragment of type IX collagen was isolated from pig hyaline cartilage after pepsin digestion (Shimokomaki *et al.,* 1980). The mole-

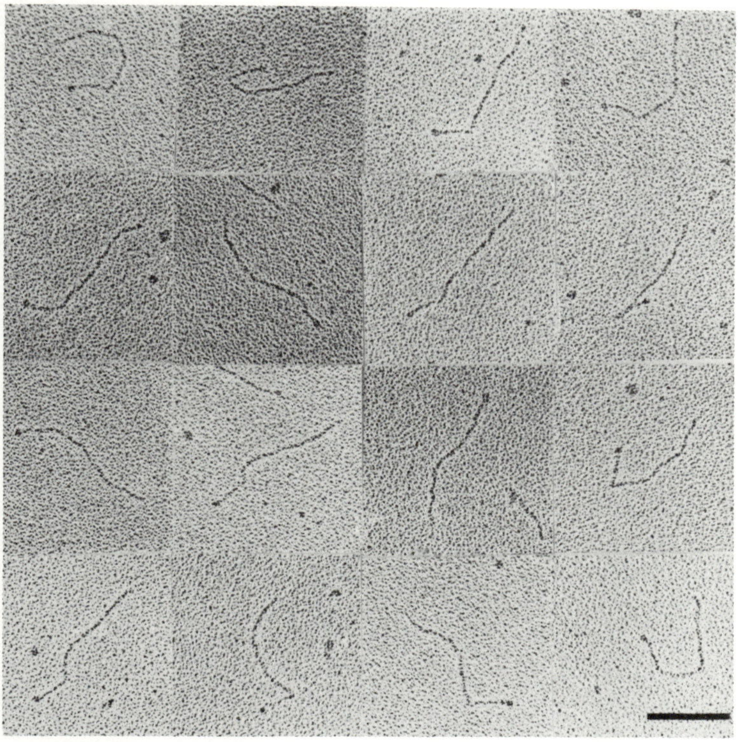

FIG. 2. Gallery of selected type IX collagen molecules visualized after rotary shadowing of the collagenous molecules secreted by a suspension culture of embryonic chick chondrocytes. Note that the NC4 domain is visualized as a small knob. In addition, a prominent kink is present in most of the molecules at the NC3 domain. Bar, 100 nm.

cule was of M_r 110,000 before reduction of interchain disulfide bridges and yielded a single band of M_r 33,000 after reduction. It was called type M collagen. The fragment was more soluble than other previously reported cartilage collagens and was not precipitated from a pepsin extract until 2.0 *M* NaCl, 0.5 *M* HAc were used. Later, additional smaller molecular-weight peptides were described by the same group and called CF1 and CF2 (Shimokomaki *et al.*, 1981).

The pepsin-resistant fragments of bovine type IX collagen were investigated independently by Ayad *et al.* (1981, 1982), using nasal cartilage, and by Ricard-Blum *et al.* (1982, 1985), using epiphyseal cartilage. The former group also described similar fractions isolated from human intervertebral disc. The separation methods used by the two groups were quite different. Ayad *et al.* (1981, 1982) found that type IX fragments were soluble in 20 m*M* Na_2HPO_4 at pH 9.2, whereas type

XI collagen (1α2α3α collagen) was precipitated. Ricard-Blum *et al.* (1982), however, used a high NaCl concentration in acetic acid to separate the type IX fragments from type II and type XI collagens which precipitated at lower salt concentrations. The high-molecular-weight component (CPS 1), as described by Ayad *et al.* (1981), was of M_r 110,000 before reduction and M_r 33,000 after reduction and, therefore, closely resembled the M fragment previously described by Shimokomaki *et al.* (1980). A low-molecular-weight component called CPS 2 (Ayad *et al.,* 1982) was subsequently isolated and separated after reduction into three components of M_r 16,000, 10,000, and 8,000. The same fragments were observed by Ricard-Blum *et al.* (1982) and called X5, X6 and X7 respectively. This latter group also obtained a number of higher molecular-weight fragments, which were called X1 to X4.

A model for the arrangement of the various pepsin-resistant fragments of bovine type IX collagen was proposed by Ricard-Blum *et al.* (1985). This model supports the view that the structural organization of mammalian type IX collagen is quite similar to the avian one to be described below, except for some differences in the location of the pepsin-sensitive sites.

Ayad *et al.* (1981, 1982) also presented electron micrographs of segment long spacing (SLS) crystallites of the CPS 1 and CPS 2 fragments. As expected from the molecular weight of these fragments, the crystallites were about one-third of the length of those obtained for type I collagen for CPS 1 and one-seventh for CPS 2.

B. *Chicken*

The pepsin fragments of chicken type IX collagen were described in previous publications as the high-molecular-weight components (HMW) and low-molecular-weight components (LMW) (Reese and Mayne, 1981; Reese *et al.,* 1982; Bruckner *et al.,* 1983: Mayne *et al.,* 1985, 1986; van der Rest *et al.,* 1985), or as M1 and M2 (von der Mark *et al.,* 1982). For the sake of consistency, we will use the former nomenclature for the rest of this chapter.

HMW and LMW were prepared by differential salt precipitation in 0.5 *M* acetic acid from pepsin-solubilized cartilage collagen. The type IX fragments were precipitated at 2.0 *M* NaCl after precipitation of type II and type XI (1α2α3α) collagen at 0.9 *M* NaCl and 1.2 *M* NaCl, respectively. During agarose gel filtration, HMW eluted as a double peak at the location of the β and α chains of type I collagen, although on SDS–PAGE it migrated as a single band between the β and α chains. After reduction, HMW migrated on SDS–PAGE as three components, called C1, C2, and C3 (with M_rs of 87,500, 51,000, and 36,400, respectively) as determined by agarose gel filtration (Reese and Mayne, 1981). Subsequently, a fourth component called C4 (M_r 14,000) was shown to participate in the assembly of HMW, although it is not disulfide bonded to the other components (Reese *et al.*, 1982). It was found that C4 will remain bound to the HMW fragment on agarose gel filtration in 1 *M* $CaCl_2$ provided the sample is not heat denatured prior to application to the column (see Fig. 3 for a model of HMW which shows the location of C4). Examination of HMW by rotary shadowing showed a prominent kink in the molecule giving rise to both a short and long arm (Reese *et al.*, 1982; von der Mark *et al.*, 1982). Removal of the C4 component by denaturation altered the rotary shadowing image of HMW, causing the short arm not to be visualized. This result indicates that C4 forms part of the triple-helical structure of the short arm (Reese *et al.*, 1982). In these conditions the long arm renatures due to the presence of interchain disulfide bridges but, with the loss of C4, the short arm cannot renature and is not visualized by rotary shadowing. In later experiments, the C2 component was resolved by reversed-phase high-performance liquid chromatography (HPLC) into equimolar amounts of two chains, which were shown to represent different primary structures by fingerprinting of their tryptic fragments (van der Rest *et al.*, 1985) and by partial amino acid sequence analysis [van der Rest *et al.* (1985) and unpublished observations]. The two chains were called C2 and C5 and are incorporated into the model of HMW as shown in Fig. 3. The C1 component probably arises from a nonreducible cross-link between C3 and either C2 or C5.

The low-molecular-weight component LMW was also present in the precipitate at 2.0 *M* NaCl, 0.5 *M* HAc, and could be separated from HMW by agarose gel filtration. LMW, after reduction and alkylation,

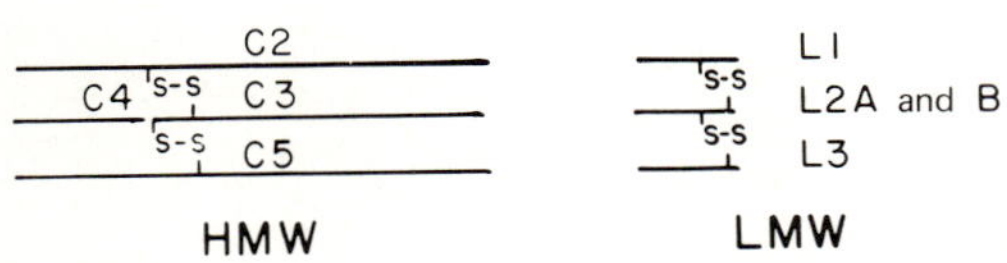

FIG. 3. Models for the structure of the pepsin-resistant fragments of chicken type IX collagen called HMW and LMW. Note that both HMW and LMW contain three different chains, and that, for HMW, one of the chains is cleaved giving rise to C3 and C4. This occurs at the NC3 domain (see Fig. 1), and in this domain the interchain disulfide bridges (S—S) of HMW are probably present. LMW also possesses interchain disulfide bridges which are present at the carboxyl-terminus within the NC1 domain (see van der Rest *et al.*, 1985).

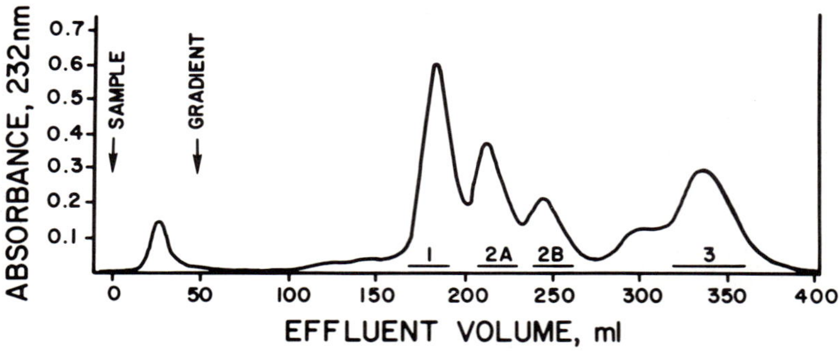

FIG. 4. Carboxymethylcellulose (CM-cellulose) chromatography of the chains of LMW obtained after reduction and alkylation followed by agarose gel filtration. The column (1.5 × 10 cm) was equilibrated at 42°C with 0.01 *M* (Na^+)sodium citrate. pH 3.6, and elution was achieved with a linear gradient of 0.0–0.2 *M* NaCl at a flow rate of 100 ml/hr over a total volume of 400 ml. Bars show the four fractions that were pooled for further analysis. Reprinted from Mayne *et al.* (1985) with permission.

was resolved by CM-cellulose chromatography into four components that were called LMW 1 (M_r 11,000), LMW 2A (M_r 10,600), LMW 2B (M_r 10,500), and LMW 3 (M_r 10,600), as is shown in Fig. 4. The molecular weights were determined by agarose gel filtration (Mayne *et al.*, 1985). By HPLC fingerprinting, using several proteolytic enzymes,

LMW 1

Leu-Arg-Lys-Pro-Leu-Ser-Pro-Gly-Met-Thr-Gly-Arg-Hyp-Gly-Pro-Ala-Gly-Pro-Hyp-Gly-Pro-Hyp-Gly-Ala-Thr-

LMW 2A

Gly-Gly-Val-Gly-Ala-Met-Gly-Pro-Hyp-Gly-Pro-Hyp-Gly-Pro-Hyp-Gly-Pro-Hyp-Gly-Glu-

LMW 2B

Ala-Lys-Arg-Ala-Ala-Leu-Gly-Gly-Val-Gly-Ala-Met-Gly-Pro-Hyp-Gly-Pro-Hyp-Gly-Pro-

Lys-Arg-Ala-Ala-Leu-Gly-Gly-Val-Gly-Ala-Met-Gly-Pro-Hyp-Gly-Pro-Hyp-Gly-Pro-Hyp

LMW 3
(Major)

Ser-Leu-Arg-Arg-Pro-Glu-Phe-Gly-Ala-Hyp-Gly-Leu-Hyp-Gly-Arg-Hyp-Gly-Pro-Hyp-Gly-

LMW 3
(Minor)

Leu-Arg-Arg-Pro-Glu-Phe-Gly-Ala-Hyp-Gly-Leu-Hyp-Gly-Arg-Hyp-Gly-Pro-Hyp-Gly-

FIG. 5. Amino-terminal amino acid sequence analyses of LMW 1, 2A, 2B, and 3. For LMW 2B, two sequences were recognized in approximately equal proportions. Both contain additional lysine and arginine residues that are not present in the sequence of LMW 2A. This probably explains the difference in elution of LMW 2A and LMW 2B during CM-cellulose chromatography. Reprinted from Mayne *et al.* (1986) with permission.

it was shown that LMW 2A and 2B are different cleavage products of the same parent chain. This was confirmed by N-terminal amino acid sequencing which showed that LMW 2B is five or six residues longer than LMW 2A and that the extra residues include a Lys-Arg sequence (Fig. 5). This will significantly increase the basic character of the fragment and probably explains the delayed retention time of LMW 2B on CM-cellulose (see Fig. 4).

LMW, after denaturation but without reduction of interchain disulfide bridges, could not be fractionated by either CM-cellulose chromatography (Mayne *et al.*, 1985) or reverse phase-HPLC (M. van der Rest and R. Mayne, unpublished observations), suggesting that the three chains are present in a single molecular assembly. LMW was also visualized after rotary shadowing as a short rod (Mayne and von der Mark, 1983) and possessed a single sharp melting curve during denaturation (Bruckner *et al.*, 1983). A model for the structure of LMW is also shown in Fig. 3.

III. The Intact Type IX Molecule

A. *The Isolation of Intact Type IX Collagen*

1. Chemical Characterization

The intact form of type IX collagen is often referred to as the precursor form of type IX collagen, although it now seems likely that the molecule is not processed extracellularly (von der Mark *et al.*, 1984; Kielty *et al.*, 1985). It was initially reported to consist of two chains with apparent M_r 68,000 and 84,000 (Bruckner *et al.*, 1983; von der Mark *et al.*, 1984), which correspond to the H and J chains earlier described by Gibson *et al.* (1983).

In the above experiments, intact type IX collagen was isolated from chondrocyte cell culture or cartilage organ cultures as a radiolabeled molecule. It can be either precipitated from the medium of cell cultures with 30% ammonium sulfate (Bruckner *et al.*, 1983; Gibson *et al.*, 1983) or extracted from the pelleted chondrocytes with 1 *M* NaCl (von der Mark *et al.*, 1984). Bruckner *et al.* (1983) further purified intact type IX collagen by immunoprecipitation with a polyclonal antibody against HMW. These authors also showed, by immunoblotting, that the antibody specifically reacted with the M_r 84,000 band and, to a lower extent, with the M_r 68,000 band. The immunoprecipitated material also showed, after reduction, sharp bands at M_r 84,000 and M_r 68,000 in addition to a fuzzy band at M_r 115,000 which was not characterized at the time. von der Mark *et al.* (1984) obtained similar results for a preparation of intact type IX collagen (called pM) purified by DEAE-cellulose chromatography. In addition, they showed that bands of the same size as the HMW and LMW components (M1 and M2 in their nomenclature) could be generated by pepsin digestion of the intact molecule. This result demonstrated for the first time that HMW and LMW are both present in a single parent molecule.

A third chain, migrating as a fuzzy band at M_r 115,000 on gel electrophoresis, was largely overlooked in these early reports. Bruckner *et al.* (1985), using a preparation purified by DEAE-cellulose chromatography, showed that intact unreduced type IX collagen could be labeled with [^{35}S]sulfate (see Fig. 6). After reduction, radioactivity was present only in the fuzzy band at M_r 115,000. The ^{35}S label was removed by both chondroitinase ABC and bacterial collagenase treat-

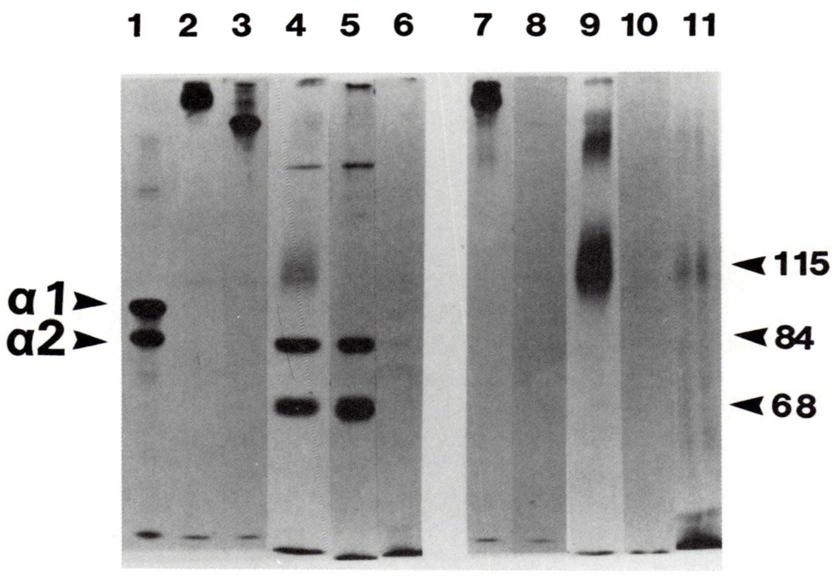

FIG. 6. Fluorogram prepared after electrophoresis of radioactively labeled type IX collagen in 6% polyacrylamide in sodium dodecyl sulfate. Type IX collagen was initially purified from other collagens by DEAE-cellulose chromatography. Lane 1, Type I collagen standard. Lanes 2–6, [^{14}C]Glycine-labeled protein. Lanes 7–11, [^{35}S]Sulfate-labeled protein. Lanes 2 and 7, Unreduced control samples. Lanes 4 and 9, Reduced control samples. Lanes 3 and 8, Unreduced samples digested with chondroitinase ABC. Lanes 5 and 10, Reduced samples digested with chondroitinase ABC. Lanes 6 and 11, Reduced samples digested with bacterial collagenase. Reprinted from Bruckner *et al.* (1985) with permission.

ment. With a [^{14}C]glycine label, the M_r 115,000 band disappeared after chondroitinase digestion while the M_r 68,000 band increased in intensity (compare lanes 4 and 5, Fig. 6). These results suggest that sulfated glycosaminoglycan chains are present in type IX collagen and are specifically located in the M_r 115,000 band. These results will be further discussed in Section III,C.

2. Rotary Shadowing Observations

Rotary shadowing was also used to recognize the intact form of type IX collagen as secreted by a suspension culture of embryonic chick chondrocytes (Irwin *et al.*, 1985). A gallery of these molecules is shown in Fig. 2. The type IX collagen molecules were originally recognized in the preparation by inclusion before rotary shadowing of a monoclonal antibody prepared against the HMW fragment. The location of the epitope for this antibody was at the end of the long arm of HMW as shown by rotary shadowing observations. The antibody was observed to bind to the intact type IX molecules at the expected location of the HMW fragment [Irwin *et al.* (1985) and see below].

In separate experiments, a large form of type IX collagen (M_r 225,000) was isolated from rat chondrosarcoma (Duance *et al.*, 1984), and its relationship with the pepsin fragment previously called type M collagen by Shimokomaki *et al.* (1980) was demonstrated by immunoassay using a polyclonal antibody raised against the type M fragment. The large form of type IX collagen was examined by rotary shadowing, and a prominent kink was observed in some of the molecules. The knob at the amino-terminus was not, however, visualized. This result suggests that the amino-terminus may have been lost either by cleavage within the tumor matrix or during the isolation procedures.

B. Comparison between Protein and cDNA Sequences

Most of the detailed structure of type IX collagen was unraveled from the sequencing of cDNA clones encoding portions of two of the constituent chains and by comparison of these sequences with selected amino acid sequences of pepsin-resistant fragments (see Fig. 7). The first cDNA studied was called pYN 1738 (Ninomiya and Olsen, 1984). It was prepared from a chicken cartilage cDNA library, and its characterization and isolation are discussed in more detail by Vasios *et al.* (this volume). The structure of the intact type IX molecule, as derived

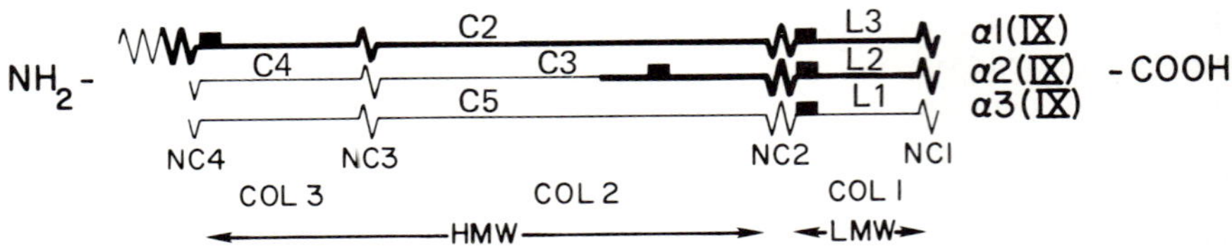

FIG. 7. A model for the structure of type IX collagen showing the relationship between the pepsin fragments HMW and LMW and the three chains of the intact molecule. The regions for which the amino acid sequences were derived from cDNA sequences are shown by heavy lines. The regions for which the amino acid sequences were established by protein sequencing are indicated by dark boxes. Light lines indicate regions of unknown sequences.

from the nucleotide sequence of this cDNA, was compatible with a model which contained both the HMW and LMW fragments (Ninomiya and Olsen, 1984). All of the HMW and LMW fragments were subjected to partial N-terminal amino acid sequence analysis. The sequences of the HMW C2 and of the LMW 3 chains were found to correspond to sequences present in the protein encoded for by the nucleotide sequence of pYN 1738. The only exception was for the prolines in the Y position of the Gly-X-Y triplets which were posttranslationally modified to hydroxyprolines (van der Rest *et al.*, 1985). In addition, all of the pepsin fragments of HMW and LMW were digested with trypsin and fingerprinted by HPLC. The amino acid composition of all of the peaks of these peptide maps was determined. These compositions were compared with the compositions predicted for the tryptic fragments from the cDNA sequence of pYN 1738. All of the cDNA-predicted tryptic peptides were recognized in the tryptic digest of LMW 3, except for the last nine residues. These were probably removed during pepsin digestion. Because of the larger size of the HMW fragments, the tryptic fingerprint of HMW after HPLC was more complex. However, seven tryptic peptides of HMW C2 could be unambiguously identified with peptides predicted from the sequence of pYN 1738. The chain encoded for by pYN 1738 was called the α1(IX) chain (van der Rest *et al.*, 1985), and it was concluded that it contains HMW C2 and LMW 3 as shown in Fig. 7.

A second cDNA clone, pYN 1731, was subsequently described which partially encodes for the α2(IX) chain (Ninomiya *et al.*, 1985). In this case, however, the cDNA did not extend to cover the N-terminus of the HMW fragments. A large tryptic fragment of HMW C3 was nevertheless recognized which, from its amino acid composition, appeared to be encoded for in the sequence of pYN 1731. A partial N-terminal amino acid sequence of this tryptic peptide was obtained and confirmed this identification. The LMW 2 fragment was also shown to be encoded for in pYN 1731 both by partial amino acid sequence analysis and by amino acid composition of the tryptic fragments. It was concluded that the $\alpha2$(IX) chain is comprised of the HMW fragments C4 and C3 and the LMW 2 fragment (see Fig. 7).

No cDNA has yet been described for the $\alpha3$(IX) chain. By elimination, it can be inferred that it contains HMW C5 and LMW 1.

C. *Type IX Collagen Is a Proteoglycan*

Noro *et al.* (1983) first isolated a collagenase-sensitive, disulfide-bonded proteoglycan from chick embryo epiphyseal cartilage. It was named PG-Lt. Subsequently, this proteoglycan was shown to be the same as type IX collagen, based on comparable chromatographic behavior and immunoblotting experiments using a polyclonal antibody prepared against the HMW component of type IX collagen (Vaughan *et al.,* 1985). For these latter experiments, the same preparation method as Noro *et al.* (1983) was used, except that chicken sterna were used as a source of cartilage. There are, however, some differences between the two reports. One difference is strictly methodological and refers to the molecular weights as determined by gel electrophoresis. In the publication of Noro *et al.* (1983), globular proteins were used as standards while Vaughan *et al.* (1985) used collagen standards. If the differences in mobilities between collagen and globular proteins are taken into account, then the two reports are completely comparable. A second difference stems from the apparent absence of the M_r 68,000 band in the paper of Noro *et al.* (1983). This is probably explained by the use of serine for protein labeling, resulting in very weak intensities of the bands after fluorography. The M_r 68,000 band might then have been missed. Finally, Noro *et al.* (1983) suggested that PG-Lt has a high content of dermatan sulfate, while Vaughan *et al.* (1985) found that the glycosaminoglycan chains were fully digested to disaccharides by chondroitinase AC, suggesting the absence of dermatan sulfate. The different cartilages that were used by the two groups may explain this discrepancy. Noro *et al.* (1983) used the epiphyses of embryonic chick tibia while Vaughan *et al.* (1985) used chicken sterna. Both groups found a ratio of four- to six-sulfated disaccharides of 3:1, which is much higher than the 1:3 ratio in the aggregating proteoglycan of chick hyaline cartilage.

The work of Vaughan *et al.* (1985), as described above, was complemented by another report from the same group [Bruckner *et al.*

(1985)]. In both papers, the existence in the intact type IX collagen molecule of a third chain of M_r 115,000 associated with glycosaminoglycan was demonstrated. These two reports establish that one or perhaps two chondroitin sulfate chains are covalently linked with a type IX chain. This is based on [^{35}S]sulfate labeling of the M_r 115,000 chain and removal of the label by chondroitinase ABC or AC digestion (see Fig. 6). Subsequently, the glycosaminoglycans were shown to be exclusively linked to the α2(IX) chain. By DEAE-Trisacryl chromatography, Huber *et al.* (1986) separated the three chains of type IX collagen (Fig. 8). By comparison of the tryptic fingerprints obtained for the pepsin fragments C3 and C4 and the isolated, intact, sulfated chain (Fig. 8, peak 3), it was concluded that the glycosaminoglycan-carrying chain is α2(IX) (Fig. 9). By the same technique, the M_r 84,000 band (Fig. 8, peak 2) was identified as α1(IX) and the M_r 68,000 band (Fig. 8, peak 1) as α3(IX) (Huber *et al.*, 1986). These results were independently confirmed by Konomi *et al.* (1986), who used polyclonal antibodies raised against synthetic peptides that correspond to known amino acid sequences for the three chains. These authors demonstrated, by immunoblotting, that there was specific labeling of the M_r 115,000 band with an antibody raised against a synthetic peptide from the α2(IX) chain (Fig. 10). However, using this antibody, immunoblotting was only successful if the migration of the M_r 115,000 band was reduced to M_r 68,000 by chondroitinase ABC treatment (Fig. 10, lanes 4 and 5). Similarly, the M_r 84,000 band was specifically labeled with an antibody against a synthetic peptide for the α1(IX) chain (Fig. 10, lanes 2 and 3) and the M_r 68,000 band with an antibody against a synthetic peptide for the α3(IX) chain (Fig. 10, lanes 6 and 7).

These results show that chondroitin sulfates are only present on the α2(IX) chain and provide further evidence that the three chains of type IX collagen are part of the same triple helix. This is demonstrated by the observation that intact and unreduced type IX migrates on SDS–PAGE as a single band whose molecular weight is reduced by chondroitinase treatment (see Fig. 6). If $\alpha1(IX)_3$ or $\alpha3(IX)_3$ existed, these molecules should be insensitive to chondroitinase treatment.

FIG. 8. Separation of the three chains of type IX collagen. (A) The reduced and alkylated chains were chromatographed on a DEAE-Trisacryl column in 7.5 *M* urea at 40°C. The α3(IX) chain (peak 1) eluted with the starting buffer, followed by the α1(IX) chain (peak 2) and the α2(IX) chain (peak 3). Fractions, pooled as shown by bars, were analyzed by SDS–PAGE and subsequent fluorography (B). Lanes a and b, Purified type IX collagen nonreduced and reduced. Lanes c, d, and e, Aliquots from peaks 1, 2, and 3, respectively. Reprinted from Huber *et al.* (1986) with permission.

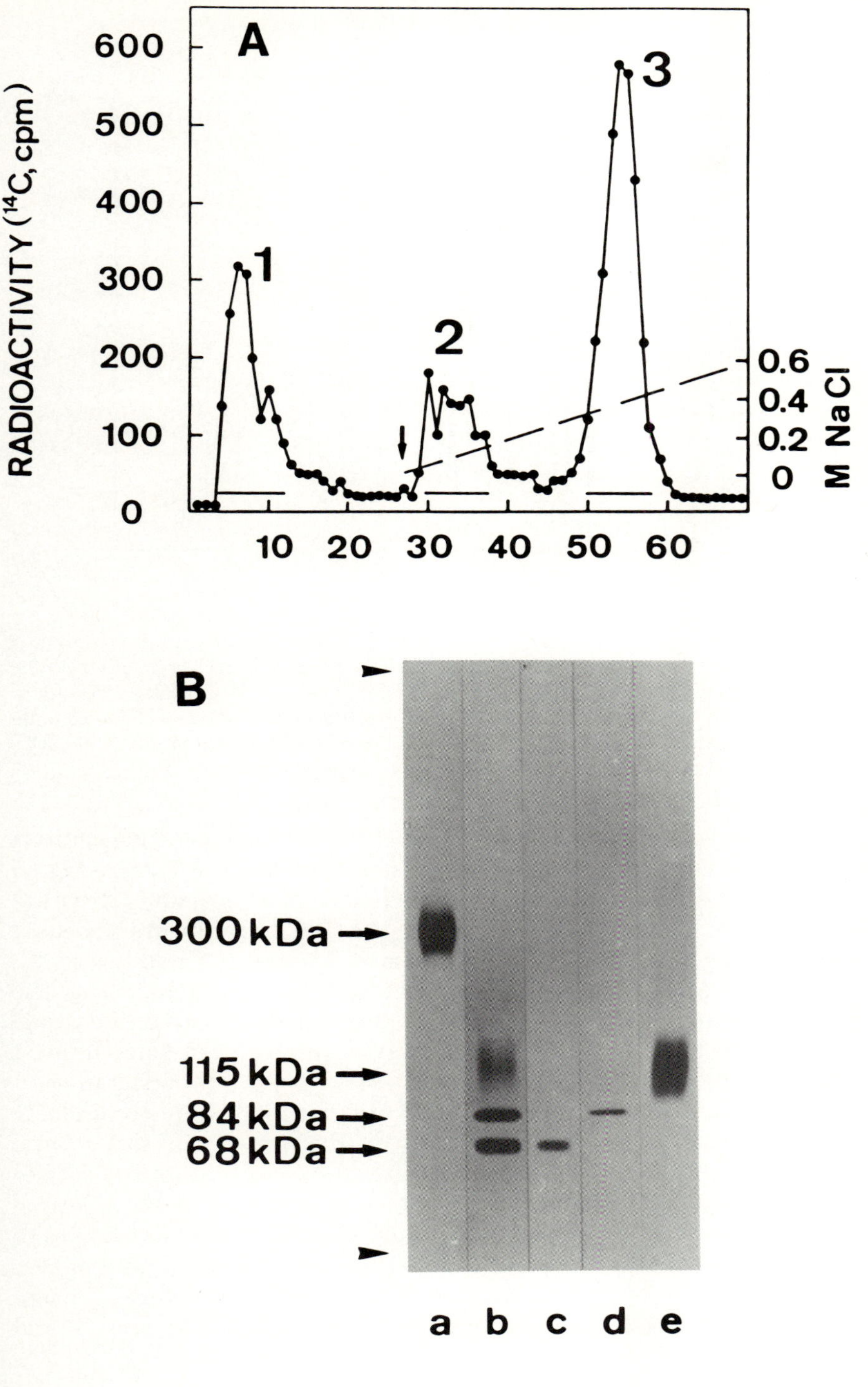

A
RADIOACTIVITY (14C, cpm)
600
500
400
300
200
100
0
1
2
3
10
20
30
40
50
60
0.6
0.4
0.2
0
M NaCl
B
300 kDa
115 kDa
84 kDa
68 kDa
a
b
c
d
e

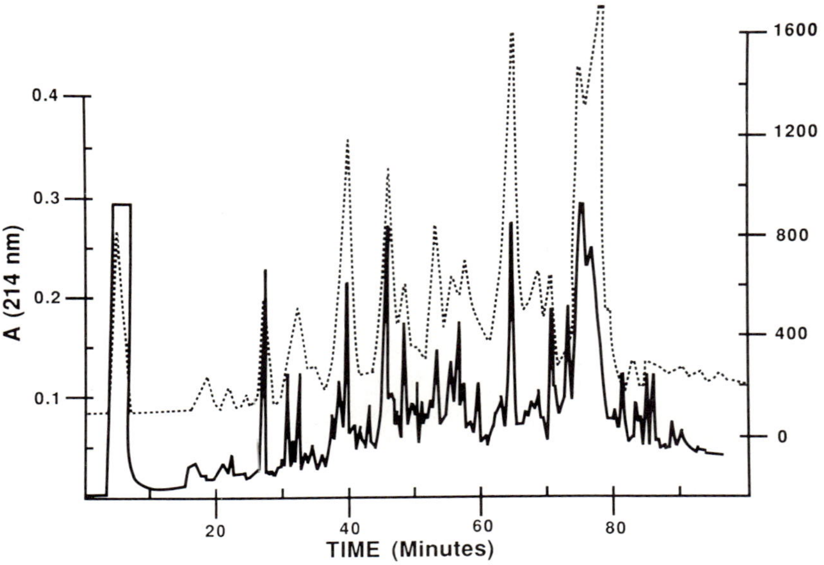

FIG. 9. Tryptic peptide profiles of an equimolar mixture of the C3 and C4 components of HMW (solid line) and of the radiolabeled 115 kDa chain (Fig. 8, peak 3) of intact type IX collagen (broken line). The tryptic digestions were performed separately. The digestion products were mixed and chromatographed together on a C18 Vydac TP 201 column, eluted in 9 m*M* trifluoroacetic acid using an aqueous acetonitrile gradient (0–32%). Fractions were collected at 1 min for radioactivity measurements.

The precise attachment site for the glycosaminoglycan side chain is not known at present, nor has the presence of a specific linkage region to serine residues been demonstrated. Based on primary structure data, the NC1 and NC2 domains of the α2(IX) chain are probably excluded, as they do not contain serine residues (Ninomiya *et al.*, 1985). The three collagenous domains can also be excluded, as the pepsin fragments HMW and LMW do not contain glycosaminoglycan chains. The possible location of the glycosaminoglycan side chain at the NC3 and/or at the NC4 domain is presently actively being investigated. Preliminary results (Irwin and Mayne, 1986) were obtained by rotary shadowing of type IX collagen in the presence of monoclonal antibodies 9A2 and 2B6 which recognize stubs of chondroitin 4-sulfate or dermatan sulfate generated after chondroitinase ABC digestion (Caterson *et al.*, 1985). Specific binding of both of these antibodies to the NC3 domain was observed, and this binding was dependent on prior digestion with chondroitinase. These results, therefore, indicate that the chondroitin sulfate may be located at the NC3 domain and

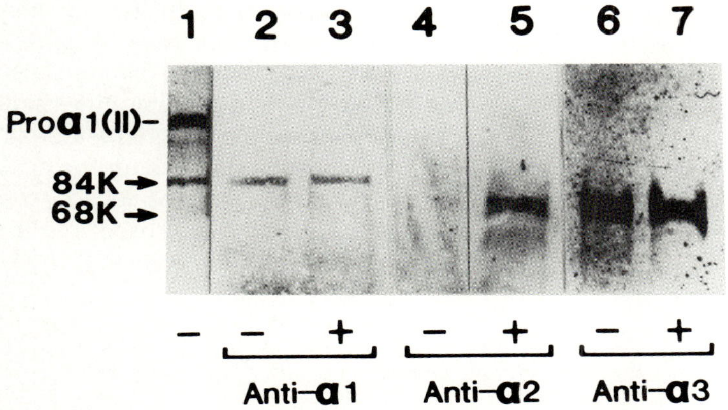

FIG. 10. Immunoblot of type IX collagen with antibodies prepared against synthetic peptides derived from nucleotide or amino acid sequences of the α1(IX), α2(IX), and α3(IX) chains. Lane 1 shows the result of staining the gel with Coomassie Brilliant Blue. Arrows indicate the location of the M_r 84,000 and M_r 68,000 bands (collagenous standards). Lanes 2–7 show the result of staining nitrocellulose blots with peptide-specific antibodies. Note that with antibodies against α1(IX) (lanes 2 and 3) or α3(X) (lanes 6 and 7) the location of the band remained unchanged after chondroitinase ABC digestion (+). For the α2(IX) chain, immunoblotting of a M_r 68,000 band only occurred after chondroitinase digestion (lanes 4 and 5). Without chondroitinase ABC digestion (−), the α2(IX) chain appears as a broad band of average M_r 115,000 (see Fig. 6, lane 9); this may explain the failure to obtain a successful immunoblot. Reprinted from Konomi *et al.* (1986) with permission.

must be present on a small peptide which is liberated from HMW during pepsin cleavage between C3 and C4.

Vaughan *et al.* (1985) demonstrated incorporation of [^{3}H]mannose into type IX collagen. This incorporation was not restricted to a single chain, and both the M_r 84,000 [α1(IX)] chain and M_r 68,000 [α3(IX)] chain were labeled. Whether the M_r 115,000 [α2(IX)] chain was also labeled is not clear from their results. Therefore, type IX collagen also apparently possesses mannose-containing oligosaccharide units.

IV. The Structure of Type IX Collagen

A. *Nomenclature*

The type IX collagen molecule is now clearly shown to be composed of seven domains that are not always of equal length in the three chains. Their triple-helical domains must, however, be in register to give the observed helical stability. In order to facilitate the identification of individual amino acids while permitting easy comparison between the three chains, we will use the following nomenclature: (1) amino acid position in a domain from the amino-terminus, (2) domain name, and (3) chain number. For example, a cysteine residue is found at position 5-NC1-α1(IX). This translates as the fifth amino acid starting from the amino end of the NC1 domain of the α1(IX) chain. In earlier publications, a different nomenclature was used: the numbering of amino acids referring to their position in cDNA pYN 1738 (Ninomiya and Olsen, 1984).

B. The Seven Domains of Type IX Collagen

1. The NC1 Domain

The NC1 domain is very short, consisting of 21 residues for α1(IX) and 15 residues for α2(IX), as determined from cDNA sequence analysis of the clones pYN 1738 (van der Rest *et al.*, 1985) and pYN 1731 (Ninomiya *et al.*, 1985). Comparison of these results with structural analyses performed on pepsin fragments indicates that very little, if any, of the NC1 domain of the α1(IX) and α2(IX) chains is removed by processing (van der Rest *et al.*, 1985). Indeed, the presence of all of the residues except the last nine amino acids was demonstrated in the pepsin fragments of the LMW 3 fragment of the α1(IX) chain. The exact size of the NC1 domain of α3(IX) is not known at present, as a cDNA has not been identified for this chain. The NC1 domain of both α1(IX) and α2(IX) contains one cysteine residue at position 5-NC1. The α3(IX) NC1 domain contains a cysteine at the same position as demonstrated by amino acid sequencing of a tryptic peptide containing the C-terminal portion of COL1 and the pepsin resistant portion of NC1 (M. van der Rest, unpublished results). In the three chains, another cysteine is found in the Y position of the last triplet of the COL1 domain, so that the three chains may be held together by pairs of interchain disulfide bridges.

2. The COL1 Domain

The COL1 domain is 115 residues long for both the α1(IX) and α2(IX) chains. It contains two imperfections in the triple helix. The first one is a Gly-X-Gly imperfection, with X being Arg, located at residues 61-COL1 to 63-COL1 for both α1(IX) and α2(IX). The second imperfection, Gly-X-Y-X-Y in both chains, is located at residues 78-

COL1 to 82-COL1 in α1(IX) and at residues 81-COL1 to 85-COL1 in α2(IX) (Ninomiya *et al.,* 1985). The 3-residue shift in the location of this imperfection may have a direct effect on the conformation of the native molecule at this site. However, further study of this conformational effect will require the sequence of the α3(IX) chain in the same region. This sequence is not yet available. The last amino acid of the COL1 domain is a cysteine in α1(IX), α2(IX), and α3(IX). The implication of this result in terms of disulfide-bond formation will be discussed in Section IV,C.

Since the COL1 domain has been the most extensively studied at the protein level, it is possible to give some information on the extent of posttranslationl modifications. In α1(IX), most prolines and all lysines in the Y position are hydroxylated. We also have evidence for only partial hydroxylation of at least one proline residue, since a tryptic fragment [37 in van der Rest *et al.* (1985)] was found in two peaks eluting 3 min apart on HPLC and differing only in the hydroxylation of one proline (M. van der Rest, unpublished observations). Glycosylation of hydroxylysine residues has not been directly demonstrated. However, the resistance to trypsin of the peptide bond next to hydroxylysines 36-COL1-α1(IX) and 48-COL1-α1(IX) and of hydroxylysines 48-COL1-α2(IX) and 54-COL1-α2(IX) indicates that they are glycosylated. No analysis for 3-hydroxyproline has been reported yet.

3. The NC2 Domain

The NC2 domain is 30 residues long in both α1(IX) and α2(IX). The size and structure of the NC2-α3(IX) domain is not known. In addition to probably providing a limited flexibility to the molecule, and also being cleaved by pepsin, this domain has no remarkable property. Sequencing of cDNA clones shows that there is a cysteine residue at position 10-NC2-α1(IX) with no equivalent in NC2-α2(IX). The three chains are, thus, not covalently held together at this domain. Since most of the NC2 domain is cleaved by pepsin, very few data are available for this domain at the protein level.

4. The COL2 Domain

The COL2 domain of α1(IX) is 339 residues long and contains no imperfections in the triple helix. Its primary structure was deduced from the sequence of the cDNA clone pYN 1738 (Ninomiya and Olsen, 1984). Several tryptic fragments from the pepsin-resistant fragment

HMW C2, which covers COL2, NC3, and COL3, were purified and analyzed by amino acid composition (van der Rest *et al.*, 1985). As was previously observed for the COL1 domain, these data showed the presence of the usual posttranslational hydroxylations of proline and lysine residues. A partial sequence of COL2 from α2(IX) was derived from a cDNA, pYN 1731, which covers the last 117 residues of the domain (Ninomiya *et al.*, 1985). N-terminal amino acid sequence analysis of a large tryptic fragment starting at residue 255-COL2-α2(IX) was obtained and confirmed the identity of the pepsin-resistant fragment containing COL2 (HMW C3) as being derived from the α2(IX) chain.

5. The NC3 Domain

The NC3 domains of the α1(IX) and α3(IX) chains are resistant to proteolysis by pepsin while this domain in the α2(IX) chain is sensitive to pepsin, resulting in a cut between COL2 and COL3 of α2(IX) (HMW C3 and C4) after pepsin solubilization (Reese *et al.*, 1982) (Fig. 3). The primary structure of NC3 of α1(IX) was derived from cDNA sequence analysis (Ninomiya and Olsen, 1984; van der Rest *et al.*, 1985). It is 12 residues long and contains two cysteine residues at positions 6-NC3 and 10-NC3. The structure of the NC3 domain of α2(IX) is presently actively being investigated (McCormick *et al.*, 1987), since immunological evidence (described above) suggests that it is the attachment site of the glycosaminoglycan side chain (Irwin and Mayne, 1986). Except that it is resistant to proteolysis by pepsin, nothing is known of the structure of NC3 from α3(IX).

6. The COL3 Domain

The COL3 domain of α1(IX) is 137 residues long and contains one Gly-X-Gly imperfection at residues 87-COL3 to 89-COL3, where X is Lys. The COL3 domain of α2(IX) is contained in the HMW C4 pepsin fragment (Reese *et al.*, 1982). It is not disulfide linked to the rest of HMW and is characterized by a very high hydroxyproline content (161 residues/1000) (Reese *et al.*, 1982). The cDNA-derived sequence of the COL1 domain of α1(IX) contains 19 prolines in position Y, which are likely sites of hydroxylation (Ninomiya and Olsen, 1984). This represents 138 residues per 1000. This high hydroxyproline content would be expected to increase the stability of the triple helix of the COL3 domain. This is probably reflected by the complex melting curve of

HMW, which contains both the COL2 and COL3 domains and only a part of which melts at a high temperature (Bruckner *et al.*, 1983). In addition, it also explains why HMW C4 remains linked to the other disulfide-bonded fragments in 1 *M* $CaCl_2$ during agarose gel filtration at room temperature (Reese *et al.*, 1982). Whether this increased stability of COL3 has any biological significance is, at present, unknown.

7. The NC4 Domain

Complete information on the size and structure of NC4 is still lacking both at the DNA and protein levels. The cDNA pYN 1738 only covers 85 residues and contains one cysteine residue. The α1(IX) chain has an apparent molecular weight of 84,000, while chondroitinase-treated α2(IX) is only 68,000, almost co-migrating with α3(IX) (Huber *et al.*, 1986; Konomi *et al.*, 1986). Since they do not differ significantly in size in the other regions, it is likely that the difference in molecular weight is due to an extended NC4 domain in α1(IX). This is in keeping with the differences in sizes of the mRNAs of α1(IX) and α2(IX) (Ninomiya *et al.*, 1985). The NC4 domain of α1(IX) was recently purified from a collagenase digest of chicken embryo sternal cartilage (Vasios *et al.*, 1987). Attempts to obtain an amino acid sequence of this peptide were unsuccessful, suggesting that the N-terminus may be blocked. However, sequence analysis of tryptic fragments confirmed the identity of the purified fragment when comparison was made with the sequence of an α1(IX) genomic DNA clone (Vasios *et al.*, 1987). The NC4 domain of α1(IX) behaves as a very hydrophobic entity and has an apparent M_r of 23,000 on SDS–PAGE. It may contain intrachain disulfide bridges but there is no evidence for interchain disulfide bridges between α1(IX) and the other two chains in this region.

C. *Disulfide Bridges in Type IX Collagen*

Type IX collagen must contain at least two sets of interchain disulfide bridges, since both HMW and LMW contain interchain disulfide bridges. A pair of cysteines are found at the C-terminal end of the molecule at residues 115-COL1 and 5-NC1 in the three chains. They must be held together by these bridges since the pepsin-solubilized C-terminal fragment (LMW) is obtained as a single, disulfide-bonded entity which is cleaved into three chains after reduction (Mayne *et al.*, 1985).

A set of cysteine residues is also found in α1(IX) in the NC3 domain (residues 6-NC3-α1(IX) and 10-NC3-α1(IX)]. The α2(IX) chain must be linked to α1(IX) by one or both of these cysteines since the only other cysteine found in α1(IX) that could contribute to the disulfide bands of the HMW pepsin fragment (Reese *et al.,* 1982) is located at position 10-NC2-α1(IX) and has no equivalent in α2(IX).

One remarkable feature of the cysteines at the C-terminus is that, in each chain, they are separated by four residues. The way the cysteines are cross-linked is not known. In type III collagen, the two cysteines of each chain mark the transition between the helical and nonhelical domain and are in adjacent positions so that the bonds can go from a cysteine occupying the Y position of the last triplet in one chain to the cysteine occupying the first nonhelical position of an adjacent chain (Glanville *et al.,* 1976). The types of conformations possible with the much more distant cysteines of type IX collagen remain to be established. A complex pattern of disulfide bonds can also be expected in the NC3 domain.

The NC4 domain of α1(IX) may contain intrachain disulfide bonds, since its apparent molecular weight increases slightly upon reduction (Vasios *et al.,* 1987).

D. Lysine-Derived Cross-Links

The only direct analysis of lysine-derived cross-links in type IX collagen was reported by Wu and Eyre (1984) on pepsin fragments isolated from bovine articular cartilage. The cross-links studied were the fluorescent hydroxypyridinium residues. They were found exclusively in the high-molecular-weight fraction at a concentration of 0.45/1000 residues, which is similar to the concentration in type II collagen (0.42/1000 residues). There was, however, a difference in the ratio of the hydroxylysyl form of pyridinoline to the lysyl form of pyridinoline (8:1 in type IX, compared to 100:1 in purified type II collagen). These numbers are only representative of the sample analyzed (articular cartilage of 2-year-old steers) and are likely to differ for other species, tissues, or age groups. Furthermore, it is possible that additional cross-links may be present in the pepsin-sensitive regions of the molecule.

Cross-linking in the chicken HMW pepsin fragment of type IX collagen is indirectly suggested by the presence of a fragment of M_r 87,500 (C1) after reduction of interchain disulfide bridges. This fragment is thought to be a cross-linked peptide between C3 and C2 or C5 (Reese *et al.,* 1982).

V. Biosynthesis and Cleavage of the Intact Molecule by Proteases

The biosynthetic form of type IX collagen appears on SDS–PAGE as a single broad band of apparent M_r 250,000 (Bruckner *et al.,* 1983, 1985; Gibson *et al.,* 1983; von der Mark *et al.,* 1984). The apparent heterogeneity of the molecule appears to arise from the presence of variable amounts of chondroitin sulfate. Kielty *et al.* (1985) recently examined explants of chick embryo cartilage for evidence of extracellular processing for type IX collagen and were unable to demonstrate any reduction in the molecular weight of this collagen before it is incorporated into the cartilage matrix.

The susceptibility of the biosynthetic form of chicken type IX collagen to chymotrypsin was investigated, and two disulfide-bonded aggregates of M_r 210,000 and 43,000 were isolated (Clark and Richards, 1985). The larger fragment gave rise on SDS–PAGE to a single band of M_r 61,000 after reduction, suggesting that cleavage by chymotrypsin does not occur at the NC3 domain of any of the three chains (see Fig. 7). Another group also observed a similar fragment of M_r 210,000 after limited *pepsin* digestion of type IX collagen, but fragments of M_r 140,000, 69,000, and 40,000 were observed after reduction (Yasui *et al.,* 1984). This result suggests that mild pepsin digestion may also cause internal cleavages in the molecule. Further pepsin digestion resulted in a smaller fragment of M_r 153,000 which appeared similar to HMW. In both of these studies, the fragments generated after protease digestion were not completely characterized. However, the results are entirely consistent with the model for type IX collagen as presented in Fig. 7.

VI. Immunofluorescent Localization of Type IX Collagen in Cartilage Matrix

Several groups have prepared polycolonal antibodies against the pepsin-resistant fragments of type IX collagen isolated from either porcine or bovine cartilages, and subsequently used the antibodies to investigate the immunofluorescent localization of type IX collagen in articular cartilages (Duance *et al.,* 1982: Ricard-Blum *et al.,* 1982) or bovine nasal septum (Evans *et al.,* 1983). In general, the results suggest that type IX collagen is predominantly present in the pericellular matrix which immediately surrounds the chondrocytes and may contain a higher proportion of proteoglycans (Poole *et al.,* 1982). The intensity of immunofluorescent staining was markedly increased after the use of testicular hyaluronidase digestion to remove "masking" proteoglycans, as previous described for type II collagen in chick cartilages (von der Mark *et al.*, 1976). Electron microscopic analyses have shown that the pericellular capsule of articular cartilage (Poole *et al.*, 1984) or the inner territorial matrix of growth-plate cartilage (Eggli *et al.*, 1985) consists of a fine "feltlike" meshwork of collagen fibrils unlike the wider, cross-banded fibrils of type II collagen found in the interterritorial matrix. In one of these studies, an attempt was also made to correlate the morphology of cartilage after hyaluronidase digestion with the observed structure of the pericellular capsule (Poole *et al.*, 1985). It was speculated that type IX collagen, and possibly type XI ($1\alpha 2\alpha 3\alpha$) collagen, may be involved in organizing or stabilizing the pericellular capsule, or possibly could form the fibrils which make up this matrix. An immunoperoxidase procedure was also used to obtain electron microscopic localization of type IX collagen in fetal calf epiphyseal cartilage (Hartmann *et al.*, 1983). The results suggested a preferential localization to the fine fibrils around the chondrocytes rather than to the larger cross-banded fibrils of type II collagen.

In other studies, a monoclonal antibody was prepared against the high-molecular-weight fragment of chicken type IX collagen and used for immunofluorescent localization of type IX collagen in a variety of cartilages (Irwin *et al.*, 1985; Mayne and Irwin. 1986). The antibody was of sufficiently high affinity that the location of the epitope along the intact type IX molecule could be viewed by rotary shadowing (Irwin *et al.*, 1985). Immunofluorescent staining of embryonic or adult chicken sterna with this antibody after hyaluronidase digestion showed codistribution with type II collagen throughout the cartilage matrix and not specifically in a pericellular location (Irwin *et al.*, 1985). Similar results were obtained with articular cartilage from 17-day chick embryos in which the type IX collagen was present throughout the cartilage matrix together with type II collagen, except on the articular surfaces where type I collagen predominated (Mayne and Irwin, 1986). Immunofluorescent staining of hypertrophic cartilage from 17-day-old chick embryo tibiotarsus also showed codistribution of type II and type IX collagens (Fig. 11). Similar results confirming the codistribution of type II and type IX collagens in several chick cartilages were found with a polyclonal antibody prepared against HMW (Müller-Glauser *et al.*, 1986).

Somewhat different results were observed when sections were not treated with hyaluronidase prior to immunofluorescent staining. In embryonic chick cartilage, some immunofluorescent staining was ob-

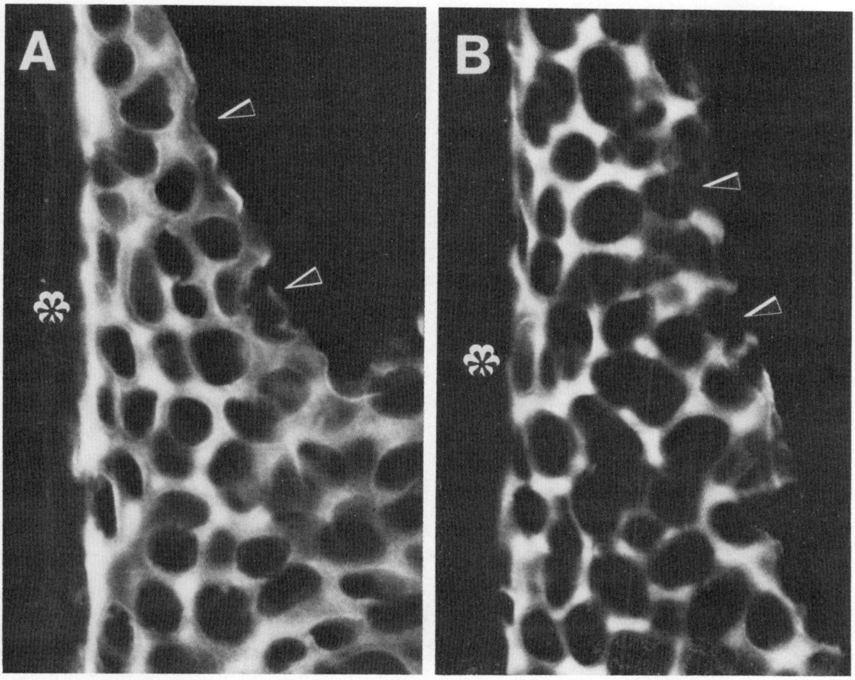

FIG. 11. Immunofluorescent staining of the diaphysis of 17-day chick embryo tibiotarsus with monoclonal antibody EB6 to type II collagen (A) and monoclonal antibody 2C2 to type IX collagen (B). Arrowheads show sites of erosion of chondrocyte lacunae and asterisks show the surface of the cartilage. ×363.

served in the center of the cartilage for type IX collagen but not for type II collagen (Irwin *et al.*, 1985), suggesting that type IX collagen may not be so extensively masked by proteoglycans. Also, in adult quail sternum without hyaluronidase digestion, some fluorescent staining occurred in a pericellular location for type IX collagen but not for type II collagen (Irwin *et al.*,1985). These results suggest that considerable caution should be exercised in interpreting immunolocatization experiments with cartilage, especially if the hyaluronidase digestion is incomplete. Recently, Poole *et al.* (1985) investigated the removal of proteoglycans from canine tibial cartilage by hyaluronidase digestion. Ruthenium red staining showed that digestion for up to 10 hr resulted in little change in proteoglycan concentration, and it was necessary to digest for 15–20 hr to remove completely all of the proteoglycan aggregates. These digestion times with hyaluronidase are far longer than are customarily employed before immunofluorescent staining. In addition, the pericellular matrix was observed to collapse after hyaluronidase digestion so that only a pericellular capsule was observed.

VII. Summary and Conclusions

Type IX collagen differs from the interstitial collagens in several important respects. The intact molecule is unusual in possessing three collagenous domains, each of which is interspersed by a short, noncollagenous domain. One of these noncollagenous domains (NC3) appears to be highly flexible when the molecule is visualized by rotary shadowing (Irwin *et al.,* 1985). In addition, the size and location of the exons in the type IX collagen genes are clearly different when compared to the fibril-forming collagens (Lozano *et al.,* 1985). The exons within the interstitial collagen genes appear to have originated from a unit of 54 base pairs [see Kühn, this volume; Kühn (1984)], but for type IX collagen no evidence was found for such a repeated structure (Lozano *et al.,* 1985). Type IX collagen may, therefore, be unrelated or only distantly related to the fibril-forming collagens, and it seems unlikely that it will be able to form collagen fibrils of the conventional type. It is, however, possible that the molecule may interact with fibrils of type II collagen and could be involved in organizing or maintaining the 3-dimensional meshwork of fine collagen fibrils found in many cartilage matrices. This possibility has been suggested several times (Wu and Eyre, 1984; Bruckner *et al.,* 1985: Mayne *et al.,* 1985), and recently some electron microscopic evidence was presented to support such a potential function (Müller-Glauser *et al.,* 1986). In cartilage matrices, type IX collagen is always present with type II collagen, and both are present in bovine vitreous humor, which also contains a fine meshwork of collagen fibrils (Ayad and Weiss, 1984). The synthesis of type IX collagen is also apparently initiated at the same time as the synthesis of type II collagen during chondrogenesis in chick limb buds (Kimura *et al.,* 1985).

The apparent preferential immunolocalization of type IX collagen to a pericellular location in several mammalian cartilages (Duance *et al.,* 1982; Ricard-Blum *et al.,* 1982) might also suggest an involvement in organizing a 3-dimensional meshwork. Only in the pericellular region of the epiphyseal cartilage of adult mammals is a fine meshwork of collagen fibrils formed, whereas in the territorial and interterritorial matrices the collagen fibrils are larger and not so extensively organized (Poole *et al.,* 1985; Eggli *et al.,* 1985). In addition, the identification of the trivalent hydroxypyridinium cross-link in some of the type

IX collagen molecules of bovine articular cartilage (Wu and Eyre, 1984) suggests a potential means of cross-linking type IX collagen to type II collagens. However, a population of type IX collagen in chicken cartilage matrix is clearly not cross-linked in this manner and can be solubilized from cartilage matrices in 4 *M* guanidine HCl (Noro *et al.*, 1983; Wu and Eyre, 1984; R. Mayne, unpublished observations).

It was initially estimated that type IX collagen constitutes only 5% of the collagen that can be solubilized from the chicken sternum with pepsin (Reese and Mayne, 1981). However, recent biosynthetic experiments with embryonic chick cartilages suggest that the synthesis of type IX collagen relative to type II collagen may be more than previously estimated and, during biosynthesis, could constitute as much as 20% of the synthesis of type II collagen (Yasui *et al.*, 1984; Kielty *et al.*, 1985; Vaughan *et al.*, 1987). Therefore, it seems highly likely that type IX collagen will play important functions in the assembly and structure of the extracellular matrix of cartilage.

Acknowledgments

The authors would like to thank Dr. B. R. Olsen, Dr. L. Vaughan, and Dr. P. Bruckner for communicating some of their results prior to publication, and Dr. P. J. Roughley for critically reading the manuscript. The authors would also like to thank Michael H. Irwin for unpublished results, and Pauline M. Mayne and Maxine Rudolph for their assistance in preparation of the manuscript. The authors acknowledge support from the Medical Research Council of Canada (Grant MT-7796) and the Shriners of North America (to M.v.d.R), and from National Institutes of Health Grant AM 30481 (to R.M.). M.v.d.R. is a chercheur-boursier sénior from the Fonds de la Recherche en Santé du Québec.

References

Ayad, S., and Weiss, J. B. (1984). *Biochem. J.* **218,** 835–840.

Ayad, S., Abedin, M. Z., Grundy, S. M., and Weiss, J. B. (1981). *FEBS Lett.* **123,** 195–199.

Ayad, S., Abedin, M. Z., Weiss, J. B., and Grundy, S. M. (1982). *FEBS Lett.* **139.** 300–347.

Bruckner, P., Mayne, R., and Tuderman, L. (1983). *Eur. J. Biochem.* **136,** 333–339.

Bruckner, P., Vaughan, L., and Winterhalter, K. H. (1985). *Proc. Natl. Acad. Sci. U.S.A.* **82,** 2608–2612.

Burgeson, R. E., and Hollister, D. W. (1979). *Biochem. Biophys. Res. Commun.* **87,** 1124–1131.

Caterson, B., Christner, J. E., Baker, J. R., and Couchman, J. R. (1985). *Fed. Proc., Fed. Am. Soc. Exp. Biol.* **44,** 386–393.

Clark, C. C., and Richards, C. F. (1985). *Collagen Relat. Res.* **5,** 205–223.

Duance, V. C., Shimokomaki, M., and Bailey, A. J. (1982). *Biosci. Rep.* **2,** 223–227.

Duance, V. C., Wotton, S. F., Voyle, C. A., and Bailey, A. J. (1984). *Biochem. J.* **221,** 885–889.

Eggli, P. S., Herrmann, W., Hunzinker, E. B., and Schenk, R. K. (1985). *Anat. Rec.* **211,** 246–257.

Evans, H. B., Ayad, S., Abedin, M. Z., Hopkins, S., Morgan, K., Walton, K. W., Weiss, J. B., and Holt, P. J. L. (1983). *Ann. Rheum. Dis.* **42,** 575–581.

Gibson, G. J., Kielty, C. M., Garner, C., Schor, S. L., and Grant, M. E. (1983). *Biochem. J.* **211,** 417–426.

Glanville, R. W., Allmann, H. and Fietzek, P. P. (1976). *Hoppe-Seyler's Z. Physiol. Chem.* **357,** 1663–1665.

Hartmann, D. J., Magloire, H., Ricard-Blum, S., Joffre, A., Couble, M.-L., Ville, G., and Herbage, D. (1983). *Collagen Relat. Res.* **3,** 349–357.

Huber, S., van der Rest, M., Bruckner, P., Rodriguez, E., Winterhalter, K. H., and Vaughan, L. (1986). *J. Biol. Chem.* **261,** 5965–5968.

Irwin, M. H., and Mayne, R. (1986). *J. Biol. Chem.* **261,** 16281–16283.

Irwin, M. H., Silvers, S. H., and Mayne, R. (1985). *J. Cell Biol.* **101,** 814–823.

Kielty, C. M., Kwan, A. P. L., Holmes, D. F., Schor, S. L., and Grant, M. E. (1985). *Biochem. J.* **227,** 545–554.

Kimura, T., Yasui, N., Ohsawa, S., and Ono, K. (1985). *Biochem. Biophys. Res. Commun.* **130,** 746–752.

Konomi, H., Seyer, J. M., Ninomiya, Y., and Olsen, B. R. (1986). *J. Biol. Chem.* **261,** 6742–6746.

Kühn, K. (1984). *Collagen Relat. Res.* **4,** 309–322.

Lozano, G., Ninomiya, Y., Thompson, H., and Olsen, B. R. (1985). *Proc. Natl. Acad. Sci. U.S.A.* **82,** 4050–4054.

McCormick, D., van der Rest, M., Goodship, J., Lozano, G., Ninomiya, Y., and Olsen, B. R. (1987). *Proc. Natl. Acad. Sci. U.S.A.,* In press.

Mayne, R., and Irwin, M. H. (1986). *In* "Articular Cartilage Biochemistry" (K. E. Kuettner, R. Schleyerbach, and V. C. Hascall, eds.), pp. 23–38. Raven, New York.

Mayne, R., and von der Mark, K. (1983). *In* "Cartilage" (B. K. Hall, ed.), Vol. 1, pp. 181–214. Academic Press, New York.

Mayne, R., van der Rest, M., Weaver, D. C., and Butler, W. T. (1985). *J. Cell. Biochem.* **27,** 133–141.

Mayne, R., van der Rest, M., Ninomiya, Y., and Olsen, B. R. (1986). *Ann. N.Y. Acad. Sci.* **460,** 38–46.

Miller, E. J. (1976). *Mol. Cell. Biochem.* **13,** 165–191.

Müller-Glauser, W., Humbel, B., Glatt, M., Sträuli, P., Winterhalter, K. H., and Bruckner, P. (1986). *J. Cell Biol.* **102,** 1931–1939.

Ninomiya, Y., and Olsen, B. R. (1984). *Proc. Natl. Acad. Sci. U.S.A.* **81,** 3014–3018.

Ninomiya, Y., van der Rest, M., Mayne, R., Lozano, G., and Olsen, B. R. (1985). *Biochemistry* **24,** 4223–4229.

Noro, A., Kimata, K., Oike, Y., Shinomura, T., Maeda, N., Yano, S., Takahashi, N., and Suzuki, S. (1983). *J. Biol. Chem.* **258,** 9323–9331.

Poole, A. R., Pidoux, I., Reiner, A., and Rosenberg, L. (1982). *J. Cell Biol.* **93,** 921–937.

Poole, C. A., Flint, M. H., and Beaumont, B. W. (1984). *J. Anat.* **138,** 113–138.

Poole, C. A., Flint, M. H., and Beaumont, B. W. (1985). *J. Ultrastruct. Res.* **91,** 13–23.

Reese, C. A., and Mayne, R. (1981). *Biochemistry* **20,** 5443–5448.

Reese, C. A., Wiedemann, H., Kühn, K., and Mayne, R. (1982). *Biochemistry* **21,** 826–830.

Ricard-Blum, S., Hartmann, D. J., Herbage, D., Payen-Meyran, C., and Ville, G. (1982). *FEBS Lett.* **146.** 343–347.

Ricard-Blum, S., Tiollier, J., Garrone, R., and Herbage, D. (1985). *J. Cell. Biochem.* **27,** 347–358.

Shimokomaki, M., Duance, V. C., and Bailey, A. J. (1980). *FEBS Lett.* **121,** 51–54.

Shimokomaki, M., Duance, V. C., and Bailey, A. J. (1981). *Biosci. Rep.* **1,** 561–570.
van der Rest, M., Mayne, R., Ninomiya, Y., Seidah, N. G., Chrétien, M., and Olsen, B. R. (1985). *J. Biol. Chem.* **260,** 220–225.
Vasios, G., Nishimura, I., Konomi, H., van der Rest, M., Ninomiya, Y., and Olsen, B. R. (1987). Submitted for publication.
Vaughan, L., Winterhalter, K. H., and Bruckner, P. (1985). *J. Biol. Chem.* **260,** 4758–4763.
Vaughan, L., Huber, S., Chiquet, M., and Wintherhalter, K. H. (1987). *EMBO J.* **6,** 349–353.
von der Mark, H., von der Mark, K., and Gay, S. (1976). *Dev. Biol.* **48,** 237–249.
von der Mark, K., van Menxel, M., and Wiedemann, H. (1982). *Eur. J. Biochem.* **124,** 57–62.
von der Mark, K., van Menxel, M., and Wiedemann, H. (1984). *Eur. J. Biochem.* **138,** 629–633.
Wu, J.-J., and Eyre, D. R. (1984). *Biochem. Biophys. Res. Commun.* **123,** 1033–1039.
Yasui, N., Benya, P. D., and Nimni, M. E. (1984). *J. Biol. Chem.* **259,** 14175–14179.

Type X Collagen

Thomas M. Schmid

Department of Biochemistry, Rush-Presbyterian–St. Luke's Medical Center, Chicago, Illinois 60612

and

Thomas F. Linsenmayer

Department of Anatomy and Cellular Biology, Tufts University Schools of Medicine, Boston, Massachusetts 02111, and The Developmental Biology Laboratory, Massachusetts General Hospital, Boston, Massachusetts 02114

I. Introduction

Type X collagen possesses a number of unique characteristics that distinguish it from the other collagens. The molecule has a very restricted tissue distribution and is synthesized predominantly by hypertrophic chondrocytes during the process of endochondral ossification. In the fetal and adolescent skeleton, the molecule is a transient intermediate in the cartilage which will be replaced by bone. In the adult, type X collagen persists in the zone of calcified cartilage which separates the hyaline cartilage from the subchondral bone.

The molecule is 138 nm in length, approximately half the length of the interstitial collagen types I, II, III, and V. Although the triple-helical domain is smaller, it has greater thermal stability than type I or type II collagens, demonstrating a T_m of 47°C. Type X collagen is susceptible to cleavage by vertebrate collagenase, but it is cleaved at two sites within its helical domain rather than at a single locus, which is characteristic of other collagen types. Another intriguing and possibly unique characteristic is the structure of its gene, which contains only one intron. The genes for other vertebrate collagen types contain numerous introns.

II. Initial Discovery

At approximately the same time, three separate laboratories working with cultures of embryonic chick chondrocytes reported the pres-

STRUCTURE AND FUNCTION OF COLLAGEN TYPES

ence of a short collagenous molecule which lacked interchain disulfide bonds (Schmid, 1980; Gibson *et al.*, 1981; Capasso *et al.*, 1982). One laboratory was studying the differentiation of chondrocytes from three different zones isolated from the tibiotarsus of 12-day chick embryos. The chondrocytes of embryonic long bones differentiate in a linear continuum from small proliferative cells (zone 1), to flattened cells with high biosynthetic capacity (zone 2), and then to hypertrophic cells (zone 3) (Stocum *et al.*, 1979). The objective of the work was to determine whether the cells within different zones were biochemically different. If so, a second objective was to determine whether monolayer chondrocyte cultures derived from the three specific zones of growth cartilage would maintain these zone-specific characteristics in culture. A variety of biochemical parameters were evaluated, including the types and rates of proteoglycan synthesis (Kim and Conrad, 1977, 1980) and the types and rates of collagen synthesis (Schmid and Conrad, 1982a,b). Matrix vesicle production was also examined (Glaser and Conrad, 1981).

Short-term organ cultures of tissue from zone 3 (hypertrophic cells) synthesized a collagenous protein with chains of lower molecular mass than α1(II) chains (Schmid and Conrad, 1982a,b). The tissues from zone 1 (proliferative cells) and zone 2 (flattened cells) did not synthesize the low-molecular-weight collagen. If the cells from each zone were dissociated and grown in monolayer culture for a week, the cells originating from both zone 2 and zone 3 now synthesized the small collagen molecule. The cells derived from all three zones, including zone 1, synthesized the small collagen molecule if the primary chondrocytes were passed into secondary cultures and maintained in monolayer culture for an additional week. The low-molecular-weight collagen appeared to be a specific marker for hypertrophic chondrocytes and showed that the chondrocyte monolayer cultures derived from different zones did not merely maintain their characteristics in culture but rather continued their program of development.

The low molecular weight collagen has chains of M_r 59,000, as analyzed by SDS–PAGE, when compared to collagen standards. It was first thought to be a degradation product of a larger collagen molecule, but further experimentation failed to reveal evidence of a precursor molecule. Analysis of the molecule by cyanogen bromide (CNBr) peptide mapping showed that it was unrelated to any known collagen. The molecule was referred to as 59K collagen or short chain (SC) collagen. The low-molecular-weight collagen has now been given the designation type X collagen.

Another laboratory reported the synthesis of type X collagen by chick sternal chondrocytes in culture (Gibson *et al.*, 1981, 1982). The chondrocytes were isolated from 18-day sterna and cultured either in native collagen gels or directly on tissue culture dishes. The major difference between the cultures was the relative amount of a short collagen molecule (type X) secreted into the medium. The chondrocytes cultured on plastic synthesized the protein in trace amounts, whereas the chondrocytes in the collagen gels synthesized the molecule as the major [^{3}H]proline-labeled product secreted into the medium. The molecule was degraded by bacterial collagenase and, by SDS–PAGE, it had chains of M_r 59,000. By CNBr peptide analysis, it was unrelated to any known collagen type. Gibson *et al.* (1982) referred to the molecule as G collagen.

A third laboratory observed the synthesis of type X collagen while studying the stability of embryonic chick chondrocytes in culture (Capasso *et al.,* 1982). This group isolated chondrocytes from the tibiotarsus of 19-day chick embryos and maintained the cells in culture as monolayers or as floating cells. Some of the attached chondrocytes were observed to spontaneously detach from the culture dishes, and the inclusion of ascorbic acid in the culture medium promoted the attachment of the floating chondrocytes. Both the floating and attached chondrocyte cultures synthesized type II collagen plus a low-molecular-weight collagenous protein with chains of M_r 64,000 (64K) when compared with globular protein standards on SDS–PAGE. The 64K chains were degraded by bacterial collagenase, and their secretion was dependent on the presence of ascorbic acid in the medium. The molecule appeared unrelated to type II collagen since it incorporated [^{35}S]methionine at a faster rate.

The above reports emphasize the importance of cell culture in the initial discovery and characterization of type X collagen. The molecule was probably not recognized by earlier investigators because of its low molecular weight, its solubility in acidic salt solutions (described later), and its restricted distribution to hypertrophic cartilage. The investigators who first observed the synthesis of type X were fortunate to have used sources of chondrocytes which were capable of producing the molecule and culture conditions which enhanced its synthesis. Type X collagen is not restricted to avian cartilage. A similar molecule has also been isolated from bovine (Remington *et al.,* 1984; Grant *et al.,* 1985) and rabbit (Remington *et al.,* 1983; Sussman *et al.,* 1984) growth-plate cartilages.

III. Source of Type X Collagen

Chondrocyte cell cultures were the first source of type X collagen, and harvesting conditioned medium from cultures continues to be the most useful means of obtaining the intact molecule in quantity. In one procedure, chondrocytes are enzymatically released from the tibiotarsi of 12-day chick embryos to prepare primary monolayer cultures (Kim and Conrad, 1977; Schmid and Conrad, 1982a; Schmid and Linsenmayer, 1983). After 1 week, the cultures reach confluence and are then passed into secondary culture. Such expanded mass cultures can be maintained and harvested for 6 weeks. There are two advantages to this method. The cells will synthesize type X collagen in increasing proportions (eventually, it constitutes the major collagen secreted into the culture medium), and type X can be harvested daily from the culture medium.

Alternate methods for the isolation of type X collagen involve the proteolytic release of type X from cartilage (Kielty *et al.,* 1984; Quarto *et al.,* 1985) or the extraction of type X from cartilage which has been made lathyritic. Both of these methods suffer from a disadvantage in that type X collagen is a minor component of hyaline cartilage. In an 18-day embryonic chick tibiotarsus, type X collagen represents approximately 2% of the collagen in the cartilage. Only a small proportion of the chondrocytes in cartilaginous tissues are hypertrophic, and even within the hypertrophic regions there exist considerable quantities of other collagen types (von der Mark and von der Mark, 1977; Schmid and Conrad, 1982b; Kielty *et al.,* 1985; Grant *et al.,* 1985). Proteolytic extraction has an additional disadvantage in that it destroys the nonhelical regions of the molecule.

IV. Isolation of Type X Collagen

Mass cultures of confluent chondrocyte monolayers (10 150-mm plates) yield greater than 50 mg of type X collagen after a month of harvesting the medium (Schmid and Linsenmayer, 1983). The type X is initially isolated from the culture medium by the addition of 30% ammonium sulfate in the presence of protease inhibitors. The precipitate is recovered by centrifugation, solubilized, and dialyzed against a neutral buffer (50 m*M* Tris, 0.4 *M* NaCl, pH 7.4). To remove residual proteoglycans and serum proteins from the preparation, the 30% ammonium sulfate precipitation is repeated.

The ammonium sulfate precipitate contains several different types of procollagen and collagen molecules. The type X collagen can be separated from most of the other collagen types by its solubility in acidic salt solutions. The types I and II collagens precipitate when samples are dialyzed against 0.5 *M* acetic acid containing 0.7 *M* NaCl.

The supernatant is enriched in type IX, type X, and type XI (1α2α3α) collagens. Type XI becomes insoluble when the NaCl concentration is raised to 1.2 *M;* the majority of the type X and type IX do not precipitate until the NaCl concentration of the acid solution is raised to 2 *M*. In practice, contamination of type X preparations with type XI and type IX collagens is slight since most of these molecules remain with the chondrocyte cell layer and are not present in the culture medium. Therefore, the collagen precipitated from the culture medium between 1.2 and 2.0 *M* NaCl in 0.5 *M* acetic acid is almost exclusively type X collagen. If completely pure type X is desired, preparations can be further purified by gel filtration or ion exchange chromatography (see Section V).

The use of cell culture as the source of type X collagen without resorting to proteolytic extraction demonstrated that the molecule contains both collagenous and noncollagenous domains (Gibson *et al.*, 1982; Schmid and Conrad, 1982a). The majority of the protein isolated by this method contains three identical chains of M_r 59,000.

The individual molecules were visualized by electron microscopy after rotary shadowing with platinum. As shown in Fig. 1A and B, two distinct domains of the molecule are evident in such preparations. One is a long, thin, rodlike region, characteristic of a collagen triple helix (Schmid *et al.*, 1984). The other is a knoblike globular region at one end of the helical domain.

The type X collagen gene has recently been isolated and sequenced (see Vasios *et al.*, this volume; Ninomiya *et al.*, 1986). The amino acid sequence of the molecule revealed a large nonhelical domain (NC1) of 170 residues at the carboxy-terminus of the molecule, a helical domain of 460 residues, and a small nonhelical domain (NC2) at the amino terminus of the molecule. The knob which is observed by rotary shadowing must, therefore, represent the carboxy-terminus of the molecule.

Digestion of type X collagen with pepsin removes the globular region of the molecule, leaving the triple-helical domain intact (Fig. 2, compare lanes 2 and 3). On SDS–PAGE the chains of the pepsin resistant domain have an M_r of 45,000 (using collagen peptides as standards). By electron microscopy (Fig. 1C), the length of this domain is 133 nm. Assuming a length of 0.286 nm per amino acid residue in a triple-helical polypeptide (Rich and Crick, 1961) and an M_r 98 per amino acid residue, the data from the gene sequence, SDS–PAGE, and electron microscopy are in agreement.

The noncollagenous domain at the carboxy-terminus can be isolated following digestion of the molecule with bacterial collagenase (Schmid

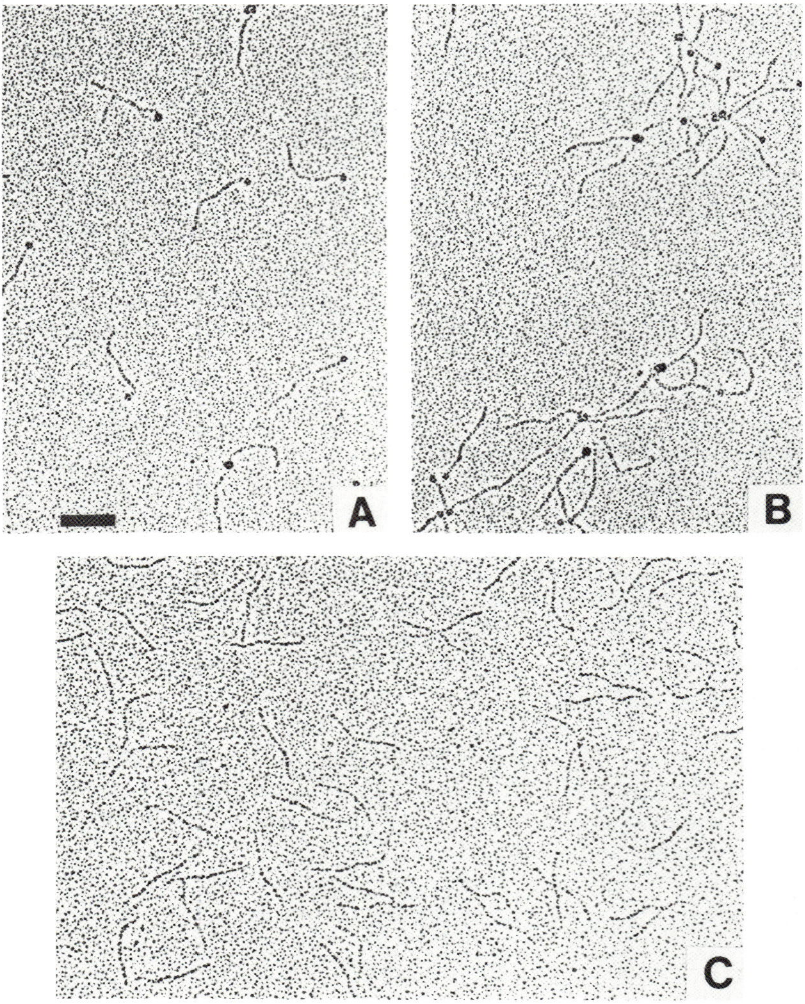

FIG. 1. Rotary-shadowed preparations of type X collagen. (A and B) Micrographs of the intact molecule (59K chains). (C) The pepsin-resistant form of type X collagen (45K chains). Bar, 100 nm. Reproduced from Schmid *et al.* (1984), *Ultrastruct. Res.* **86,** 186–191, by copyright permission of Academic Press, Orlando, Florida.

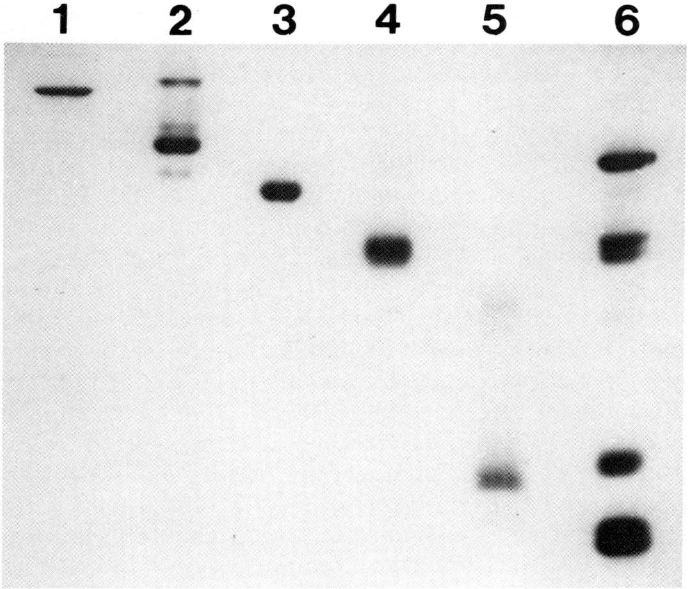

FIG. 2. SDS–PAGE (12% gel). Lane 1 contains α1(II) chains. Lane 2 contains the chains of intact type X collagen (59K). Lane 3 contains the pepsin-resistant chains (45K) of type X. Lane 4 contains a bacterial collagenase digestion of type X, but the sample was not heated before loading the gel. Lane 5 contains a sample which was treated like that in lane 4, except the mixture was heated to 100°C before loading the gel. Lane 6 contains the following protein mixture: bovine serum albumin (67K), ovalbumin (45K), soybean trypsin inhibitor (21.5K), and myoglobin (17K). Reprinted with permission from Schmid, T. M., and Lisenmayer, T. F. (1984). *Biochemistry* **23,** 553–558. Copyright (1984) American Chemical Society.

and Linsenmayer, 1984). As shown in Fig. 2, lane 5, the products of such a digestion contain chains of M_r 15,000–20,000. If the SDS–PAGE is performed without heating the samples (lane 4), the products migrate with an M_r of 45,000 (using noncollagenous proteins as standards). Therefore, the presence of SDS alone is not sufficient to separate the chains of the noncollagenous domain; both heat and detergent are necessary, suggesting the chains are tightly associated through strong noncovalent interactions.

V. Chromatography of Type X Collagen

Gel filtration of collagen molecules on Sephacryl S-500 (Schmid and Linsenmayer, 1983) or agarose (Quarto *et al.*, 1985) is a convenient method for separating the short type X molecules from small amounts of larger collagen molecules like type II. Sephacryl S-500 has certain characteristics which make it useful for such purifications. The elution position of the type X molecule can be modified to maximize the desired separation. The type X molecules, with 59K chains, have a slight affinity for the gel, which appears to be due to a hydrophobic interaction with the globular domain of the molecule. Thus, this form of the molecule elutes from a column slightly after pepsin-treated molecules composed of 45K chains. The affinity of the globular domain for the gel can be eliminated if the polarity of the solvent is reduced by the addition of 25% ethylene glycol to the buffer. Under such conditions, both forms of the molecule elute in the same position.

Ion exchange chromatography was also used to purify both the 59K and 45K chains of type X collagen (Schmid and Linsenmayer, 1983; Cancedda *et al.,* 1985). In one method, (carboxymethyl) cellulose was equilibrated at 42°C in 20 m*M* sodium acetate buffer containing 1 *M* urea, pH 4.8, and the two forms of type X eluted with a linear salt gradient (Schmid and Linsenmayer, 1984). The 45K chains eluted in the salt gradient (conductivity, 9 mmho) between the $\alpha1(I)$ and $\alpha2(I)$ chains, while the 59K chains of type X eluted (conductivity, 12.5 mmho) after the $\alpha2(I)$. The recoveries from these columns were in the range of 30–35%.

Attempts to use DEAE-cellulose chromatography for the purification of type X have not been successful using chromatographic conditions which have been successful for the purification of procollagen molecules (Linsenmayer and Little, 1978). Type X preparations bound so tightly to the resin that the material could not be recovered with high salt concentrations or extreme changes in pH (T. M. Schmid and T. F. Linsenmayer, unpublished results).

VI. Amino Acid Composition of Type X Collagen

Table I shows a comparison of the amino acid composition of chicken type X collagen reported from three different laboratories. In addition, the composition as deduced from the nucleotide sequence of the type X collagen gene (Ninomiya *et al.,* 1986) is also presented. Comparison can therefore be made of the amino acid composition from the sequence data with that of the molecule containing its full complement of posttranslational modifications. Examination of the type X helical domain, i.e., the pepsin-resistant domain with 45K chains, reveals an amino acid composition typical of a collagen triple helix. The molecule

TABLE I

AMINO ACID COMPOSITION OF TYPE X COLLAGEN[a]

Residue	45K[b]	45K[c]	45K[d]	45K[e]	59K[b]	59K[c]	NC1[b]
4-Hyp	ND	112	101	126	ND	80	ND
Asp	9	35	41	39	13	49	24
Thr	9	11	18	13	27	17	59
Ser	20	16	30	20	43	38	94
Glu	46	85	93	85	37	90	24
Pro	247	136	113	117	186	118	65
Gly	333	316	312	333	249	259	71
Ala	46	53	65	58	48	53	53
Cys	0	0	0	0	6	0	18
Val	28	25	20	24	40	36	65
Met	17	23	14	16	30	24	41
Ile	24	28	22	25	37	31	71
Leu	59	56	56	48	68	57	88
Tyr	2	3	2	3	24	25	76
Phe	13	11	13	12	22	26	35
Hyl	ND	35	31	33	ND	30	ND
Lys	52	20	26	20	54	25	59
His	9	7	7	8	12	12	18
Arg	28	28	37	29	30	30	29
Gln	42	ND	ND	ND	46	ND	53
Asn	17	ND	ND	ND	25	ND	47
Trp	0	ND	ND	ND	3	ND	12
Glc	ND	ND	ND	29	ND	ND	ND
Gal	ND	ND	ND	23	ND	ND	ND

[a]All values are reported as residues/1000 and are not corrected for loss or incomplete release of the amino acids during hydrolysis. Regions of the type X molecule: 59K, entire molecule (672 residues); 45K, pepsin resistant domain (460 residues); NC1, nonhelical region at carboxyterminus (170 residues); NC2, nonhelical region at amino-terminus (42 residues); ND, not determined.

[b]Composition derived from sequence of the type X gene (Ninomiya *et al.*, 1986) and does not include posttranslational modifications of the protein.

[c]Composition derived from protein hydrolysate (Schmid and Linsenmayer, 1983).

[d]Composition derived from protein hydrolysate (Kielty *et al.*, 1984).

[e]Composition derived from protein hydrolysate (Quarto *et al.*, 1985).

has 33% of its residues as glycine, 25% as imino acids, and high levels of hydroxyproline and hydroxylysine.

The total content of imino acids is higher than that of other collagens, and this may contribute to the high thermal stability of the triple helix (see Section VIII). About half of the proline residues and two-thirds of the lysine residues are hydroxylated. Based on the glucose and galactose content of the molecule (Quarto *et al.*, 1985), a large proportion of the hydroxylysine residues must be glycosylated.

The overall amino acid composition of type X collagen is more similar to type IX collagen than to other collagen types. Like type IX collagen, the bovine type X molecule contains interchain disulfide bonds in the helical and globular domains of the molecule (Remington *et al.*, 1984; Grant *et al.*, 1985). However, avian and rabbit (Remington *et al.*, 1983; Sussman *et al.*, 1984) type X collagen lack interchain disulfide bonds. Type X contains higher levels of imino acids and methionine than does type IX.

A comparison of the amino acid composition of the triple-helical and the noncollagenous regions of the type X molecule reveals some striking differences. The most dramatic difference is the high content of aromatic residues in the noncollagenous regions (NC1 and NC2). Of the tyrosine residues, 94% (15) are in these regions, as are 60% (9) of the phenylalanine residues, and all of the tryptophan residues (2). More than half (12) of the methionine residues are found in the nonhelical regions, which account for only 25–30% of the total mass. The regions are also rich in serine, threonine, valine, isoleucine, and leucine.

From the amino acid sequence derived from analysis of the chick type X gene, four cysteine residues are present in the noncollagenous regions. One cysteine is in the small NC2 domain at the amino-terminus of the molecule, while the other three are found in the larger NC1 domain at the carboxy-terminus. In chick type X, as mentioned above, the cysteine residues do not appear to form interchain disulfide bonds, but the possibility exists that they may form intrachain bonds.

It should be pointed out that using the gene sequence to derive the amino acid compositions of the different forms of the molecule is complicated by our lack of knowledge of the potential processing of the molecule. A definitive relationship has not been established between the translation product of the type X gene and the form in which the molecule is secreted and ultimately deposited in the extracellular matrix. During these processes, there may or may not be small proteolytic cleavages of the molecule near its amino- and/or carboxy-terminus. Thus, the data derived by the different methods must be interpreted and compared with some degree of caution.

VII. Cyanogen Bromide Cleavage of Type X Collagen

The cleavage of the type X collagen with cyanogen bromide produces a large number of peptides, as would be predicted from its high content of methionine. Compared to the peptides generated from type I or type II collagen, the CNBr peptides of type X are all low molecular weight. A separation of the CNBr peptides of type X collagen by SDS–PAGE is shown in Fig. 3. The majority of the peptides, which are fixed and stained in this gradient gel, have molecular weights between 6000 and 13,000. These results compare favorably with the sizes of the peptides predicted from the location of methionine residues in the sequence of the molecule (see Vasios *et al.,* this volume, Fig. 5). The largest CNBr peptides of type X should have M_r values of approximately 12,000, 8000, and 7000.

The pattern of type X collagen CNBr peptides is easily recognizable when compared to the peptides of other collagen types, most of which are considerably larger (Schmid and Conrad, 1982a; Gibson *et al.,* 1982; Remington *et al.,* 1984; Bornstein and Sage, 1980). However, the relatively large number of peptides which migrate in the low-molecular-weight region makes comparisons less useful in complex collagen mixtures; it also makes the comparison of minor differences in type X preparations more difficult. A 2-dimensional separation procedure, such as that of Benya (1981), would probably be more effective for the detection of differences in type X samples.

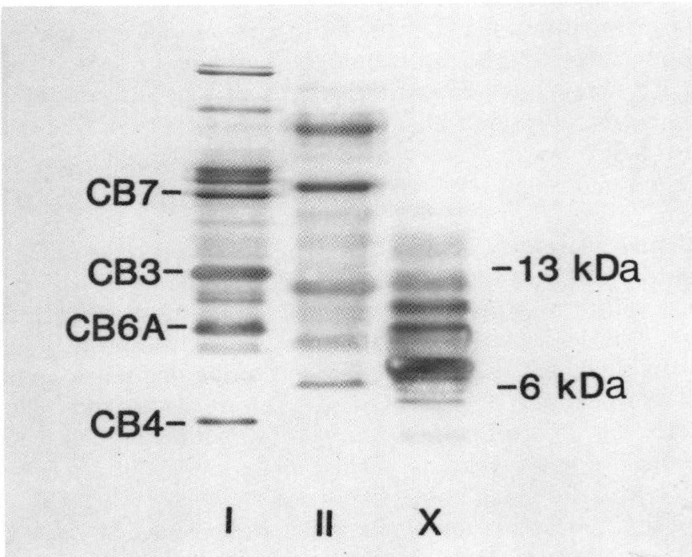

FIG. 3. CNBr peptides of types I, II, and X collagens separated by SDS–PAGE (8–20% gradient gel). CNBr peptides of type I collagen are labeled on the left side of the gel. The molecular weights of two collagenous CNBr peptides are shown for comparison on the right side of the gel.

Other methods of cleaving type X might also be advantageous, such as the specific cleavage of the Asn-Gly bonds with hydroxylamine (Bornstein and Balian, 1977). From the sequence of the type X molecule, six bonds should be cleaved, resulting in seven peptides with M_rs of approximately 18,000, 10,000, 8000, 7000, 6000, and 4000.

VIII. Thermal Stability of Type X Collagen

Several methods for the chromatographic separation of other collagen chains involve the thermal denaturation of the collagen triple helix and subsequent chromatography under denaturing conditions. In initial attempts to purify the denatured chains of type X collagen, the molecule was heated to temperatures greater than 70°C and chromatography was performed on a agarose column (BioGel A-5m) in the presence of 1 *M* $CaCl_2$. The majority of the type X material eluted in a position expected for a molecule larger than an α1(II) chain. This observation suggested that individual type X chains were not separated by this procedure, which prompted an examination of the thermal stability of the type X molecule.

The thermal stability of the type X triple helix was analyzed by circular dichroic (CD) spectropolarimetry. The intact type X molecule and its pepsin-resistant helical domain showed CD spectra typical of collagenous molecules (Schmid and Linsenmayer, 1984). They both exhibited a small peak of positive ellipticity at 222 nm and a very strong negative ellipticity at 197 nm. The ellipticity of the samples at both wavelengths was greatly reduced at 55°C, which suggested that the collagen had undergone a conformational change from triple helix to random coil.

The temperature-dependent change in the helical conformation was examined by monitoring the ellipticity of the samples at 222 nm as they were raised to progressively higher temperatures. Both the type X molecule and its pepsin-resistant domain exhibited a thermal denaturation midpoint (T_m) of 47°C. Similar determinations on chick collagen types I and II generated T_m values of 42–44°C. Thus, the triple helix of type X collagen is more stable than that of types I and II.

The noncollagenous regions of the molecule did not affect the thermal stability of the triple helix. Their presence, however, did affect the ability of the helical portion of the molecule to renature after thermal denaturation. If the type X triple helix was completely denatured by a 30-min incubation at 55°C and then the preparation was rapidly cooled to 25°C, the preparation regained greater than 60% of the helical signal within 40 min. An identical experiment performed with the pepsin-resistant domain of the molecule resulted in little, if any, reformation of the triple-helical structure. This was true even when a much longer time period was allowed for potential renaturation to occur. It appears that some aspect of the noncollagenous domains of type X collagen is quite refractory to thermal denaturation. Thus, it keeps the chains in "register," facilitating a rapid and correct reformation of the helical structure in the adjacent domain. Similar observations on the efficiency and completeness of renaturation have been made for procollagen vs collagen molecules (Hayashi *et al.,* 1979). In type I procollagen, the molecule has interchain disulfide bonds within its nonhelical domain at the carboxy-terminus, which account for the stability of the molecule. Since interchain disulfide bonds have not been detected in chicken type X collagen, noncovalent bonds, probably of a hydrophobic nature, may stabilize the nonhelical domains of the molecule. As described above, the noncollagenous domains of type X are enriched in aromatic and other hydrophobic amino acids. The thermal stability of the interchain interactions in the noncollagenous regions and the necessity for both heat and detergent to disrupt the interactions are consistent with the nature of hydrophobic interactions.

The return of 60% of the helical signal upon cooling could be the result of each molecule regaining this percentage of its structure or of 60% of the molecules regaining their complete helical structure. These possibilities were tested by determining the size distribution of the renatured molecules. If all molecules renatured to the same extent, one would expect them to elute as a single peak. If some had renatured completely and others had not, the elution profile should be more heterogeneous. The results obtained by gel filtration on Sephacryl S-500 (Schmid and Linsenmayer, 1984) showed that the majority of the renatured type X molecules migrated to a position indistinguishable from that of native molecules, but approximately one-third of the material eluted as a smaller species. The smaller molecular-weight material, which eluted as a shoulder on the peak of native-sized molecules, was larger than the size of an individual denatured type X chain, suggesting that some structure had reformed. However, they had not reformed native molecules.

Similar results were obtained by immunochemical analyses using a monoclonal antibody directed against type X collagen. This antibody recognizes a specific triple-helical epitope and will not bind to the denatured chains of the molecule. The complete characterization of the antibody will be described in Section XII. However, the conformational dependence of this reagent provides a rigorous test of the ability of the type X molecule to reform an exact triple-helical conformation (at least at the site of the epitope). Denatured type X molecules were cooled to 25°C, and the extent of renaturation was determined by employing the antibody in a competition enzyme-linked immunosorbent assay (ELISA). Again, approximately two-thirds of the molecules regained the conformation necessary to be recognized by the antibody (Schmid and Linsenmayer, 1985a). When the assay was used to examine possible renaturation of the isolated collagenous domain of type X (45K chains), none was detected.

IX. Vertebrate Collagenase Susceptibility of Type X Collagen

The vertebrate collagenases are a class of enzymes that produce a scission through the triple helix, thereby reducing the thermal stability of the resulting products well below that of physiological temperature (Gross, 1981). It was originally hypothesized by Gross and coworkers that the thermal stability of collagenase reaction products was of paramount importance, since this would render the collagen chains susceptible to cleavage by other proteolytic enzymes, thus completing the degradation process. As has been discussed elsewhere in this volume, different collagen types exhibit different susceptibilities to the vertebrate collagenases. For example, within the extracellular matrix of cartilage, type II collagen is cleaved by the enzyme, while the minor collagens including type IX (Liotta *et al.*, 1982) and type XI (Eyre *et al.*, 1984) are not. Type X collagen is susceptible to cleavage by the enzyme (Schmid and Conrad, 1982a; Gibson *et al.*, 1983). It displays the unique property of being cleaved at two sites within its triple helix rather than at a single locus as is typical of other susceptible collagens (Schmid *et al.*, 1986a).

Identification of the cleavage products, and some idea as to their location within the type X molecule, was achieved by analyzing digests of radiolabeled type X with purified human skin collagenase. For comparison, digests of radiolabeled type II collagen were also analyzed. One type X substrate was preferentially radiolabeled in its helical domain by [^{3}H]proline incorporation (90% of the proline residues are found in this region). A second type X preparation was preferentially radiolabeled by [^{3}H]tyrosine incorporation in its nonhelical domain (80% of tyrosine residues occur there). Digestions of both substrates were performed with various concentrations of enzyme at 25°C. Since this temperature is approximately 22°C below the T_m of the molecule, it should insure the integrity of the triple helices of the substrate and reaction products. SDS–PAGE analysis of the reaction products is shown in Fig. 4. As can be seen in lane 10, a complete digestion of the 59K substrate chains generated three products with M_rs of 32,000, 18,000, and 9000. The sum of these three products accounts for the entire molecular mass of the substrate chains.

A comparison of the collagenase cleavage products derived from [^{3}H]proline-labeled (lane 10) and [^{3}H]tyrosine-labeled (lane 15) type X substrates shows that the 18K fragment must contain the tyrosine-

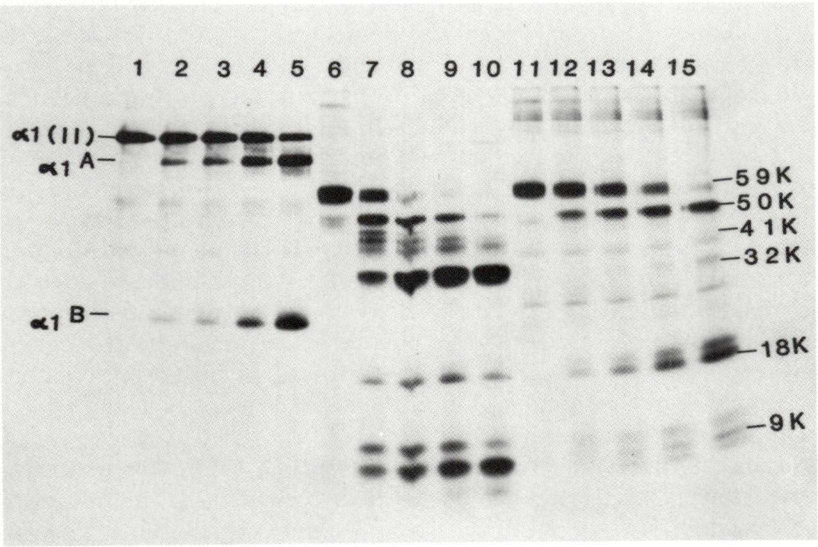

FIG. 4. SDS–PAGE of the human skin collagenase digestion of type X and type II collagens. Each lane contained 10,000 cpm of collagen substrate. Lanes 1–5 contained [^{3}H]proline-labeled type II collagen substrate. Lanes 6–10 contained [^{3}H]proline-labeled type X collagen substrate. Lanes 11–15 contained [^{3}H]tyrosine-labeled type X collagen substrate. The human skin collagenase was activated with trypsin and all the assays were performed at 25°C for 22 hr. Lanes 1, 6, and 11 contained no enzyme. Lanes 2, 7, and 12 contained 0.5 μl of enzyme. Lanes 3, 8, and 13 contained 1 μl of enzyme. Lanes 4, 9, and 14 contained 2 μl of enzyme. Lanes 5, 10, and 15 contained 4 μl of enzyme. The vertebrate collagenase products of type II collagen are marked on the left side of the gel, and the relative molecular mass of the collagenase products of type X are shown on the right side of the gel. Reproduced from Schmid *et al.* (1986a), *J. Biol. Chem.* **261,** 4184–4189, by permission of the copyright owner The American Society of Biological Chemists, Inc.

rich globular domain at the carboxy-terminus of the molecule. The 32K and 9K fragments are enriched in [^{3}H]proline and are probably derived from the helical domain of type X. The position of the 32K and 9K fragments in the type X molecule can be determined from an analysis of the single cleavage intermediates shown in lane 7. The largest intermediate has M_r 50,000 chains and is probably generated by the loss of the 9K fragment from the 59K substrate. The 50K intermediate also contains the tyrosine-rich globular region (see lane 15). The 9K fragment must, therefore, be derived from the amino-terminus of the helical domain, since this is the only possibility for generation of a 50K intermediate.

The location of the vertebrate collagenase cleavage sites in type X collagen was also established by electron microscopic visualization of rotary shadowed preparations (Schmid *et al.*, 1986a). To serve as a point of reference for the sites, a monoclonal antibody (X-AC9) was included with the cleaved preparations. The monoclonal antibody is known to bind within the triple-helical domain, at a locus approximately 20 nm from the globular domain. Figure 5 shows a preparation of the major collagenase cleavage product (32K chains) of type X collagen to which the monoclonal antibody is bound. For comparison, intact type X molecules with bound antibodies (from a separate preparation shown in black rectangles) have been superimposed such that the monoclonal antibodies on the cleaved and native preparations are aligned. It can be seen that two cleavages are made by the enzyme. One cleavage removes the carboxy-terminal globular region plus a short portion of the adjacent helix; the second removes a short portion of the helix at the amino-terminus.

Similar products were observed after type X collagen was digested with a variety of sources of vertebrate collagenase, including ones from human skin, synovium, neutrophil, and rabbit cornea. The exact bonds cleaved by the enzyme are unknown. Assuming the cleavages occur between Gly-Leu or Gly-Ile residues, the standard sites of cleavage for the enzyme in other collagen types (Woolley, 1984), there are several potential sites at the predicted locations within the type X molecule. As deduced from the genomic DNA sequence, two of the most likely candidates are a Gly-Leu bond 92 residues from the amino-terminus of the helical domain and a Gly-Ile bond 40 residues from the carboxy-terminus of the triple helix. Each of these bonds is close to an interrupted site along the triple helix in which a Gly-X-Y triplet is separated by two additional amino acids from the next Gly-X-Y triplet. These may be potential sites of instability within the molecule. Each interruption is followed by a sequence deficient in imino acid residues,

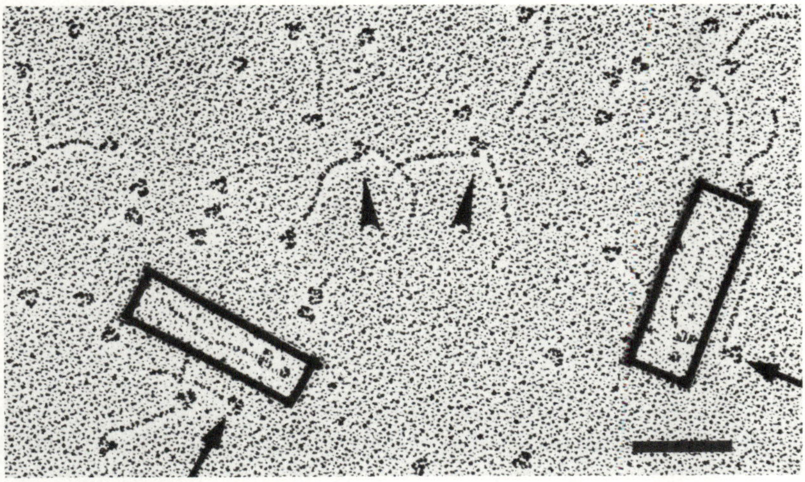

FIG. 5. Electron micrograph of rotary-shadowed preparations of a vertebrate collagenase digestion of type X collagen and the X-AC9 monoclonal antibody. After the digestion of type X collagen with human skin collagenase, the X-AC9 antibody was added and the mixture was rotary shadowed. The long arrows point to the molecules that represent the major vertebrate collagenase cleavage products (32K) of type X. These molecules retain the epitope recognized by the X-AC9 antibody at one end of the molecule. For comparison, native type X molecules with a bound antibody have been superimposed on the electron micrograph (outlined in the black rectangles) and aligned with respect to the bound antibody. The arrowheads mark X-AC9 molecules which have cross-linked two 32K collagenase cleavage products of type X collagen. Bar, 100 nm. Reproduced from Schmid *et al.* (1986a), *J. Biol. Chem.* **261,** 4184–4189, by permission of the copyright owner The American Society of Biological Chemists, Inc.

which might further compromise the helical stability of these regions (see sequence of type X molecule in Vasios *et al.*, this volume, Fig. 5).

Type X collagen appears to be degraded more rapidly than type II collagen using *in vitro* solution assays. Preparations of type X and type II collagens were isolated from the same chondrocyte cultures after labeling with [^{3}H]proline. A comparison of similar digestions by human skin collagenase, with equal amounts of radioactive type X and type II collagen substrates, showed that the type X substrate (Fig. 4, lane 9) is more rapidly degraded than the type II substrate (Fig. 4, lane 4).

Since vertebrate collagenase is thought to facilitate the degradation of collagenous molecules by generating products with reduced thermal stability, the thermal stability of the major collagenase product (32K chains) of type X collagen was examined. The thermal denaturation curve for the molecule by CD spectropolarimetry gave a T_m of 43°C in neutral solutions (Schmid *et al.,* 1986a). Thus, the denaturation temperature for the fragment with 32K chains is only slightly reduced from the T_m of the native molecule (47°C) and should not undergo spontaneous denaturation at the physiological temperature of the chick. Instead, the triple-helical fragment with 32K chains may be released *en bloc* from the center of the type X molecule. Whether it persists *in vivo* and has any physiological function is at present unknown.

X. Molecular Structure of Type X Collagen

The most complete information on the structure of type X collagen has been derived from the nucleotide sequence of the type X collagen gene (see Vasios *et al.,* this volume, Fig. 5; Ninomiya *et al.,* 1986). A cDNA library was constructed from poly(A)$^+$ RNA isolated from long-term cultures of chick embryonic tibiotarsus chondrocytes. One cDNA clone (600 nucleotides) hybridized to mRNA (2500 nucleotides) from long-term chondrocyte cultures and mRNA isolated from the cephalic region of chick embryo sterna. The cDNA clone did not hybridize to mRNA from a 17-day chick embryo sterna (which would synthesize very low levels of type X collagen). The cDNA clone was used to screen a chicken *Eco*RI genomic library. A genomic clone was isolated which contained the cDNA sequence. The amino acid sequence predicted from the genomic DNA sequence was identical to the amino acid sequence of two CNBr peptides of type X collagen. The coding sequence of the type X collagen gene was not interrupted by introns.

The largest form of the molecule isolated from the medium of chondrocyte cultures is similar in size to the molecule which can be inferred from the DNA sequence of the type X collagen gene. However, the exact size of the molecule, as transported into the extracellular milieu, is not known. The molecule contains two noncollagenous domains, NC1 and NC2. NC1, the largest noncollagenous domain, is located at the carboxy-terminus and contains 170 amino acid residues. The NC2 domain is a short telopeptide-like region located at the amino-terminus of the molecule.

The collagenous domain contains 460 amino acid residues and, in general, exhibits the repetitive Gly-X-Y sequence characteristic of a collagen molecule. There are, however, two general types of deviation from the Gly-X-Y sequence. Four positions in the molecule have a Gly-X-Gly sequence; three others have a Gly-X-Y which is separated from the next triplet by two additional amino acids. As mentioned above, two of these positions may be sites for cleavage by vertebrate collagenase. These deviations from the usual Gly-X-Y sequence do not produce sufficient defects in the helical structure to make it susceptible to cleavage by pepsin (Schmid and Conrad, 1982a; Gibson *et al.*, 1982), chymotrypsin (Gibson and Flint, 1985), or trypsin (T. M. Schmid, unpublished results).

The sequence of the type X molecule (Vasios et al., this volume, Fig. 5) also reveals the presence of a hydrophobic sequence in the nonhelical domain (NC1) towards the carboxyl end of the molecule (Ninomiya *et al.*, 1986). A hydrophobic sequence of 29 amino acids is bounded by two lysine residues and contains two cysteinyl residues. The hydrophobic nature and length of this sequence are characteristic of polypeptides that span a lipid bilayer and suggest the molecule may associate with a membrane.

The intact type X collagen molecule purified from cell culture has been rotary shadowed and visualized in the electron microscope. As stated earlier, both the collagenous domain and the major globular domain (NC1) can be clearly visualized, the molecule having an overall length of 138 ± 5 nm (Schmid *et al.*, 1984, 1986a; Kielty *et al.*, 1985). Frequently, the molecules associate at their globular domains to form aggregates (Fig. 1B). It is not clear whether such aggregation occurs *in vivo* and has physiological significance or whether it is simply a preparational artifact. Pepsin digestion removes the globular domain from the carboxy-terminus of the molecule; the remaining collagenous domain (133 ± 8 nm) accounts for most of the length.

Type X collagen will form segment long spacing (SLS) crystallites when dialyzed against ATP (Schmid *et al.*, 1984; Kielty *et al.*, 1984), but this only occurs in molecules from which the noncollagenous domain has been removed. The crystallites often occur as end-to-end dimers and, when positively stained, exhibit a banding pattern different from that of other collagen types (i.e., types I and II). The length determined for the crystallites (133 nm) is equivalent to that of the individual pepsinized molecules viewed in rotary shadowed preparations. Preparations of type X molecules with intact noncollagenous domains formed masses of thin, fibrous material which lacked discernable periodicity (T. M. Schmid, R. R. Bruns, and T. F. Linsenmayer, unpublished results).

At present, we do not know whether type X collagen becomes assembled into a supramolecular structure *in vivo*. Studies are presently in progress to localize type X *in vivo* employing immunoelectron microscopy with monoclonal antibody X-AC9.

XI. Biosynthesis of Type X Collagen

Little information is available on the intracellular steps involved in the biosynthesis of type X collagen. Presumably, the molecule is synthesized on the rough endoplasmic reticulum, but specific information about the organization of the ribosomes or the sites of chain association and triple helix formation are unknown. Cell-free translation studies indicate that the newly synthesized chains of type X migrate close to the 59K chains of the fully hydroxylated and glycosylated molecules secreted by cells (Cancedda *et al.,* 1985; Ninomiya *et al.,* 1986; Balian and Morris, 1986). Castagnola *et al.* (1986) have recently shown that type X contains a signal sequence of M_r 1500–2000 which is removed by dog pancreas membranes in cell-free translation experiments. The NC2 domain at the amino-terminus of type X collagen contains a group of hydrophobic amino acids resembling a typical signal sequence (see sequence in Vasios *et al.,* this volume, Fig. 5). Therefore, approximately half of the NC2 domain is probably removed before type X collagen is secreted. Type X collagen requires ascorbate for the proper hydroxylation of the molecule and, in the absence of ascorbate, at least some under-hydroxylated molecules are secreted at 37°C (Capasso *et al.,* 1982).

Type X collagen was detected in several different sizes of intracellular vesicles by immunofluorescent histochemistry with monoclonal antibody X-AC9 (Linsenmayer *et al.,* 1986b). In certain cartilages undergoing hypertrophy, the intracellular type X-containing vesicles are readily detected within chondrocytes just preceding the boundary at which the type X begins to be deposited as extracellular matrix. Figure 6 shows an example of such antibody-reactive intracellular vesicles within chondrocytes in a developing vertebral body. The detection of intracellular type X within cells which have no type X in their surrounding matrix suggests the molecule may be stored intracellularly for a period of time before it is secreted.

The thermal stability of the intracellular type X molecules was ex-

---

Fig. 6. Fluorescence micrographs of sections reacted at different temperatures with anti-type X collagen monoclonal antibody. Sections were equilibrated (30 min), stained (30 min), and washed at various temperatures (A, 45°C; B, 50°C; C, 60°C. D was incubated at 60°C, (30 min) and cooled to 40°C before reacting with antibody at the lower temperature. In C, the white arrows point to cells which are nonreactive. In D, the arrowheads point to two cells in which the intracellular reactivity has returned upon cooling to the lower temperature. All are ×420. Reproduced from Linsenmayer *et al.* (1986b), *Exp. Cell Res.,* in press, with copyright permission of Academic Press, Orlando, Florida.

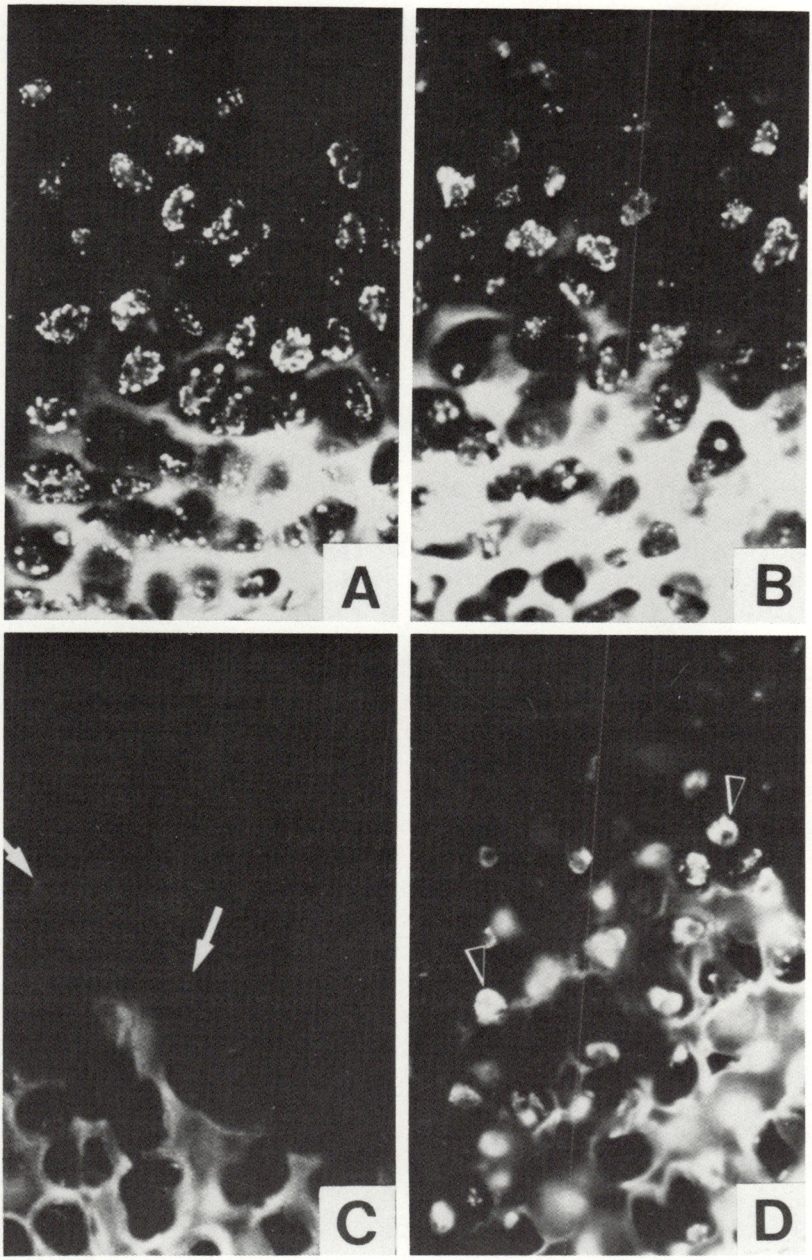
A
B
C
D

amined with monoclonal antibody X-AC9. When unfixed cryostat tissue sections were incubated with the antibody at various temperatures, the intracellular vesicles still retained their antibody reactivity at 55°C, but lost it by 60°C. This is considerably higher than one would deduce from *in vitro* measurements. The reactivity within the surrounding extracellular matrix was still retained at 67.5°C. This is the highest temperature we could examine, since the antibody itself is thermally denatured above this temperature. It is tempting to speculate that the increased thermal stability of the type X collagen, when deposited in the extracellular matrix, is the result of its assembly into a supramolecular form following its movement from intracellular to extracellular compartments. Extracellular events might include proteolytic processing, fibril formation, and cross-link formation.

Type X collagen is synthesized in a form that resembles a procollagen molecule, i.e., it contains both a collagenous triple-helical domain and a sizable noncollagenous domain at one end of the molecule. However, any experiments suggesting that the 59K chains of type X are processed to some other form must be evaluated with caution, since the globular domain is sensitive to general proteolytic activity (Schmid and Conrad, 1982a; Gibson *et al.,* 1982, Gibson and Flint, 1985). A recent report presented data suggesting the cleavage of the 59K chains of type X to 50K chains (Kielty *et al.,* 1985). The loss occurred from the globular domain of the molecule. These experiments were performed using organ cultures of hypertrophic cartilage from the 17-day chick tibiotarsus. The cartilage pieces were pulsed with [^{3}H]proline for 1 hr and chased with unlabeled proline. After 6 hr, the majority of the 59K chains had been converted to 50,000 chains. Another pulse–chase organ culture study using rabbit wrist and costochondral growth cartilage reported that type X was lost after a 20 hr chase (Sussman *et al.,* 1984). A third study reported a rapid degradation of type X collagen in cell culture (Capasso *et al.,* 1984a).

Other pulse–chase experiments have produced different results. Pulse–chase experiments using organ cultures of cartilage from the fetal bovine costochondral junction showed no detectable processing of type X, even after the chase was extended to 20 hr (Grant *et al.,* 1985). In other experiments, type X was not processed in the medium of chondrocyte monolayer cultures during a time period in which type II procollagen was almost completely converted to molecules composed of α1(II) chains (Schmid and Conrad, 1982b). It is possible that the different conditions of cell and organ culture employed may have contributed to at least some of the variability in these results.

The majority of type X molecules isolated from cartilage *in vivo* are not processed (T. M. Schmid, unpublished observations). Chick embryos were made lathyritic with an injection of β-aminoproprionitrile (β-APN) into their airsacs on day 15 of development. After 3 days, the embryos were sacrificed and the type X extracted from their long bones with 4 *M* guanidinium chloride in the presence of protease inhibitors. The type X was isolated with an affinity gel made of monoclonal antibody X-AC9 coupled to Sepharose. The type X molecules isolated by the affinity gel migrated on SDS–PAGE at a position indistinguishable from the 59K chains of molecules isolated from chondrocyte cell cultures. The types I and II collagens extracted from these same tissues have been completely processed to α1 and α2 chains. Thus, little if any type X is processed beyond the stage of 59K chains *in vivo*.

The presence of the globular domain on the form of the mature type X molecule which is deposited in the extracellular matrix may greatly affect the ability of the molecule to form macromolecular aggregates. For example, type VII collagen appears to be present in tissue as SLS-like aggregates (Bentz *et al.*, 1983). However, the globular domain of type X appears to inhibit SLS crystallite formation, at least under *in vitro* conditions. Rotary shadowed preparations of type X show aggregation of the molecules by their globular domains, but as stated earlier, it is not known if this is physiologically important.

In chicken cartilage, type X collagen within the extracellular matrix appears to be covalently cross-linked. The extractability of type X collagen from cartilage *in vitro* (Reginato *et al.*, 1986) and *in vivo* (T. M. Schmid, unpublished observations) was increased if the cartilage was incubated with the lathyrogen β-APN. Experiments employing reducing agents suggest the cross-links do not involve disulfide bonds. Direct evidence for the involvement of specific amino acid residues has not been determined. SDS gels of purified type X collagen preparations isolated from chondrocytes cultured in the presence of β-APN sometimes show a collagenous band of M_r 120,000 (Schmid and Linsenmayer, 1983). This protein most likely represents a dimeric form (β chain) of type X collagen chains, since it is precipitated from type X preparations with the anti-type X monoclonal antibody. The β chains seem to be covalently cross-linked since their size does not change after boiling the molecules in the presence of a reducing agent and detergent. The β chains also appear to be susceptible to proteolytic cleavage with pepsin, generating the 45,000 chains typical of the type X helical domain. Thus type X appears to contain covalent cross-links in a pepsin-sensitive region of the molecule.

A quantitative comparison of the extractability of type X collagen from radiolabeled 17-day chick sterna showed that the molecule was not easily extracted unless a lathyrogen was present. In the presence of β-APN, most of the radiolabeled type X collagen was extracted in either a 1 *M* NaCl, 50 m*M* Tris, pH 7.4 buffer or 0.5 *M* acetic acid (Reginato *et al.,* 1986). In the absence of β-APN, only about 20% of the type X was extracted. The remainder of the type X collagen could be recovered only after pepsin treatment followed by extraction in 0.15 *M* NaCl, 50 m*M* Tris, 20 m*M* dithiothreitol, pH 7.4 (Jimenez *et al.,* 1986). The presence of the reducing agent was necessary to extract the type X collagen. Since the molecule recovered from this procedure contained 45K chains which lacked cysteine residues, it was concluded that other cysteine-rich molecules must hold the type X collagen in the tissue.

XII. Immunohistochemical Localization of Type X Collagen with a Monoclonal Antibody

Several monoclonal antibodies have been produced which recognize type X collagen, and one antibody, X-AC9, has been extensively characterized. The X-AC9 antibody has been useful as an immunohistochemical reagent to determine the location of type X in embryonic and adult tissue (Schmid and Linsenmayer, 1985a,b; Linsenmayer *et al.,* 1986a), as a molecular marker in ultrastructural studies (Schmid *et al.,* 1986a), as a probe for the thermal stability of the type X triple helix (Linsenmayer *et al.,* 1986b), and as a preparative reagent for the quantitative isolation of type X from complex mixtures of proteins (Solursh *et al.,* 1986).

To produce this antibody, female SJL/J mice were hyperimmunized with native type X collagen (59K chains) and their splenocytes fused with NS-1 myeloma cells. The resulting hybridomas were selected in hypoxanthine, aminopterin and thymidine-containing medium, and those producing the desired antibody were identified by ELISA using plates that were coated with intact type X (59K chains) or the triple-helical domain (45K chains) of type X.

The antibody produced by the X-AC9 hybridoma gave the strongest reaction in ELISA and, therefore, the cells were cloned and the antibody extensively characterized. The X-AC9 antibody is an IgG_1 molecule which binds in the triple-helical domain of the type X molecule. In competition assays, it did not react with the chicken collagen types I, II, III, IV, V, IX, or XI (1α2α3α). The antibody also did not bind to the thermally denatured chains of type X collagen. Examination of rotary shadowed preparations showed that the antibody binds to a single epitope on the type X molecule, located approximately 20 nm from the carboxy-terminal end of the triple-helical domain (Fig. 5).

The antibody was used for indirect immunofluorescent staining of unfixed frozen tissue sections. The embryonic chick tibiotarsus was examined first, since the earlier biochemical analyses had shown that the molecule is restricted to a defined region within this structure (Schmid and Conrad, 1982b). In the tibiotarsus, the earliest detectable reactivity for type X collagen was found at 7.5 days of development (stage 32 of Hamburger and Hamilton, 1951) in the cartilage matrix of the middiaphysis (Schmid and Linsenmayer, 1985b). This distribution is restricted compared to that of type II collagen, which is found throughout the entire cartilaginous anlage. At this stage, type I collagen is present in the sheath of periosteal bone that surrounds the middiaphysis and defines the primary center of ossification. At 9.5 days of development (stage 35.5), the early formation of the marrow cavity can be observed as a middiaphyseal focus of cartilage erosion. At this time, the reactivity for type X collagen extends considerably outwards from the center of the diaphysis, occupying the middle 40% of its length.

In Fig. 7, the location of type X collagen is compared with the distribution of types I and II collagen in serial sections of a 13-day tibiotarsus (Schmid and Linsenmayer, 1985a). Figure 7C shows that type II collagen is distributed throughout the entire cartilaginous anlage. The marrow cavity (mc) now fills approximately the middle half of the structure, and the sheath of periosteal bone, which stains brightly for type I collagen (Fig. 7A), has thickened and progressed toward each end of the tibiotarsus. The type X (Fig. 7B) is localized in the cartilage matrix near the marrow cavity.

The regions showing reactivity with the anti-X antibody at 7.5, 9.5, and 13 days of development are all composed of hypertrophic cartilage. Other investigators have shown that the hypertrophic chondrocytes within the tibiotarsus are first observed in the tibiotarsus, at approximately 6.5 days of development, in the center of the shaft (Scott-Savage and Hall, 1979). These large chondrocytes generally correspond to, and are demarcated by, the overlying sleeve of periosteal bone (identified in Fig. 7 with arrowheads). Thus, the deposition of type X collagen lags slightly behind the hypertrophy of the chondrocytes, a result which is in complete agreement with our biochemical analyses.

The distribution of type X collagen at 19 and 21 days of incubation is also consistent with its synthesis solely by the hypertrophic region of chondrocytes. By 19 days of development, most of the cartilage re-

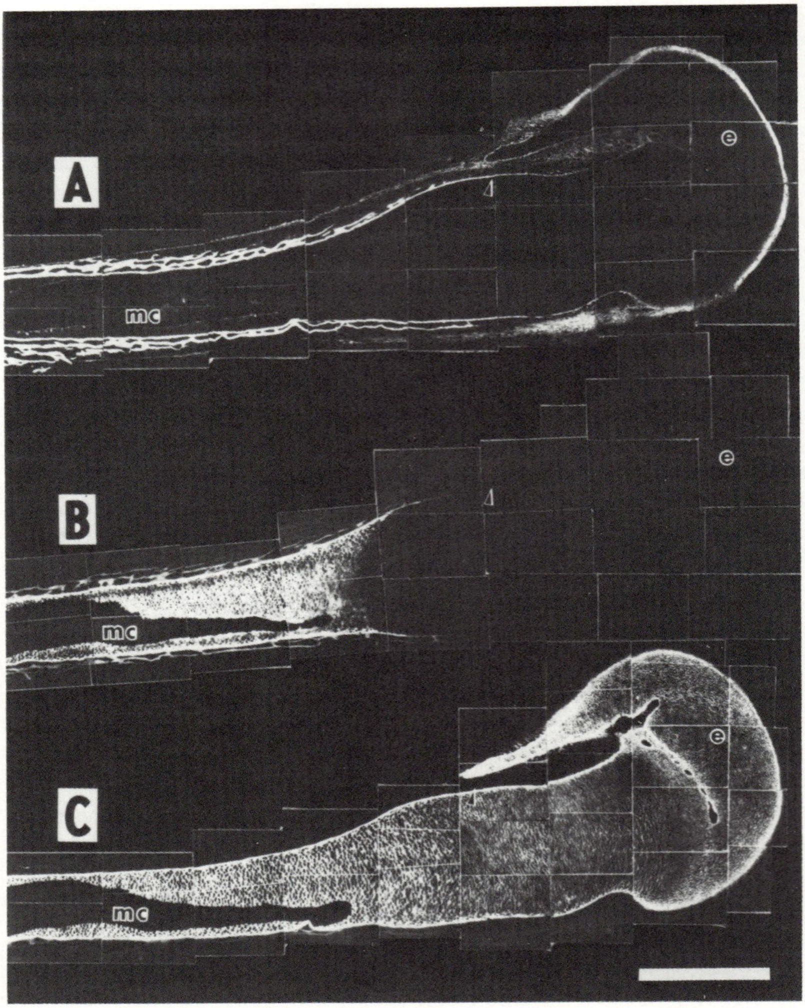

FIG. 7. Fluorescence micrographs of distal half of 13-day chick tibiotarsus treated with monoclonal antibodies directed against collagen types I (A), X (B), and II (C). Arrowheads mark the distal boundary of hypertrophic chondrocytes. The marrow cavity (mc) and epiphysis (e) are labeled. Bar, 1 mm. Reproduced from *The Journal of Cell Biology*, 1985, **100,** 598–605 by copyright permission of The Rockefeller University Press.

maining in the diaphyseal region now shows reactivity for type X. However, the region adjacent to the epiphysis is still negative. Most of the epiphysis itself is still negative, except for a small perivascular region of hypertrophic cells demarcating the region in which a secondary center of ossification characteristically forms. At a higher magnification (shown in Fig. 8), this region shows the most intense reactivity for type X collagen in the matrix directly surrounding the blood vessel channels, with a zone of progressively decreasing matrix reactivity extending outward in a radial fashion. Frequently, however, isolated foci of strong cellular/pericellular reactivity (circular profiles) exist some distance from the region of strong matrix staining. The matrix surrounding numerous other epiphyseal blood vessels is negative for type X, suggesting that blood-borne stimuli are not solely responsible for the initiation of type X synthesis, as has been previously suggested (Capasso *et al.*, 1984b).

In the 1-day-old chick, the relationships of collagen type II (Fig. 9A) and type X (Fig. 9B) to the center of hypertrophy within the epiphysis

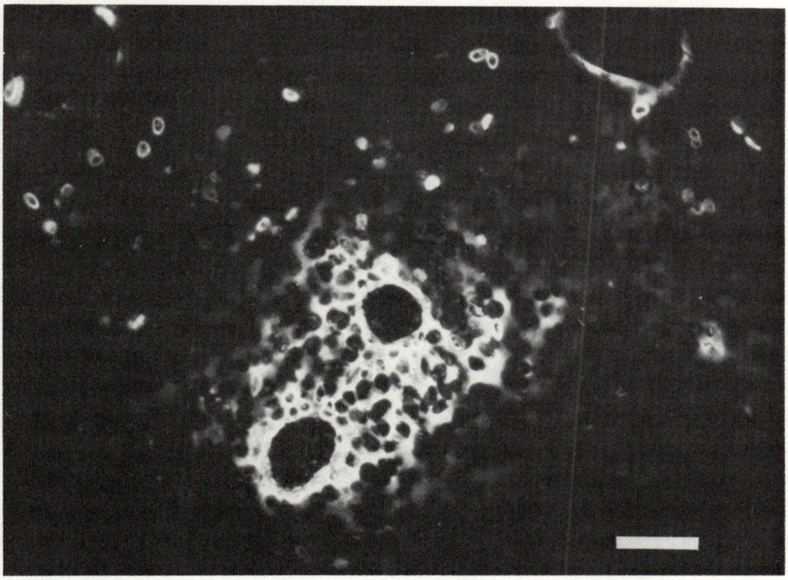

FIG. 8. Fluorescence micrograph of center of anti-type X reactivity in the epiphysis of a 19-day chick tibiotarsus. Type X reactivity surrounds two blood vessels at the future site of the secondary center of ossification. Bar, 200 μm. Reproduced from Schmid and Linsenmayer (1985b), *Dev. Biol.* **107,** 373–381, with copyright permission of Academic Press, Orlando, Florida.

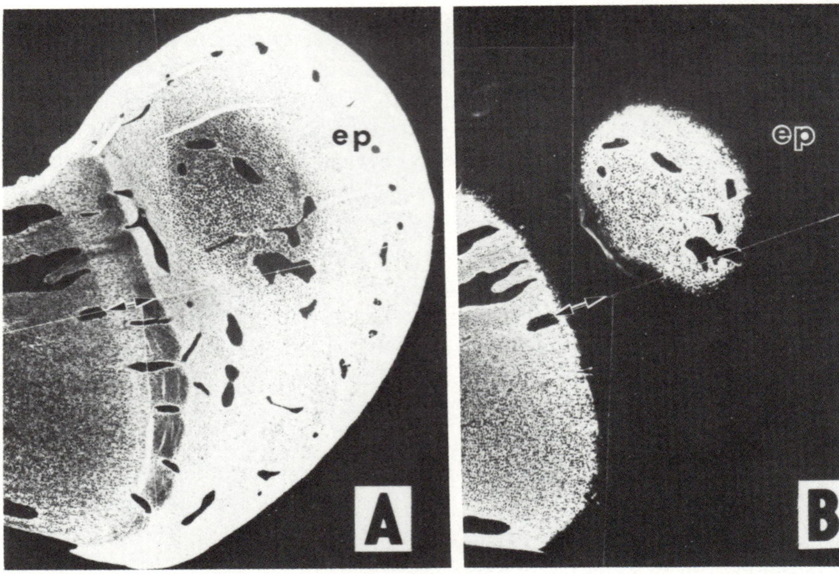

FIG. 9. Fluorescence micrograph of the epiphysis (ep) of a 21-day (1 day posthatch) chick tibiotarsus treated with anti-type II (A) and anti-type X (B) antibodies. Type X reactivity is present throughout most of the tibia and also in the epiphysis at the presumptive site of the secondary center of ossification. The double-pointed arrows mark a zone (0.4 mm) of growth cartilage which does not react with the anti-type X antibody. This region separates the tibia shaft from the epiphysis. Reproduced from Schmid and Linsenmayer (1985b), *Dev. Biol.* **107,** 373–381, with copyright permission of Academic Press, Orlando, Florida.

have now become more definitive. Extensive reactivity for type II collagen is found throughout the epiphysis (ep) and the adjacent portions of the diaphysis. Within the epiphysis there is a loss of reactivity in the midepiphyseal zone of hypertrophic cells. In a serial section reaction with the anti-X antibody (Fig. 9B), this zone is strongly fluorescent and has considerably increased in diameter from what it was 2 days earlier.

In the sections reacted for type II collagen, there is a 0.4 mm band (demarcated by the double-headed arrow in Fig. 9A) between the epiphysis and the diaphysis. This band consists of a weakly reactive zone (nearest the epiphysis) and a strongly reactive zone (nearest the diaphysis). In the section reacted for type X collagen (Fig. 9B), the corresponding region (again demarcated with arrowheads) is completely negative. Strong reactivity for type X begins abruptly on the diaphyseal side of this band, where the hypertrophic cells are now clearly evident. By contrasting the fluorescent patterns obtained with the two antibodies, it can be concluded that this band plus the adjacent hypertrophic region of the diaphysis comprise the characteristic regions of an epiphyseal growth zone (see Bloom and Fawcett, 1975).

Preliminary experiments examining the distribution of the molecule in tibiotarsi from more mature chickens (7–9 weeks) show that type X remains in the region of calcified cartilage which separates the articular hyaline cartilage from the subchondral bone. In the chicken, the most superficial layer of the epiphysis is a fibrous layer containing type I collagen (von der Mark *et al.,* 1976). A region of hyaline cartilage containing a type II collagen underlies the fibrous layer. Type X is present a few cell layers above the tidemark, the boundary between the uncalcified and calcified cartilage. Type X is also present in the remnant cartilage at the core of spicules of endochondral bone (Schmid and Linsenmayer, 1985a).

The brightest staining intensity of the X-AC9 antibody was always in regions of hypertrophic cartilage, while regions of bone also showed very light but positive reactivity. The sheath of periosteal bone of the developing tibiotarsus in Fig. 7 also reacted lightly, as did calvarial bone. Appropriate controls suggested that this light reactivity of bone was specific; however, further biochemical experiments were unable to detect the presence of type X collagen in chick calvaria. At this time, we assume the antibody is cross reacting with some other substance in bone or that type X, if present, is there in exceedingly small amounts.

Type X collagen is also present in the notochord (Linsenmayer *et al.,* 1986a). The notochord is a cartilage-like skeletal tissue which has been shown to have an inductive influence on the formation of the cartilaginous vertebral bodies (Holtzer, 1961; Lash, 1968). Temporally, the appearance of type II collagen in the notochord precedes the detectable appearance of type II collagen in the surrounding vertebral tissue (Linsenmayer *et al.,* 1973). Conversely, however, the temporal appearance of type X collagen in the notochord is subsequent to its appearance in the surrounding vertebral bodies. It is detectable first in the hypertrophic cartilage of the vertebral bodies, and several days later it is detectable in the adjacent notochordal tissue. Also, the spatial distribution of type X within the notochord differs from that of type II. The embryonic notochord and adjacent vertebral bodies show a segmental distribution in which type X (Fig. 10A) is preferentially deposited in the midvertebral regions. Conversely, the type II collagen (Fig. 10B) accumulates in the intervertebral regions. This inverse distribution of type X and II within the notochord seems to correlate with the different developmental fates of these regions. The type X-rich

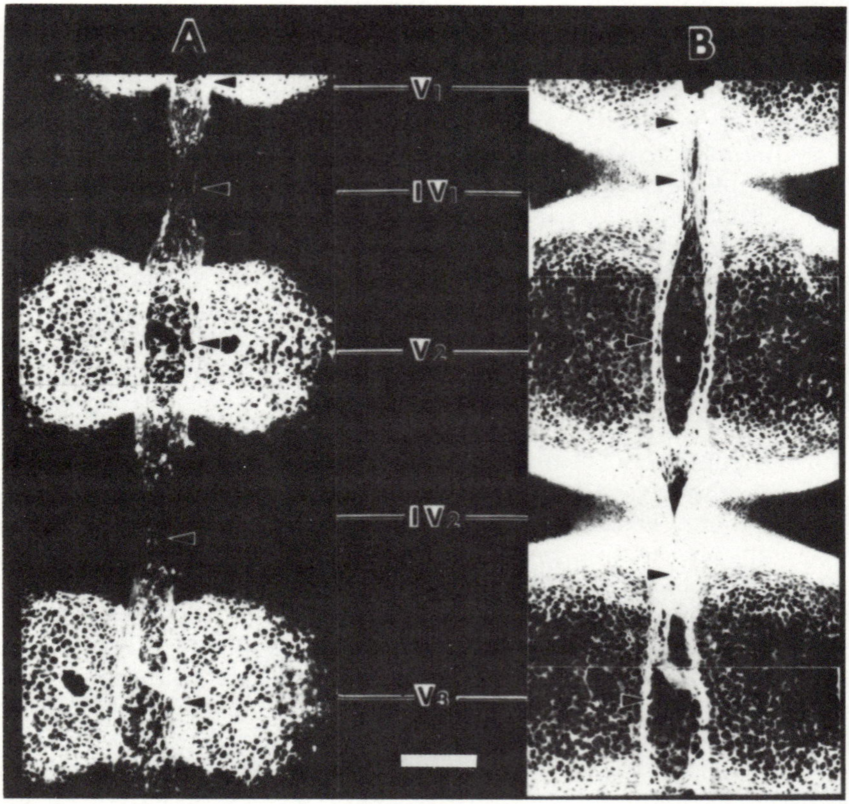

FIG. 10. Fluorescence micrograph of a longitudinal section through three sequential vertebrae of a 19-day embryo. The vertebrae are designated V_1, V_2, V_3, and the intervertebral regions are designated IV_1 and IV_2. (A) Section has been reacted with an anti-type X antibody; (B) section has been reacted with an anti-type II antibody. In both A and B, the notochord is indicated by arrowheads. In A, the notochord is sectioned such that it runs uninterrupted through all the vertebral bodies and intervertebral regions; in B, it is uninterrupted except at site IV_2 where, due to the plane of section, it appears to be constricted. Bar, 200 μm. Reproduced from Linsenmayer *et al.* (1986a), *Dev. Biol.* **113,** 467–473, with copyright permission of Academic Press, Orlando, Florida.

region will become part of the bone and marrow of the centrum of the vertebrae (Williams, 1908); the type II-rich regions will give rise to the nucleus pulposus of the intervertebral discs (Wolfe *et al.,* 1965).

Type X was not detectable by immunohistochemical methods in embryonic tissue outside the skeletal system, nor was it detectable in permanent cartilaginous tissues, such as those of the trachea and sclera. Several noncartilaginous tissues which synthesize type II collagen tested negative [i.e., the chick primary corneal stroma (Linsenmayer *et al.*, 1977) and the vitreous body (Smith *et al.*, 1976)]. However, it is always possible that a lack of immunohistochemical reactivity is the result of masking of a particular epitope as occurs for type V collagen (Linsenmayer *et al.*, 1983).

XIII. Biochemical Localization of Type X Collagen

Several laboratories have analyzed the distribution of type X collagen within skeletal tissue by biochemical methods. Capasso *et al.* (1984b) examined the distribution of type X collagen during several stages of embryonic chick tibiotarsus development. The first detectable synthesis of type X was in the diaphysis at 9 days of development and in the epiphysis by day 17. Immunohistochemical analysis seems to be more sensitive, since an anti-X antibody was able to detect the synthesis of type X in the diaphysis at 7.5 days of development (Schmid and Linsenmayer, 1985b).

Grant *et al.* (1985) analyzed collagen synthesis by bovine chondrocytes derived from the growth plate at the costochondral junction of a bovine fetus. Type X collagen was synthesized only in the zone of cartilage degeneration and calcification, which included the first 400 μm of tissue between the bone and adjacent growth cartilage. Type X was not synthesized by the zones of maturation or proliferation, the reserve zone, or other structural cartilages.

Several laboratories (Gibson *et al.*, 1984; Gibson and Flint, 1985; Reginato *et al.*, 1986; Jimenez *et al.*, 1986) have analyzed the distribution of type X in the chick sternum by biochemical methods. Whereas the epiphyseal growth plate of long bones exhibits a linear continuum of chondrocyte maturation, the sternum is a more complex structure which undergoes endochondral bone formation in three dimensions. The cephalic half of the sternum begins to ossify just prior to the hatching of the chick. The caudal aspect of the sternum remains cartilaginous for a long period of time (>8 weeks), but eventually it also begins to ossify. Type X is synthesized by embryonic chondrocytes from the cephalic half of the tissue but is not synthesized by embryonic chondrocytes from the caudal half of the sternum. Before 16 days of development, sternal type X collagen is not detectable by biochemical methods, but between day 16 and 17 the synthesis of type X increases suddenly (Gibson and Flint, 1985). This is concomitant with the hypertrophy of the cells in the region and consistent with the immunohistochemical observations of the tissue (Schmid *et al.*, 1986b).

The hypertrophic cells in the cephalic aspect of the sternum were recognized as a translucent zone (Gibson and Flint, 1985; Reginato *et al.*, 1986). Gibson and Flint (1985) isolated these regions and radiolabeled them in organ culture. Their experiments showed type X to represent 2% of the newly synthesized collagen on day 16 of development, 60% of the collagen by day 17, and approximately 80–90% of the collagen after 18 days of development. On day 19, the translucent region of cartilage begins to calcify (Gibson and Flint, 1985; Reginato *et al.*, 1986). Thus, although type X represents a quantitatively minor collagen when an entire cartilaginous tissue is examined, it appears to be the major collagenous product of chondrocytes during their terminal stage of maturation. The accumulation of type X collagen precedes the calcification of the tissue.

XIV. Functions of Type X Collagen

Cartilaginous tissues contain at least three collagen types: type II, type XI (1α2α3α), and type IX collagen. Type X collagen is only present in a cartilage matrix containing hypertrophic chondrocytes and becomes the major type synthesized by the most mature chondrocytes involved in endochondral bone formation. The molecule, therefore, bridges the spatial and temporal transition from cartilage to bone. This is true for the process of endochondral ossification during skeletal growth, for the maintenance of the articular cartilage–bone interface of adult bones, and for the endochondral repair of fractured bones (Grant *et al.*, 1986). During long bone development and fracture repair, most of the cartilaginous tissue forms a temporary model which will eventually be removed and replaced by bone. Therefore, the majority of type X synthesized by an organism will be removed within a relatively short period of time.

One possible function of type X collagen may be to facilitate the removal of the hypertrophic cartilage. We would speculate this could be accomplished if type X collagen were more susceptible to degradation than type II. Our preliminary experiments suggest that type X collagen is indeed more rapidly degraded by vertebrate collagenase than is type II collagen (Schmid *et al.*, 1986a), and recent results by Dean *et al.* (1985) suggest that vertebrate collagenase activity is present within hypertrophic cartilage. The degradation of type X collagen may “open up” the hypertrophic cartilage matrix to other degradative enzymes and to the invasion of other cells.

The structural characteristics of an extracellular matrix composed primarily of type X collagen may be very different from those of a cartilage matrix composed of type II collagen. The matrix surrounding the most hypertrophic chondrocytes may consist primarily of type X collagen. The different length of the collagen molecule or the size of its molecular aggregate may significantly affect the interaction with other macromolecular constituents and the hydrodynamic properties of the matrix (Hukins and Aspden, 1985).

The second possible function of type X may be in the formation of an extracellular matrix which is amenable to calcification. In some cartilaginous tissues, the deposition of type X collagen either slightly precedes or is concomitant with the calcification of hypertrophic cartilage. During ossification of the cephalic portion of the embryonic chick sternum, the region of hypertrophic cartilage becomes calcified soon after the appearance of type X. Within the organization of adult articular cartilage, the superficial hyaline layer is separated from the subchondral bone by a layer of calcified cartilage. Type X is present in this region and a few cell diameters above the tidemark, the boundary between the calcified and uncalcified cartilage.

Matrix vesicles seem to be involved in the initiation of cartilage calcification (Anderson, 1967; Bonucci, 1967), and type X collagen is known to be associated with such vesicles (Habuchi *et al.,* 1985). The vesicles are released from the plasma membrane of chondrocytes. They have a unique phospholipid composition, are rich in alkaline phosphatase, and contain elevated levels of calcium (Anderson, 1967; Bonucci, 1967; Ali, 1983). The release of matrix vesicles increases with the maturation of the chondrocyte (Ali, 1976; Glaser and Conrad, 1981). Habuchi *et al.* (1985) showed that radiolabeled type X collagen was present in matrix vesicle preparations isolated from chondrocyte culture medium by ultracentrifugation. They also showed that the synthesis of collagen and alkaline phosphatase were coordinately regulated. These results are particularly interesting, considering the presence of a hydrophobic region at the carboxy-terminus of type X which might span the lipid bilayer of a matrix vesicle (see sequence of type X in Vasios *et al.,* this volume, Fig. 5). Thus, type X may be an integral component of matrix vesicles. Further experimentation is necessary to understand the relationship of type X collagen to the calicification process.

XV. Unresolved Questions about Type X Collagen

The list of unanswered questions concerning type X collagen is long. What are the biosynthetic steps involved in its synthesis and transport? What information directs the association of the three type X chains, rather than allowing the assembly of a mixed molecule containing, for example, chains of both type IX and type X collagens? Are any interchain or intrachain disulfide bonds formed during the synthesis of type X? Are hydrophobic interactions between the chains important for chain assembly? Are such interactions a prerequisite for, or a result of, triple-helical formation? How are type X molecules transported to the extracellular matrix?

Although the tissue distribution of type X is known from immunohistochemical and biochemical studies, we do not know if type X collagen forms aggregates within the tissue *in vivo*. Immunoelectron microscopic studies are needed to define this. Is type X assembled in a lateral or linear aggregate? Are the molecules staggered? What kind of cross-links occur and where are they located?

Does the accumulation of type X in the tissue facilitate the turnover of the tissue? What are the quantitative relationships of type X to the other cartilage collagens during endochondral bone formation? Are the synthesis and/or removal of several different cartilage collagens coordinately regulated? What regulates the levels of type X collagen in tissues? What events initiate the synthesis of type X? Do the cytoskeletal changes which accompany cell hypertrophy trigger the synthesis of type X? Is there active vertebrate collagenase in the area of hypertrophic cartilage which will be removed? What other enzymes are necessary for the removal of the extracellular matrix in this region? What cells are responsible for the turnover of cartilage matrix?

Is type X involved in the calcification of cartilage? Does type X bind calcium? Do any other macromolecules bind to type X and are such interactions physiologically significant?

What is the evolutionary relationship of type X collagen to other collagen types? Why does the type X gene have only one intron? Is type X present in organisms whose skeleton remains cartilaginous throughout life?

How is human type X collagen related to chick or bovine type X collagen? How would abnormal type X molecules affect the development of the skeleton? Is type X collagen necessary for the endochondral repair of fractured bones? Might degradation products of type X collagen in body fluids be indicative of calcified cartilage erosion or endochondral bone repair? Would assays for these molecules be useful in monitoring the progress of osteoarthritis, rheumatoid arthritis, or bone fracture repair?

The answers to these questions should provide some insight into the structural function of type X collagen. Such knowledge will provide a rational basis for future experimentation related to the diseases and development of the skeletal system.

Acknowledgments

This work was supported by the National Arthritis Foundation, Illinois chapter of the Arthritis Foundation, and NIH Grants AM 03564 and HD 20715. This is publication #1002 of the Robert W. Lovett Memorial Group for the Study of Diseases Causing Deformities.

References

Ali, S. Y. (1976). *Fed. Proc., Fed. Am. Soc. Exp. Biol.* **35,** 135–142.

Ali, S. Y. (1983). *In* "Cartilage" (B. K. Hall, ed.), Vol. 1, pp. 343–378. Academic Press, New York.

Anderson, H. C. (1967). *J. Cell Biol.* **35,** 81–101.

Balian, G., and Morris, J. B. (1986). *Trans. Annu. Meet.–Orthop. Res. Soc.* **11,** 75.

Bentz, H., Morris, N. P., Murray, L. W., Sakai, L. Y., Hollister, D. W., and Burgeson, R. E. (1983). *Proc. Natl. Acad. Sci. U.S.A.* **80,** 3168–3172.

Benya, P. D. (1981). *Collagen Relat. Res.* **1,** 17–26.

Bloom, W., and Fawcett, D. W. (1975). "A Textbook of Histology," pp. 233–287. Saunders, Philadelphia.

Bonucci, E. J. (1967). *J. Ultrastruct. Res.* **20,** 33–50.

Bornstein, P., and Balian, G. (1977). *Methods Enzymol.* **47,** 132–145.

Bornstein, P., and Sage, H. (1980). *Annu. Rev. Biochem.* **49,** 957–1003.

Cancedda, R., Capasso, O., Castagnola, P., Descalzi-Cancedda, F., and Quarto, N. (1985). *J. Cell. Biochem.* **28,** 159–166.

Capasso, O., Gionti, E., Pontarelli, G., Ambesi-Impiombato, F. S., Nitsch, L., Tajana, G., and Cancedda, R. (1982). *Exp. Cell Res.* **142,** 197–206.

Capasso, O., Quarto, N., Descalzi-Cancedda, F., and Cancedda, R. (1984a). *EMBO J.* **3,** 823–827.

Capasso, O., Tajana, G., and Cancedda, R. (1984b). *Mol. Cell. Biol.* **4,** 1163–1168.

Castagnola, P., Descalzi-Cancedda, F., and Cancedda, R. (1986). *Biochem. Biophys. Res. Commun.* **136,** 109–115.

Dean, D. D., Muniz, O. E., Berman, I., Pita, J. C., Carreno, M. R., Woessner, J. F., Jr., and Howell, D. S. (1985). *J. Clin. Invest.* **76,** 716–722.

Eyre, D. R., Wu, J. J., and Woolley, D. E. (1984). *Biochem. Biophys. Res. Commun.* **118,** 724–729.

Gibson, G. J., and Flint, M. H. (1985). *J. Cell Biol.* **101,** 277–284.

Gibson, G. J., Schor, S. L., and Grant, M. E. (1981). *Biochem. Soc. Trans.* **9,** 550–551.

Gibson, G. J., Schor, S. L., and Grant, M. E. (1982). *J. Cell Biol.* **93,** 767–774.

Gibson, G. J., Kielty, C. M., Garner, C., Schor, S. L., and Grant, M. E. (1983). *Biochem. J.* **211,** 417–426.

Gibson, G. J., Beaumont, B. W., and Flint, M. H. (1984). *J. Cell Biol.* **99,** 208–216.

Glaser, J. H., and Conrad, H. E. (1981). *J. Biol. Chem.* **256,** 12607–12611.

Grant, W. T., Sussman, M. D., and Balian, G. (1985). *J. Biol. Chem.* **260,** 3798–3803.

Grant, W. T., Wang, G. J., and Balian, G. (1986). *Trans. Annu. Meet.–Orthop. Res. Soc.* **11,** 76.

Gross, J. (1981). *In* "Cell Biology of the Extracellular Matrix" (E. Hay, ed.), pp. 217–258. Plenum, New York.

Habuchi, H., Conrad, H. E., and Glaser, J. H. (1985). *J. Biol. Chem.* **260,** 13029–13034.

Hamburger, V., and Hamilton, H. L. (1951). *J. Morphol.* **88,** 49–92.

Hayashi, T., Curran-Patel, S., and Prockop, D. J. (1979). *Biochemistry* **18,** 4182–4187.

Holtzer, H. (1961). *In* "Synthesis of Molecular and Cellular Structure" (D. Rudnick, ed.), pp. 35–87. Ronald Press, New York.

Hukins, D. W. L., and Aspden, R. M. (1985). *Trends Biochem. Sci.* **10,** 260–264.

Jimenez, S. A., Yankowski, R., and Reginato, A. M. (1986). *Biochem. J.* **233,** 357–367.

Kielty, C. M., Hulmes, D. J. S., Schor, S. L., and Grant, M. E. (1984). *FEBS Lett.* **169,** 179–184.

Kielty, C. M., Kwan, A. P. L., Holmes, D. F., Schor, S. L., and Grant, M. E. (1985). *Biochem. J.* **27,** 545–554.

Kim, J. J., and Conrad, H. E. (1977). *J. Biol. Chem.* **252,** 8292–8299.

Kim, J. J., and Conrad, H. E. (1980). *J. Biol. Chem.* **255,** 1586–1597.

Lash, J. W. (1968). *In* "The Stability of the Differentiated State" (H. Ursprung, ed.), pp. 17–24. Springer-Verlag, New York.

Linsenmayer, T. F., and Little, C. D. (1978). *Proc. Natl. Acad. Sci. U.S.A.* **75,** 3235–3239.

Linsenmayer, T. F., Trelstad, R. L., and Gross, J. (1973). *Biochem. Biophys. Res. Commun.* **53,** 39–45.

Linsenmayer, T. F., Smith, G. N., Jr., and Hay, E. D. (1977). *Proc. Natl. Acad. Sci. U.S.A.* **74,** 39–43.

Linsenmayer, T. F., Fitch, J. M., Schmid, T. M., Zak, N. B., Gibney, E., Sanderson, R. D., and Mayne, R. (1983). *J. Cell Biol.* **96,** 124–132.

Linsenmayer, T. F., Gibney, E., and Schmid, T. M. (1986a). *Dev. Biol.* **113,** 467–473.

Linsenmayer, T. F., Gibney, E., and Schmid, T. M. (1986b). *Exp. Cell Res.* **166,** 15–22.

Liotta, L. A., Kalebic, T., Reese, C. A., and Mayne, R. (1982). *Biochem. Biophys. Res. Commun.* **104,** 500–506.

Ninomiya, Y., Gordon, M., van der Rest, M., Schmid, T., Linsenmayer, T., and Olsen, B. R. (1986). *J. Biol. Chem.* **261,** 5041–5050.

Quarto, N., Cancedda, R., and Descalzi-Cancedda, F. (1985). *Eur. J. Biochem.* **147,** 397–400.

Reginato, A. M., Lash, J. W., and Jimenez S. A. (1986). *J. Biol. Chem.* **261,** 2897–2904.

Remington, M. C., Bashey, R. I., Brighton, C. T., and Jimenez, S. A. (1983). *Collagen Relat. Res.* **3,** 271–278.

Remington, M. C., Bashey, R. I., Brighton, C. T., and Jimenez, S. A. (1984). *Biochem. J.* **224,** 227–233.

Rich, A., and Crick, F. H. C. (1961). *J. Mol. Biol.* **3,** 483–506.

Schmid, T. M. (1980). Ph.D. thesis. University of Illinois, Urbana.

Schmid, T. M., and Conrad, H. E. (1982a). *J. Biol. Chem.* **257,** 12444–12450.

Schmid, T. M., and Conrad, H. E. (1982b). *J. Biol. Chem.* **257,** 12451–12457.

Schmid, T. M., and Linsenmayer, T. F. (1983). *J. Biol. Chem.* **258,** 9504–9509.

Schmid, T. M., and Linsenmayer, T. F. (1984). *Biochemistry* **23,** 553–558.

Schmid, T. M., and Linsenmayer, T. F. (1985a). *J. Cell Biol.* **100,** 598–605.

Schmid, T. M., and Linsenmayer, T. F. (1985b). *Dev. Biol.* **107,** 373–381.

Schmid, T. M., Mayne, R., Bruns, R. R., and Linsenmayer, T. F. (1984). *J. Ultrastruct. Res.* **86,** 186–191.

Schmid, T. M., Mayne, R., Jeffrey, J. J., and Linsenmayer, T. F. (1986a). *J. Biol. Chem.* **261,** 4184–4189.

Schmid, T. M., Gibney, E., and Linsenmayer, T. F. (1986b). *Ann. N.Y. Acad. Sci.* **460,** 497–499.

Scott-Savage, P., and Hall, B. K. (1979). *J. Morphol.* **162,** 453–464.

Smith, G. N., Jr., Linsenmayer, T. F., and Newsome, D. A. (1976). *Proc. Natl. Acad. Sci. U.S.A.* **73,** 4420–4423.

Solursh, M., Jensen, K. L., Reiter, R. S., Schmid, T. M., and Linsenmayer, T. F. (1986). *Dev. Biol.* **117,** 90–101.
Stocum, D. L., Davis, R. M., Leger, M., and Conrad, H. E. (1979). *J. Embryol. Exp. Morphol.* **54,** 155–170.
Sussman, M. D., Ogle, R. C., and Balian, G. (1984). *J. Orthop. Res.* **2,** 134–142.
von der Mark, K., and von der Mark, H. (1977). *J. Bone Joint Surg., Br. Vol.* **59B,** 458–464.
von der Mark, K., von der Mark, H., and Gay, S. (1976). *Dev. Biol.* **53,** 153–170.
Williams, L. W. (1908). *Am. J. Anat.* **8,** 251–284.
Wolfe, H. J., Putschar, W. G. J., and Vickery, A. L. (1965). *Clin. Orthop.* **39,** 205–212.
Woolley, D. E. (1984). *In* "Extracellular Matrix Biochemistry" (K. A. Piez and A. H. Reddi, eds.), pp. 119–157. Elsevier, New York.

Type XI or 1α2α3α Collagen

David Eyre

Departments of Orthopaedics and Biochemistry, University of Washington, Seattle, Washington 98195

and

Jiann-Jiu Wu

Department of Orthopaedics, University of Washington, Seattle, Washington 98195

I. Introduction

This minor collagen was first discovered by Burgeson and Hollister (1979) during differential salt fractionation of pepsin digests of hyaline cartilage. The protein is best recognized by the separation of three distinct α chains on SDS–polyacrylamide electrophoresis, all of which migrate slightly slower than α1(II) chains. The chains were assigned the provisional names, 1α, 2α, and 3α. The collagen resembled type V collagen in the following properties: (1) both collagens precipitated at 1.2 *M* NaCl from acid solution; (2) each accounted for 5% or less of total tissue collagen; (3) their α chains migrated in the same distinctive region on SDS–polyacrylamide electrophoresis; and (4) they shared characteristic features of amino acid composition that set them apart from types I, II, and III collagens. However, the 1α and 2α chains gave unique cyanogen bromide (CNBr) peptide patterns on electrophoresis and chromatography, consistent with each being a genetically distinct product (Burgeson and Hollister, 1979; Reese and Mayne, 1981; Burgeson *et al.*, 1982). The 3α chain, however, gave an almost identical CNBr peptide map to that of the α1(II) chain, except that all the CNBr peptides of 3α ran slightly slower than their α1(II) counterparts. From these results, it was suggested that the 3α and α1(II) chains have the same primary sequence, or at least are very close genetic variants, but differ posttranslationally in their degree of lysine hydroxylation and glycosylation.

II. Nomenclature

Although first described several years ago, no type number has so far been assigned to this collagen. The main reasons for this have been, probably, a suspicion that this collagen fraction lacks any established molecular formula and consists of a mixture of several different homo- and heterotrimeric molecules, and the unknown significance of the apparent type II collagen gene product in the system. Indeed, in an earlier review Miller (1985) referred to the protein as type K collagen until its chemistry could be better resolved.

Because a consensus of recent data now point to a primary heterotrimeric parent molecule of composition 1α2α3α and establish, at the very least, that the 3α chain exists in heterotrimeric molecules with combinations of 1α and 2α chains, it now seems appropriate to give 1α2α3α collagen a type number. The next available number is type XI, which was already suggested for 1α2α3α collagen in a recent review of the collagen genes (Cheah, 1985). We propose, therefore, that the term type XI collagen be adopted. For consistency when using the new nomenclature, the 1α, 2α, and 3α chains will therefore become α1(XI), α2(XI), and α3(XI), respectively. However, the original names will probably persist for a while and, indeed, are used in this article in most figures. It remains possible that type XI collagen includes native molecules with various combinations of α1(XI), α2(XI), and α3(XI) chains, but this sort of complexity applies equally to the type V collagen system (Bornstein and Sage, 1980; Fessler *et al.*, 1985; see Fessler and Fessler, this volume). Clearly, types V and XI collagens should be thought of as close relatives that, together, constitute a specialized subclass of the collagen family of proteins.

III. Isolation and Analysis

A. *Pepsin Digestion and Salt Fractionation*

Originally, Burgeson and Hollister (1979) prepared type XI collagen from growing human (fetal and 4-year postnatal rib and femoral head) cartilages and from young adult bovine (18-month rib) cartilage. The homogenized tissue was extracted in 4 *M* guanidine HCl to remove proteoglycans. The washed residue was treated with pepsin (0.5–1.0

mg/ml) at 4°C for 24 hr, the digest neutralized to give a water-clear solution, then dialyzed against water; the precipitate that formed was sedimented and redissolved in 1 *M* NaCl at neutral pH. The type XI collagen was resolved from type II collagen by dialyzing this neutral solution against 0.1 *M* acetic acid at 4°C and centrifuging hard (6.2×10^6 *g*-min) to recover type XI in the pellet (Burgeson and Hollister, 1979; Burgeson *et al.*, 1982). This procedure was adopted after initial experiments showed that the protein tended to be insoluble in dilute acetic acid. Later, others (Shimokomaki *et al.*, 1980; Furuto and Miller, 1983a; Sussman *et al.*, 1984; Eyre *et al.*, 1984b) have reported the isolation of type XI collagen by precipitating at 1.2 *M* NaCl directly from the acidic pepsin digest of cartilage, having removed type II collagen at 0.7–0.9 *M* NaCl. Milder centrifugation was used to clarify the digest initially and to sediment the type II collagen, e.g., 18,000 rpm for 60 min in a Sorvall RC5C, SS-34 rotor (Eyre *et al.*, 1984b). Possibly, the sedimentation of type XI from dilute acetic acid requires ultracentrifugation, perhaps needing or being aided by the presence of proteoglycan digestion products to which type XI may preferentially bind (Smith *et al.*, 1985; see later).

B. Chromatography and Electrophoresis

Figure 1 shows the resolution by carboxymethylcellulose (CM-cellulose) column chromatography of the three α chains of type XI collagen prepared after pepsin digestion of neonatal human cartilage as reported by Burgeson *et al.* (1982). The α3(XI) chain coelutes with α1(II), the α1(XI) chain elutes just later than α1(V), and the α2(XI) chain comes last and later than all other α chains of the common collagen types, consistent with it being the most basic by amino acid composition (see Table II).

Figure 2 shows how the chains migrate on SDS–polyacrylamide electrophoresis using the buffer system of Laemmli (1970). The α3(XI) chain runs just above and merging with α1(II) in the position of α2(V), and the α1(XI) chain runs in the position of α1(V). Despite being resolved by SDS–electrophoresis, all three chains coeluted on agarose sieve chromatography in the position of α1(I), indicating that all are of similar length (Burgeson *et al.*, 1982).

C. Peptide Mapping

Each α chain gives a distinct profile of CNBr peptides on SDS–polyacrylamide electrophoresis (Burgeson and Hollister, 1979; Reese

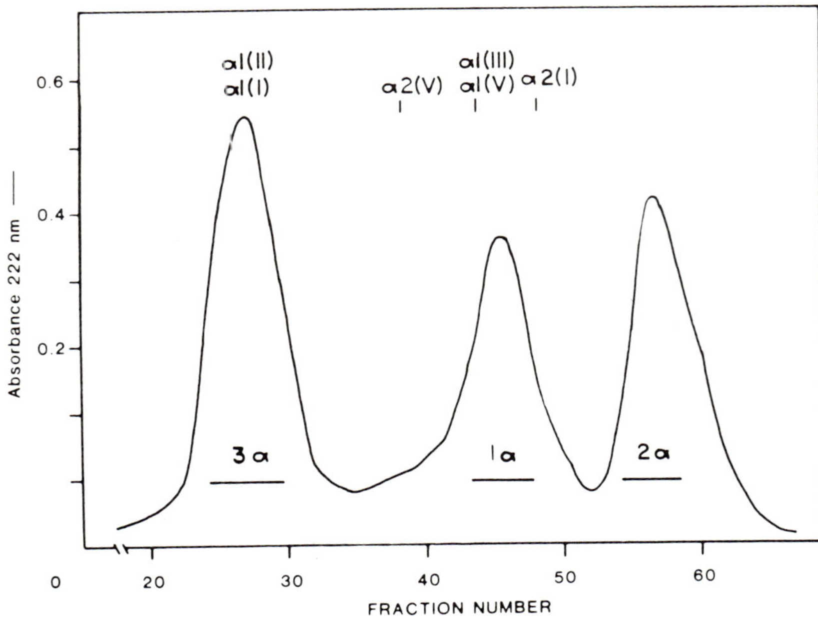

FIG. 1. Elution profile of 1α [α1(XI)], 2α [α2(XI)], and 3α [α3(XI)] chains on CM-cellulose chromatography. The column was eluted at 60°C with a linear gradient of 0–0.1 *M* LiCl in 0.05 *M* Li acetate, pH 4.8, containing 1% (v/v) isopropanol. [Reproduced with permission from Burgeson *et al.* (1982)].

and Mayne, 1981; von der Mark *et al.,* 1982; see Fig. 3) and CM-cellulose chromatography (Burgeson *et al.,* 1982). The peptide profiles of α1(XI) and α2(XI) are unique, differing significantly from those of α1(V) and α2(V), the chains to which α1(XI) and α2(XI) seem most closely related based on their other chemical properties. The α1(XI) chain was distinguished from α1(V) by giving one major peptide band (Fig. 3), which may be a doublet of peptides of about 400 residues each judging by their electrophoretic mobility (unpublished observations). All the CNBr peptides of the α2(XI) chain from the rat chondrosarcoma were isolated by HPLC molecular sieving and ion-exchange chromatography, and their amino acid compositions determined (Furuto and Miller, 1983b). The belief that α1(XI), α2(XI), α1(V), and α2(V) are each distinct gene products was reinforced by the marked differences in their peptide maps on electrophoresis after digestion by staphylococcal V8 protease (Reese and Mayne, 1981; Burgeson *et al.,* 1982). The α3(XI) chain, however, gave a very similar CNBr peptide pattern to that of the α1(II) chain, except that the major fragments of

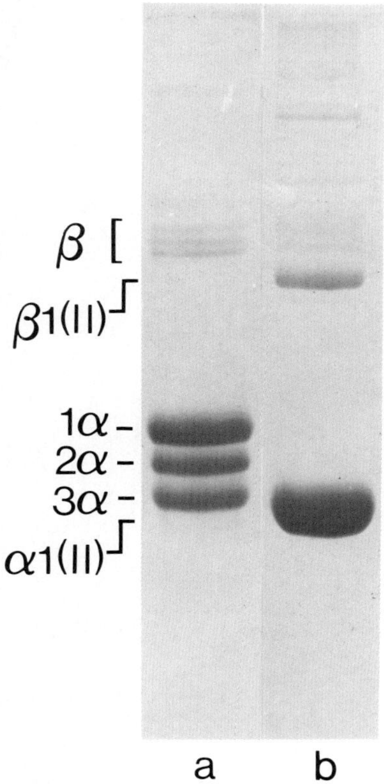

FIG. 2. Resolution of 1α [α1(XI)], 2α [α2(XI)], and 3α [α3(XI)] chains by SDS–5% polyacrylamide slab gel electrophoresis. The collagen was extracted from calf growth cartilage by pepsin digestion. a, Type XI precipitated by 1.2 *M* NaCl at pH 3; b, type II collagen precipitated from the same digested by 0.7 *M* NaCl at pH 3.

α3(XI) ran slower, suggesting they differed posttranslationally (Burgeson and Hollister, 1979). Furuto and Miller (1983a) have isolated the individual CNBr peptides of 3α by column chromatography and compared them with their α1(II) counterparts. Amino acid analyses confirmed that α3(XI) was richer in hydroxylysine and hydroxylysine glycosides than α1(II) prepared from the same tissue sample (Burgeson and Hollister, 1979; Furuto and Miller, 1983a; also see Table II).

D. Tissue Distribution

Table I lists tissues in which type XI collagen has been found. It appears as a constant minor ingredient of all types of cartilage in

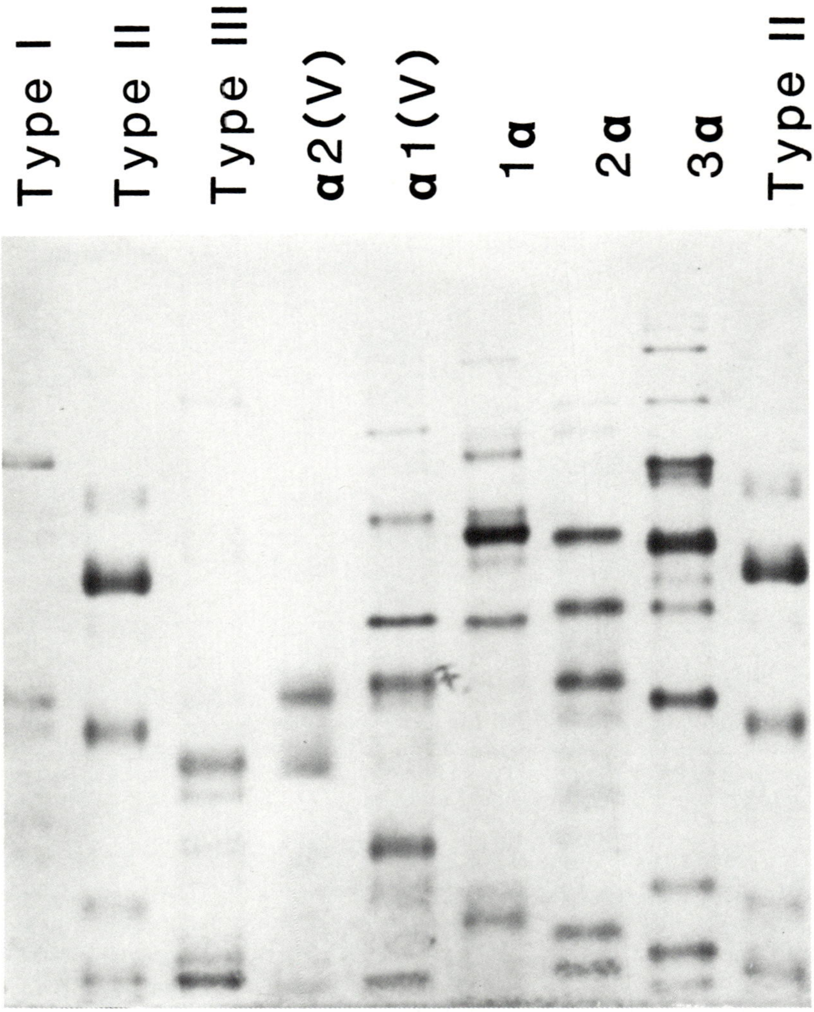

FIG. 3. Peptide profiles (CNBr) of the individual type XI collagen α-chains on SDS–10% polyacrylamide slab gel electrophoresis compared with other collagen chain types. [Adapted and reproduced with permission from Burgeson and Hollister (1979), *Biochem. Biophys. Res. Commun.* **87,** 1124–1131.]

TABLE I

Specific Tissues in Which Type XI Collagen Has Been Identified

Tissue	Species	Age	References
Hyaline cartilages			
Epiphyseal growth	Human	Fetal, infant	Burgeson and Hollister (1979), Burgeson *et al.* (1982)
	Pig	Neonatal	Shimokomaki *et al.* (1980)
	Chick	Embryo	Kimura *et al.* (1985)
Articular	Human	Adult	Burgeson and Hollister (1979), Németh-Csóka and Mészáros (1984)
	Bovine	Adult	Eyre *et al.* (1984b)
	Rabbit	Adult	Unpublished results
	Dog	Adult	Smith *et al.* (1985)
Costosternal	Human	Fetal, infant	Burgeson and Hollister (1979)
	Bovine	18 months	Burgeson and Hollister (1979)
	Chicken	Adult, 6 weeks	Reese and Mayne (1981), von der Mark *et al.* (1982)
Growth plate	Bovine		Grant *et al.* (1985), Jimenez *et al.* (1986)
Nasal septum	Bovine	Adult	Ayad *et al.* (1981), Ayad and Weiss (1984)
Laryngeal and tracheal	Bovine	Adult	Unpublished results
Chondrosarcoma	Rat	—	Furuto and Miller (1983a,b), Duance *et al.* (1984)
Fibrocartilage			
Intervertebral disc	Human, bovine	Neonatal and adult	Ayad *et al.* (1981), Ayad and Weiss (1987), unpublished results
	Ovine	Adult	Unpublished results
Elastic cartilage			
Ear	Rabbit		Madsen *et al.* (1984)
Vitreous humor[a]	Bovine	Adult	Ayad and Weiss (1984)
Notochordal sheath	Lamprey	—	Sheren *et al.* (1986)

[a]In our hands, however, this tissue yields type V, not type XI, collagen chains (see text).

which type II collagen forms the main collagen framework. Its ratio to type II collagen (about 1:30 in adult cartilage) is similar to the ratio of type V to type I collagen presented by most noncartilaginous connective tissues. An exception to this rule may be vitreous humor, which has been reported to contain 1α, 2α, and 3α chains in addition to the main collagen ingredient, type II collagen (Ayad and Weiss, 1984). Using similar tissue (2-year steer vitreous humor), type V collagen was found (in addition to type II collagen) and not 1α, 2α, and 3α chains (unpublished data). Thus, the anomalously fast electrophoretic mobility noted by Ayad and Weiss (1984) for their vitreous 2α may be explained by this chain being α2(V).

Although type V collagen apparently can be expressed with type II collagen in tissues that lack type I collagen (vitreous and also articular cartilage, see below), the parallel situation of type XI collagen being expressed in a tissue that lacks type II collagen has not been observed. Fibrocartilage of the knee meniscus, in which type I collagen forms the main collagen framework with only traces of type II collagen, contains type V collagen with no type XI detected (Eyre and Wu, 1983). Type XI collagen is present in the intervertebral disc (Table I) and appears to be most concentrated where type II collagen predominates in the nucleus pulposus.

E. Articular Cartilage

A mixture of types V and XI collagens appears to be present in the 1.2 *M* NaCl precipitated collagen fraction from adult bovine articular cartilage (Eyre, unpublished work). Moreover, the tissue samples were of normal hyaline cartilage, with no signs of fibrous degeneration or of type I collagen in the pepsin digests. On electrophoresis of the collagen fraction that precipitates at 1.2 *M* NaCl, pH3, from a pepsin digest of 18-month to 2-year steer cartilage, the ratio of 1α to 2α was obviously greater than 1:1 and nearer to 2:1, whereas the ratio in such fractions from fetal human or fetal bovine cartilage appeared to be about 1:1. Moreover, on CM-cellulose chromatography of the 1.2 *M* NaCl fraction from adult bovine articular cartilage, the 1α peak was asymmetrical. The trailing shoulder gave a CNBr peptide profile of 1α and the leading half of α1(V) (not shown). Using reverse phase HPLC, the suspected α1(V) and 1α chains could be fully resolved (Fig. 4), each giving a distinctive CNBr peptide profile (not shown). These HPLC/SDS–PAGE profiles of whole chains from articular cartilage also resolve a component that runs in the position of authentic α2(V) (Fig. 4).

The above results, therefore, indicate that type XI and V collagens

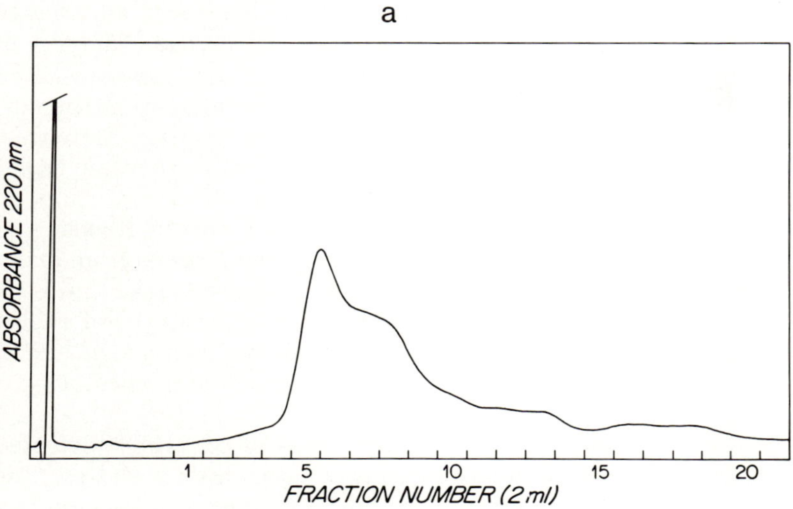

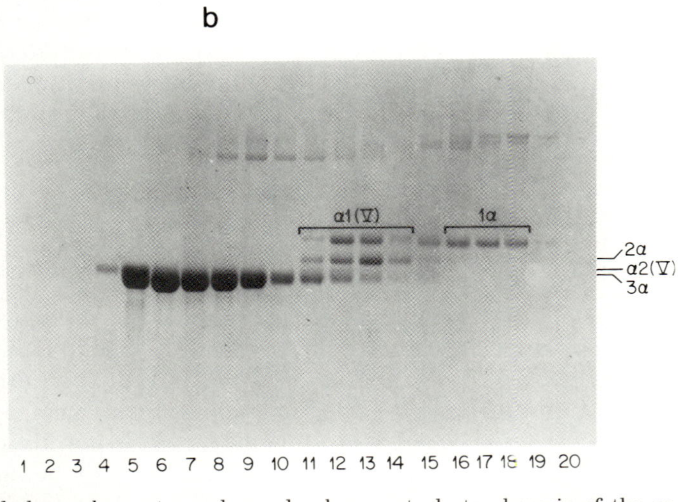

FIG. 4. Reversed-phase chromatography and subsequent electrophoresis of the resolved minor collagen α chains from mature bovine articular cartilage. (a) Elution profile of 200 μg of collagen (1.2 *M* NaCl, pH 3, precipitate from a pepsin digest) on an Aquapore RP-300 column (4.6 mm × 25 cm, Brownlee) using a composite gradient of acetonitrile:*n*-propanol (3:1 v/v) in aqueous 0.1% trifluoroacetic acid (Eyre and Wu, 1983). (b) Collected 2-ml fractions from (a) were dried and then run in consecutive lanes of a SDS–5% polyacrylamide slab gel. This sample still contained type II collagen which coeluted from the column with 3α and overlapped with 3α on electrophoresis.

coexist in adult bovine articular cartilage, with type V apparently having accumulated with age after the neonatal period. Analyses of serial slices (200 μm) taken from surface to deep cartilage showed no obvious enrichment of type V relative to type XI collagen at the surface or in any region of the tissue (unpublished results). Moreover, type II collagen was the only major collagen species present; no type I collagen was detected.

These observations may have a bearing on the report by Rhodes and Miller (1978) that a type V-like fraction could be prepared from articular cartilage which appeared to consist exclusively of α1(V) chains. It seems likely, also, that the collagen identified in cultures of rabbit articular chondrocytes and called X and Y chains (Cheung *et al.*, 1976; Benya *et al.*, 1977), consisted of a mixture of type V and XI collagen α chains.

By immunolocalization experiments, Evans *et al.* (1983) detected type V collagen in bovine nasal cartilage. Later, they withdrew this conclusion when they discovered that their anti-type V antiserum recognized type VI collagen, which then was presumed to be the source of the observed immunoreactivity in cartilage (Ayad *et al.*, 1984). Biochemical data, however, indicate that normal adult articular cartilage can contain both types V and VI collagens in addition to type XI, type IX, and the bulk type II collagen (Wu and Eyre, unpublished results).

IV. Structure

A. *Amino Acid and Carbohydrate Composition*

Types V and XI collagens prepared after pepsin digestion share distinctive features of amino acid composition (Table II). All four chains, α1(XI), α2(XI), α1(V), and α2(V), contain half the number of alanine residues and are much richer in hydroxylysine compared with the α chains of types I, II, and III collagens. In subtle details, α1(XI) matches α1(V) and α2(XI) matches α2(V), the latter pair being richer in histidine and threonine. In contrast, the α3(XI) chain is indistinguishable from α1(II) in amino acid composition, except in degrees of lysine hydroxylation and hydroxylysine glycosylation (Table II).

B. *Chain Composition of the Native Molecule*

Although growing evidence points to a principal native molecule of composition, α1α2α3(XI), it is still possible that type XI collagen prep-

TABLE II
AMINO ACID AND HYDROXYLYSINE GLYCOSIDE COMPOSITIONS OF TYPE XI COLLAGEN CHAINS FROM ADULT BOVINE ARTICULAR CARTILAGE[a]

	α(XI) (1α)	α2(XI) (2α)	α3(XI) (3α)	α1(II)	α1(V)[b]	α2(V)[b]
3-Hyp	2	2	2	1–2	ND[a]	ND[a]
4-Hyp	100	96	104	104	107	109
Asp	49	49	45	48	48	51
Thr	20	25	21	21	19	26
Ser	28	25	26	26	33	31
Glu	101	94	88	86	91	84
Pro	121	112	118	116	118	97
Gly	326	334	332	333	334	341
Ala	51	51	100	100	46	52
Cys	0	0	0	0	0	0
Val	22	23	17	18	23	24
Met	8	8	7	10	8	11
Ile	17	14	9	10	18	16
Leu	36	38	25	25	38	35
Tyr	3	4	2	2	1	2
Phe	11	11	13	13	11	14
Hyl	37	40	21	17	35	24
Lys	19	17	15	18	20	18
His	7	11	3	3	8	11
Arg	42	46	52	50	44	50
Molar ratio, glcgalHyl/galHyl	8.3	7.5	2.7	1.6	1.5[c]	2.5[c]
Glycosylation (percentage)	77	75	74	61	86[c]	58[c]

[a]D. R. Eyre, unpublished data. The chains of types XI and II collagens were purified by salt fractionation and CM-cellulose column chromatography from a single tissue sample. ND, Not determined.

[b]Data from Burgeson *et al.* (1982).

[c]Calculated from the glucose (glc) and galactose (gal) contents, whereas in columns 1–4 the values are based on hydroxylysine glycosides in a base hydrolysate by ion-exchange chromatography and ninhydrin color yield.

arations include a mixture of various heterotrimeric combinations of chains such as $(\alpha 1)_2\alpha 2$(XI), $(\alpha 2)_2\alpha 3$(XI), $\alpha 1_3$(XI) etc.

The following are the points of evidence for a heterotrimeric molecule that combines the three different chains.

1. Highly purified preparations of type XI that are not contaminated with type II or type V collagens (for example, from fetal car-

tilage) give a ratio of the three α chains of 1:1:1 by densitometry after SDS–polyacrylamide electrophoresis.

2. Three β bands of roughly equal prominence are seen in such preparations (see Fig. 2) that have mobilities consistent with their being the three possible heterodimers from intramolecular cross-linking in 1α2α3α molecules. However, monomeric native molecules were not isolated for this analysis, so the dimers could result from either intramolecular or intermolecular cross-linking. This evidence is therefore suggestive, but not proof. It is clear, nevertheless, that β dimers of 3α are absent from pure type XI preparations; if present, they would be resolved on electrophoresis running close to the β1(II) chain position (see Fig. 2).

3. Attempts to fractionate potential subpopulations of native type XI molecules by differential salt precipitation or chromatography under nondenaturing conditions (Burgeson and Hollister, 1979, Ayad and Weiss, 1984) have not succeeded.

4. Native type XI collagen molecules extracted from lathyritic chick cartilage consist of α1(XI), α2(XI), and α3(XI) chains in equal amounts, all three of which are about 10,000–15,000 MW larger than the chains of the pepsin-treated protein. The peptide extensions are apparently the retained amino-propeptides (Broek, 1984; G. Balian, personal communication). This is consistent with the native functioning form of type V collagen which appears to be a molecule retaining a portion of its propeptide sequences (Fessler *et al.,* 1985; Broek *et al.,* 1985).

5. Finally, unpublished observations that type XI collagen forms only one species of segment long spacing (SLS) crystallite as judged by electron microscopy (S. Ayad and B. Brodsky, personal communications) provide further evidence for a single molecular species of composition α1α2α3(XI).

The evidence is still inconclusive whether α3(XI) and α1(II) have identical sequences. Furuto and Miller (1983a) isolated all the 3α [α3(XI)] CNBr peptides and could find no basic differences with α1(II) in their chromatographic properties or amino acid compositions. Though data and opinions are, therefore, leaning toward posttranslational differences only, subtle distinguishing features of their CNBr peptide electrophoretic profiles remain puzzling and leave room for localized sequence differences (von der Mark *et al.,* 1982; Eyre *et al.,* 1984b). Even if α3(XI) is a product of the α1(II) gene, as seems likely, there remains the question of how some proα1(II) chains are selected to associate with proα1(XI) and proα2(XI) chains to form type XI molecules. One explanation might be alternative splicing of the primary transcript in the formation of functional mRNA. Perhaps, for example, the triple-helical body of the α3(XI) chain is identical to that of the α1(II) chain, but one or both extension sequences might be spliced differently for α1(II) and α3(XI), analogous to the alternative splicing that produces fibronectin protein variants from the single fibronectin gene (Tamkun *et al.*, 1984). A firm answer, most probably, must await sequence data on 3α and α1(II) from the same animal source at the protein level. Although protein sequence data on the bovine α1(II) chain showed signs of microheterogeneity consistent with a major and minor form (Butler *et al.*, 1977), more recent gene cloning work indicates only one human copy of the α1(II) gene (Sheffield and Upholt, 1985; Sangiorgi *et al.*, 1985), which implies that these sequence variants were due to allelic differences.

Neither protein sequences nor molecular genetic data have yet been reported for the α1(XI) or α2(XI) collagen chains.

C. *Polymeric Form*

Based on its molecular size and conformation, type XI collagen fits best in the class of Group 1 fibril-forming collagens (Miller, 1985). The actual polymeric form and location of type XI in tissues is still unknown. For example, it remains to be established if it exists as banded fibrils in its own right similar to those formed by types I, II, and III collagens. The pepsin-solubilized native protein molecules can be induced to form banded thin fibrils (of 60–67 nm periodicity) *in vitro* (Smith *et al.*, 1985; B. Doyle, personal communication) using phosphate-buffered saline or low-ionic-strength phosphate and conditions similar to those needed to form banded fibrils from type V collagen (Chiang *et al.*, 1980; Adachi and Hayashi, 1985). However, there is a growing suspicion that type V, and by implication type XI, collagens function in the extracellular matrix not as homopolymers in their own right but copolymerized within the fibrils that form the bulk fibrillar network, i.e., type V hybridized with type I and type XI hybridized with type II. Indirect evidence for such a codistribution of types I and V collagens in the extracellular collagen framework of cornea was presented by Linsenmayer *et al.* (1985).

D. *Cross-Linking*

Type XI collagen contains similar cross-linking amino acid residues (Eyre, 1987; Wu and Eyre, 1985; Eyre, unpublished) to those formed on the hydroxylysine-aldehyde pathway (Eyre *et al.,* 1984a). Thus, type XI collagen, isolated from a pepsin digest of adult bovine cartilage, was treated with $NaBH_4$, acid hydrolyzed, and the hydrolysate run on the cross-linking amino acid program of a Glenco amino acid analyzer (Eyre, 1987; unpublished results). Roughly equal amounts of dihydroxylysinonorleucine (DHLN, 0.38 mol per mol of collagen in leucine equivalents of ninhydrin color) and hydroxylysyl pyridinoline (HP, 0.37 mol per mol of collagen in leucine equivalents) were present. In contrast, type II collagen from the same digest contained mainly HP (1.5 mol/mol) with only a trace of dihydroxylysinonorleucine (0.04 mol/mol). It would seem, therefore, that cross-linking in type XI collagen tends to stay at the stage of divalent reducible bonds, perhaps because retained propeptides inhibit further cross-linking maturation reactions. Nevertheless, this common basic mechanism of cross-linking further supports the notion that type XI collagen should be considered a member of the same class of fibrillar collagens as types I, II, and III collagens, along with type V collagen (Eyre, 1987). Furthermore, it seems that as a general rule the chemical type of cross-linking exhibited by these fibril-forming collagens (i.e., whether predominantly based on lysine aldehydes or hydroxylysine aldehydes) tends to be tissue specific rather than collagen-type specific (Eyre, 1987).

V. Biosynthesis

Type XI procollagen chains have been identified as products of cultures of chondrocytes and cartilage tissue in several laboratories (Schmid and Conrad, 1982; Remington *et al.,* 1983; Sussman *et al.,* 1982, 1984; Clark and Richards, 1985; Jimenez *et al.,* 1986; Reginato *et al.,* 1986). They are present as disulfide-bonded trimeric procollagen molecules, the individual proα chains of which resolve above proα1(II) on SDS electrophoresis (Clark and Richards, 1985). In contrast to the pepsin-trimmed chains, the individual proα chains appear slightly larger than the proα chains of type I collagen as judged by their elution position on agarose sieve chromatography (Clark and Richards, 1985). However, with current data it is not yet clear whether a single form of procollagen molecule, of composition proα1,proα2,proα3(XI), is present. Further analyses of the kind designed to answer this question for type V procollagen molecules (Kumamoto and Fessler, 1980) should give definitive answers.

It was noted that type XI procollagen, in contrast to type II procollagen, was preferentially associated with the cell layer rather than being put out into the culture medium (Mayne *et al.*, 1983; Clark and Richards, 1985; G. Balian, personal communication). This might be taken as evidence that type XI is concentrated and functions pericellularly but could equally reflect differences in the solubility or aggregation properties of the two collagens under the culture conditions used.

Chondrocytes can be induced to change their phenotype in culture under a variety of conditions to produce type I collagen instead of type II collagen. It has been shown that this switch in bulk collagen expression is accompanied by a change in synthesis from type XI to type V collagen using chick chondrocytes under various culture conditions (Mayne *et al.*, 1983, 1984; Yasui *et al.*, 1986).

Synthesis of type XI collagen chains has also been demonstrated in cultures of chondrocytes and whole tissue from rabbit ear elastic cartilage (Madsen *et al.*, 1984).

VI. Protease Susceptibility

None of the chains of bovine type XI collagen is cleaved by synovial or gastric mucosal collagenase at 25 or 37°C under conditions that cleave types I, II, and III collagens into three-quarter TC^A and one-quarter TC^B fragments (Eyre *et al.*, 1984b). One study had concluded that the 3α chain of chick type XI collagen was degraded by skin collagenase while the 1α and 2α chains were not (Liotta *et al.*, 1982). This evidence of cleavage probably resulted from residual contamination of the substrate by type II collagen, for no cleavage of α3(XI) was observed in other studies on pure preparations of type XI (Fig. 5) (Furuto and Miller, 1983a; Eyre *et al.*, 1984b). Why the α3(XI) chain fails to cleave is unknown. The reasons may be, for example, because it differs in amino acid sequence and so lacks the specific cleavage site present in α1(II), or because this sequence is inaccessible when folded in heterotrimeric molecules together with α1(XI) and α2(XI) chains. This issue goes back to the fundamental question, whether α3(XI) and α1(II) are posttranslational variants of the identical gene product or differ in primary sequence.

Unhydroxylated type XI collagen made by chondrocyte cultures is unusually sensitive to pepsin, trypsin, and chymotrypsin (Clark and Richards, 1987). Chondrocytes isolated from chick embryo sterna were incubated with 2,2′-dipyridyl to inhibit prolyl and lysyl hydroxylation. All three chains, α1(XI), α2(XI), and α3(XI), in the disulfide-bonded type XI precursor molecules were extensively degraded even at low temperatures (0–15°C) under conditions when type II precursor mole-

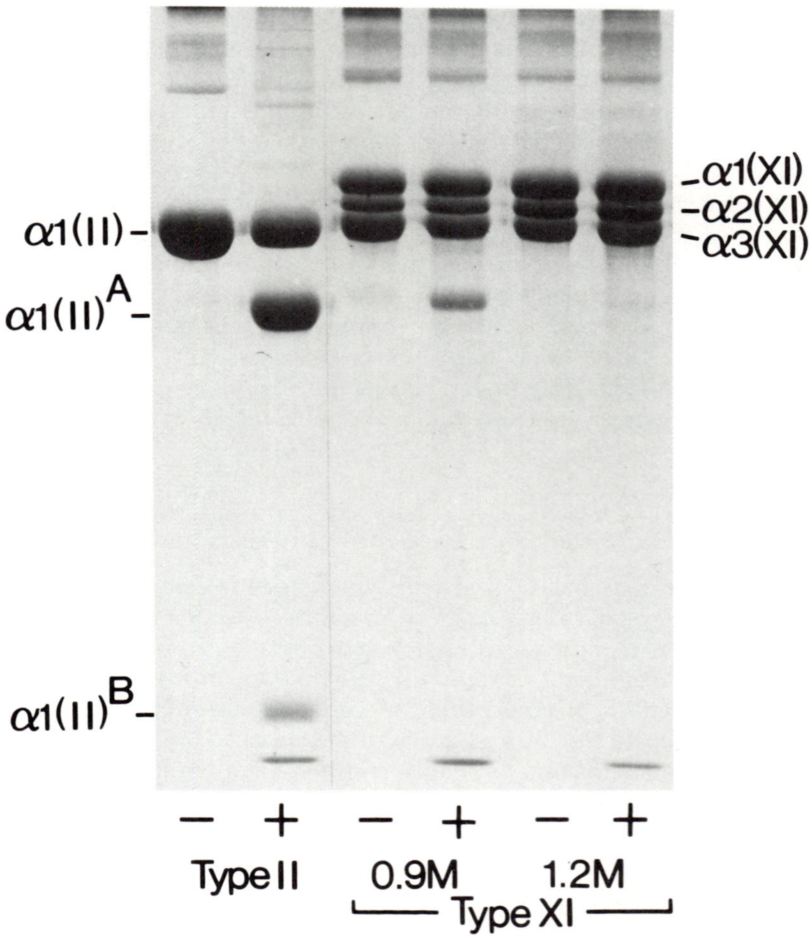

FIG. 5. SDS–7.5% polyacrylamide electrophoresis of type XI collagen showing the failure of human synovial collagenase to cleave any of the chains. Type II and type XI were prepared by pepsin from young bovine cartilage. The 0.9 *M* NaCl fraction of type XI still contained type II molecules which were cleaved giving α1(II)^A fragments, whereas no such cleavage is evident for the purer 1.2 *M* NaCl fraction. −, Control incubations with buffer alone; +, collagenase incubations.

cules were not cleaved within the triple-helical domains of the unhydroxylated molecules. The results imply that unhydroxylated type XI procollagen fails to adopt a stable triple helix under conditions that promote unhydroxylated type II collagen to do so. The physiological significance of this behavior is unknown.

Type XI collagen resembles type V collagen in its lack of susceptibility to the animal collagenases that can cleave types I, II, and III collagens. It also resembles type V collagen in being cleaved in its α1(XI) and α2(XI) chains by thrombin at 36°C, but not at 33°C, and by a tumor metalloproteinase at 30°C that also degrades type V collagen at 25–30°C (Liotta *et al.*, 1982). The susceptibility of the pepsin-soluble substrate to thrombin at 36°C reflects denaturation of the collagen substrate. The cleavage of native type XI chains into specific large fragments by a metalloproteinase that has a similar activity against native type V collagen molecules (Liotta *et al.*, 1982) invites comparison to the specificity of the collagenases that degrade types I, II, and III collagens. However, only the α1(XI) and α2(XI) chains (1α and 2α, not 3α) seemed to be cleaved (Liotta *et al.*, 1982), though contaminating type II collagen may have obscured any 3α cleavage in these experiments.

In summary, it seems unclear whether any of the type V-degrading metalloproteinases are collagen-type specific, in the manner of the known collagenases, or represent nonspecific gelatinases (Mainardi *et al.*, 1980; Murphy *et al.*, 1985) that can attack partially unfolded regions of types V and XI collagen molecules at 25–30°C. On the other hand, failure to be cleaved by specific collagenases that degrade types I, II, and III collagens might be a desirable biological feature in types V and XI collagens.

VII. Immunochemistry

An antiserum raised in rabbits against fetal bovine type XI collagen (pepsin digested, 1.2 *M* NaCl, pH 3 precipitate) showed staining by indirect immunofluorescence concentrated as a diffuse halo only around chondrocyte lacunae in epiphyseal cartilage (Ricard-Blum *et al.*, 1982). Similarly, antibodies raised against bovine type XI collagen in guinea pigs diffusely stained the territorial matrix around the chondrocytes of bovine nasal cartilage (Evans *et al.*, 1983). Although the latter antiserum was purified by serial absorptions on affinity columns prepared from types I, II, III, IV, V, and IX collagens, type VI collagen at that time was unavailable. This is important as a later report (Ayad *et al.*, 1984) noted that an anti-type V collagen antiserum was strongly positive against type VI collagen and so raises similar questions on the specificity of anti-type XI polyclonal antibodies since type VI collagen is widely distributed.

The molecule isolated from cartilage, and thought to consist of

$\alpha 1_3$(V), was localized pericellulary using immunofluorescence and a polyclonal antibody to α1(V) (Gay *et al.,* 1981).

When type XI collagen was injected into rats, it induced an arthritis in one strain (Alderley Park) similar to that induced by type II collagen (Morgan *et al.,* 1983). In Sprague–Dawley rats, however, only type II collagen induced arthritis, though both collagen types were immunogenic in both rat strains.

In unpublished studies (A. R. Poole and D. R. Eyre), rabbit antibodies raised against bovine type XI collagen reacted strongly with α1(XI) and α2(XI) chains on immunoblotting after SDS–polyacrylamide electrophoresis. This was evident even after absorbing the antiserum against types II and V collagens, showing that α1(XI) and α2(XI) contain epitopes not present on types II and V collagens and providing further evidence that these chains are genetically distinct from the type V collagen chains. However, it was not possible with this antiserum to absorb all reactivity against type IX collagen, which by immunoblotting after SDS electrophoresis could be detected as a contaminant of the type XI antigen.

VIII. Pathology

Among normal tissues, type XI collagen has been found only in those that contain type II collagen as a major matrix constituent. However, type XI was found recently in the abnormal cartilage of a child who died at birth with a very rare heritable dwarfing syndrome, the Langer–Saldino variant of achondrogenesis (Eyre *et al.,* 1986). No type II collagen could be detected in any cartilaginous tissue of this infant, despite types XI and IX collagens making up about 10% each of the total collagen (Eyre *et al.,* 1986; unpublished results) and type I collagen replacing type II collagen. In the future, such cases might give insight on the biological function of the minor cartilage collagens and help decide whether the α3(XI) and α1(II) chains are identical gene products or, perhaps, the result of alternative splicing of mRNA.

IX. Possible Functions

Although no function has been described for type XI collagen, some suspected functions include (1) regulating the diameter of type II collagen fibrils; (2) being a specific building block of the perichondrocyte basket of fibrils; and (3) mediating collagen–proteoglycan interactions.

The possibility that type XI collagen is involved in regulating fibril diameters is supported by the finding that type XI can be induced to form only very thin fibrils *in vitro*. This was observed using the pepsin-solubilized protein (Smith *et al.*, 1985) or the intact protein with retained N-propeptides that was extracted by salt from lathyritic tissues (Broek, 1984). Therefore, if type XI collagen is naturally distributed within the same fibrils as type II collagen, it may be able to control the diameters of the hybrid polymer, either by forming a coating layer on the surface of the fibril or through its quantitative ratio to type II collagen in the fibril.

The notion that type XI collagen may be restricted to the pericellular domain rests on immunofluorescence light microscopy using polyclonal antibodies to type XI collagen (see above) and by extrapolation of similar early findings with antibodies to type V collagen (Gay *et al.*, 1981). For type V collagen, at least, this concept of a restricted pericellular location seems to be changing. Data which are accumulating using very well characterized monoclonal antibodies to chick type V collagen show a uniform distribution throughout the extracellular matrix (Linsenmayer *et al.*, 1985).

Smith *et al.* (1985) have provided evidence that type XI collagen, in the form of native molecules in solution or reconstituted fibrils, exhibits a firm binding to cartilage proteoglycans. The binding was relatively nonspecific in that free sulfated polysaccharides, including chondroitin sulfate, heparin and dextran sulfate, competed very effectively in the binding assay. However, the binding was specific in that, under the same conditions, type II collagen did not bind at all. It was speculated that such binding might function *in vivo* to anchor the collagen in its fibrillar form to cartilage matrix proteoglycans or, perhaps, to sulfated glycosaminoglycans on the cell membrane. It should also be considered that if type XI collagen *in vivo* is uniformly distributed throughout the matrix as a copolymer within the same fibrils as type II collagen, then one of its functions may be to mediate physical interactions between the fibrils and proteoglycans.

Acknowledgments

Work described in the authors' laboratory was supported, in part, by PHS Grants AR37318, AR36794, and AM34078. We thank Drs. S. Ayad, G. Balian, C. C. Clark, and B. Brodsky for sharing information and manuscripts prior to publication, and Ms. Ginny Apone for preparing the manuscript.

References

Adachi, E., and Hayashi, T. (1985). *Collagen Relat. Res.* **5,** 225–232.

Ayad, S., and Weiss, J. B. (1984). *Biochem. J.* **218,** 835–840.

Ayad, S., and Weiss, J. B. (1987). *In* "The Lumbar Spine and Back Pain" (M. I. V. Jayson, ed.), 3rd Ed. Pitman, London.

Ayad, S., Abedin, M. Z., Grundy, S. M., and Weiss, J. D. (1981). *FEBS Lett.* **123,** 195–199.

Ayad, S., Evans, H., Weiss, J. B., and Holt, L. (1984). *Collagen Relat. Res.* **4,** 165–168.

Benya, P. D., Padilla, S. R., and Nimni, M. E. (1977). *Biochemistry* **16,** 865–872.

Bornstein, P., and Sage, H. (1980). *Annu. Rev. Biochem.* **49,** 957–1003.

Broek, D. L. (1984). Ph.D. Dissertation. Rutgers University, The State University of New Jersey, and the University of Medicine and Dentistry of New Jersey, Piscataway.

Broek, D. L., Madin, J., Eikenberry, E. F., and Brodsky, B. (1985). *J. Biol. Chem.* **260,** 555–562.

Burgeson, R. E., and Hollister, D. W. (1979). *Biochem. Biophys. Res. Commun.* **87,** 1124–1131.

Burgeson, R. E., Hebda, P. A., Morris, N. P., and Hollister, D. W. (1982). *J. Biol. Chem.* **257,** 7852–7856.

Butler, W. T., Finch, J. E., Jr., and Miller, E. J. (1977). *J. Biol. Chem.* **252,** 639.

Cheah, K. S. E. (1985). *Biochem. J.* **229,** 287–303.

Cheung, H. S., Harvey, W., Benya, P. D., and Nimni, M. E. (1976). *Biochem. Biophys. Res. Commun.* **68,** 1371–1378.

Chiang, T. M., Mainardi, C. L., Seyer, J. M., and Kang, A. H. (1980). *J. Lab. Clin. Med.* **95,** 99–107.

Clark, C. C., and Richards, C. F. (1985). *Collagen Relat. Res.* **5,** 205–223.

Clark, C. C., and Richards, C. F. (1987). Submitted.

Duance, V. C., Wotton, S. F., Voyle, C. A., and Bailey, A. J. (1984). *Biochem. J.* **221,** 885–889.

Evans, H. B., Ayad, S., Abedin, M. Z., Hopkins, S., Morgan, K., Walton, K. W., Weiss, J. B., and Holt, P. J. L. (1983). *Ann. Rheum. Dis.* **42,** 575–581.

Eyre, D. R. (1987). *In* "Methods in Enzymology" (L. W. Cunningham, ed.), Vol. 144, pp. 115–139. Academic Press, Orlando.

Eyre, D. R., and Wu, J. J. (1983). *FEBS Lett.* **158,** 265–270.

Eyre, D. R., Paz, M. A., and Gallop, P. M. (1984a). *Annu. Rev. Biochem.* **53,** 717–748.

Eyre, D. R., Wu, J. J., and Woolley, D. E. (1984b). *Biochem. Biophys. Res. Commun.* **118,** 724–729.

Eyre, D. R., Upton, M. P., Shapiro, F. D., Wilkinson, R. H., and Vawter, G. F. (1986). *Am. J. Hum. Genet.* **39,** 52–67.

Fessler, J. H., Shigaki, N., and Fessler, L. I. (1985). *Ann. N.Y. Acad. Sci.* **460,** 181–186.

Furuto, D. K., and Miller, E. J. (1983a). *Arch. Biochem. Biophys.* **226,** 604–611.

Furuto, D. K., and Miller, E. J. (1983b). *Collagen Relat. Res.* **3,** 433–443.

Gay, S., Rhodes, R. K., Gay, R. E., and Miller, E. J. (1981). *Collagen Relat. Res.* **1,** 53–58.

Grant, W. T., Sussman, M. D., and Balian, G. (1985). *J. Biol. Chem.* **260,** 3798–3803.

Jimenez, S. A., Yankowski, R., and Reginato, A. M. (1986). *Biochem. J.* **233,** 357–367.

Kimura, T., Yasui, N., Ohsawa, S., and Ono, K. (1985). *Biochem. Biophys. Res. Commun.* **130,** 746–752.

Kumamoto, C. A., and Fessler, J. H. (1980). *Proc. Natl. Acad. Sci. U.S.A.* **77,** 6434–6438.

Laemmli, U. K. (1970). *Nature (London)* **227,** 680.

Linsenmayer, T. F., Fitch, J. M., Gross, J., and Mayne, R. (1985). *Ann. N.Y. Acad. Sci.* **460,** 232–245.

Liotta, L. A., Kalebic, T., Reese, C. A., and Mayne, R. (1982). *Biochem. Biophys. Res. Commun.* **104,** 500–506.
Madsen, K., von der Mark, K., van Menxel, M., and Friberg, U. (1984). *Biochem. J.* **221,** 189–196.
Mainardi, C. L., Seyer, J. M., and Kang, A. H. (1980). *Biochem. Biophys. Res. Commun.* **97,** 1108–1115.
Mayne, R., Reese, C. A., Williams, C. C., and Mayne, P. M. (1983). *In* "Limb Development and Regeneration" (R. O. Kelley, P. F. Goetinck, and J. A. MacCabe, eds.), pp. 125–135. Liss, New York.
Mayne, R., Elrod, B. W., Mayne, P. M., Sanderson, R. D., and Linsenmayer, T. F. (1984). *Exp. Cell Res.* **151,** 171–182.
Miller, E. J. (1985). *Ann. N.Y. Acad. Sci.* **460,** 1–13.
Morgan, K., Evans, H. B., Firth, S. A., Smith, M. N., Ayad, S., Weiss, J. B., and Holt, P. J. L. (1983). *Ann. Rheum. Dis.* **42,** 680–683.
Murphy, G., McAlpine, C. G., Poll, C. T., and Reynolds, J. J. (1985). *Biochim. Biophys. Acta* **831,** 49–58.
Németh-Csóka, M., and Mészáros, T. (1984). *Acta Biol. Hung.* **35,** 163–180.
Reese, C. A., and Mayne, R. (1981). *Biochemistry* **20,** 5443–5448.
Reginato, A. M., Lash, J. W., and Jimenez, S. A. (1986). *J. Biol. Chem.* **261,** 2897–2904.
Remington, M. C., Bashey, R. S., Brighton, C. T., and Jimenez, S. A. (1983). *Collagen Relat. Res.* **3,** 271–278.
Rhodes, K., and Miller, E. J. (1978). *Biochemistry* **17,** 3442–3448.
Ricard-Blum, S., Hartmann, D. J., Herbage, D., Payen-Meyron, C., and Ville, G. (1982). *FEBS Lett.* **146,** 343–347.
Sangiorgi, F. O., Benson-Chanda, V., de Wet, W. J., Sobel, M. E., Tsipouros, P., and Ramirez, F. (1985). *Nucleic Acids Res.* **13,** 2207–2225.
Schmid, T. M., and Conrad, H. E. (1982). *J. Biol. Chem.* **257,** 12444–12450.
Sheffield, V. C., and Upholt, W. B. (1985). *Collagen Relat. Res.* **5,** 1–8.
Sheren, S. B., Eikenberry, E. F., Broek, D. L., van der Rest, M., Doering, T., Kelly, J., Hardt, T., and Brodsky, B. (1986). *Comp. Biochem. Physiol.* **85B,** 5–14.
Shimokomaki, M., Duance, V. C., and Bailey, A. J. (1980). *FEBS Lett.* **121,** 51–54.
Smith, G. N., Jr., Williams, J. M., and Brandt, K. D. (1985). *J. Biol. Chem.* **260,** 10761–10767.
Sussman, M. D., Ogle, R. C., Hubbard, S. L., and Balian, G. (1982). *In* "Extracellular Matrix" (S. P. Hawkes and J. L. Wang, eds.), pp. 35–44. Academic Press, New York.
Sussman, M. D., Ogle, R. C., and Balian, G. (1984). *J. Orthop. Res.* **2,** 134–142.
Tamkun, J. W., Schwarzbauer, J. E., and Hynes, R. O. (1984). *Proc. Natl. Acad. Sci. U.S.A.* **81,** 5140–5144.
von der Mark, K., van Menxel, M., and Wiedemann, H. (1982). *Eur. J. Biochem.* **124,** 57–62.
Wu, J. J., and Eyre, D. R. (1985). *Ann. N.Y. Acad. Sci.* **460,** 520–523.
Yasui, N., Benya, P. D., and Nimni, M. E. (1986). *J. Biol. Chem.* **261,** 7997–8001.

Analysis of Collagen Structure by Molecular Biology Techniques

George Vasios,* Yoshifumi Ninomiya, and Bjorn Reino Olsen

Department of Anatomy and Cellular Biology, Harvard Medical School, Boston, Massachusetts 02115

I. Introduction

The isolation and characterization of cDNA and genomic DNA clones for collagenous molecules are essential for understanding how collagen genes are regulated. In addition, nucleotide sequence analysis of DNA provides a powerful approach to primary structure analysis of collagens. In this article, we will discuss how molecular biology techniques are being used to generate structural information about collagen types. We will first describe some of the general strategies for constructing cDNA and genomic DNA clones. This will be followed by a discussion of how such clones are identified. Finally, we will examine the kinds of structural information that can be easily obtained from cloned DNA, illustrated by some recent examples from our own studies.

II. Cloning Strategies

A. cDNAs

1. Purification and Characterization of mRNA [poly(A)$^+$ RNA]

Synthesis of cDNA requires isolation of mRNA. Isolation of intact mRNA that is enriched for a specific mRNA of interest is probably the

* Present address: Dana-Farber Cancer Institute, Boston, Massachusetts 02115.

most critical step in a cDNA cloning project and, therefore, warrants considerable attention. Subsequent steps, such as cDNA synthesis, construction of recombinant plasmid (or phage), transformation, and cloning, are now quite routine and can be performed readily with the use of commercial kits. The successful isolation of long cDNAs, however, depends critically on high quality, well-characterized RNA. Extra weeks spent on isolation of RNA could well save months of subsequent cloning work.

Several methods are available for the isolation of RNA. A popular method is extraction with SDS in the presence of proteinase K (Burnett and Rosenbloom, 1979; see also Maniatis *et al.,* 1982). This has been used by several investigators for the isolation of RNA from both cultured cells and tissues. The method has been used in our own laboratory with chick calvarial bone tissue with good results. However, we did not find the method satisfactory for cartilage tissues. Extraction with guanidine hydrochloride (Adams *et al.,* 1977) gave better yields of high-molecular-weight (and presumably intact) RNA from chick embryo sternal cartilage than did the proteinase K technique. The best method for collagen mRNA extraction, however, is the use of guanidine thiocyanate (Chirgwin *et al.,* 1979). We now use this method almost exclusively to isolate high-molecular-weight RNA in high yields with a minimum of degradation from a variety of tissues.

Having obtained total cellular or tissue RNA by one of the above methods, the enrichment of poly(A)$^+$ RNAs (mRNAs) over rRNAs is accomplished by oligo(dT) cellulose (see Maniatis *et al.,* 1982) or paper affinity chromatography [messenger affinity paper (mAP)] (Werner *et al.,* 1984). If size fractionation of mRNAs is desired, one may use a 5–20% sucrose gradient with a denaturant such as methyl mercury hydroxide (Payvar and Schimke, 1979; Myers *et al.,* 1981). Or, as in the case of α1(I) and α2(I) collagen mRNAs, size fractionation can be accomplished using Sepharose-4B chromatography (Frischauf *et al.,* 1978).

Characterizing the RNA before it is used for cDNA synthesis is as important as choosing the best method for its extraction from cells and tissues. Cell-free translation (Pelham and Jackson, 1976) with a commercially available lysate and analysis of the translation products by SDS–polyacrylamide gel electrophoresis is now relatively simple and provides important information about the complexity of the mRNA. The use of highly purified bacterial collagenase provides, of course, a convenient probe for collagenous polypeptides.

The characterization of mRNA by cell-free translation analysis can also provide data that are essential for selecting an efficient method for screening cDNAs constructed from the mRNA. For example, the strategy used for the recent cloning of type X collagen cDNA in our laboratory (Ninomiya *et al.,* 1986) was based on cell-free translation analysis of RNA isolated from long-term cultured chondrocytes and sternal cartilage. Gel electrophoresis of translation products synthesized with chondrocyte RNA showed a prominent, collagenase-sensitive band comigrating with the bovine serum albumin molecular-weight maker, while products synthesized with RNA from 17-day-old chick embryo sternal cartilage only showed a very weak band (Fig. 1A). On this basis, we constructed a cDNA library from the chondrocyte RNA and screened the library using a differential screening method (Ninomiya *et al.,* 1984a). Clones were hybridized with alkali-fragmented, ^{32}P-labeled chondrocyte RNA (positive probe) and alkali-fragmented, ^{32}P-labeled sternal cartilage RNA (negative probe). One of the clones that hybridized strongly to chondrocyte RNA, but did not hybridize to sternal cartilage RNA, was selected and shown to code for type X collagen (Ninomiya *et al.,* 1986; Fig. 1B).

Finally, the quality of the RNA may be assessed by Northern blot analysis using an existing collagen DNA probe to probe the nitrocellulose blot. For example, in analyzing cartilage mRNA, we have found it useful to examine whether type II collagen mRNA is intact or degraded. Usually, extensive degradation of one mRNA (type II), indicated by a smear of hybridization in a Northern blot, also indicates that other mRNAs are degraded.

2. cDNA Synthesis

A variety of methods are available for synthesis and cloning of cDNA with poly(A)$^+$ RNA as a template. These are described in detail in several excellent handbooks and laboratory manuals (Maniatis *et al.,* 1982; Perbal, 1984; Glover, 1985; see also Grossman and Moldave, 1980; Wu, 1979; Wu *et al.,* 1983a,b). In addition, good commercial kits are available containing all the necessary enzymes and buffers as well as detailed protocols. Since cloning vectors and transformation-competent host cells also can be purchased, the whole process of cDNA synthesis and cloning is becoming rather routine. In fact, some companies even supply ready-made cDNA libraries from a variety of sources, and it is likely that such libraries will soon become available for an increasing number of tissues and animal species.

There are, however, several basic components to the overall procedure of cDNA synthesis, and these will be briefly presented. In-

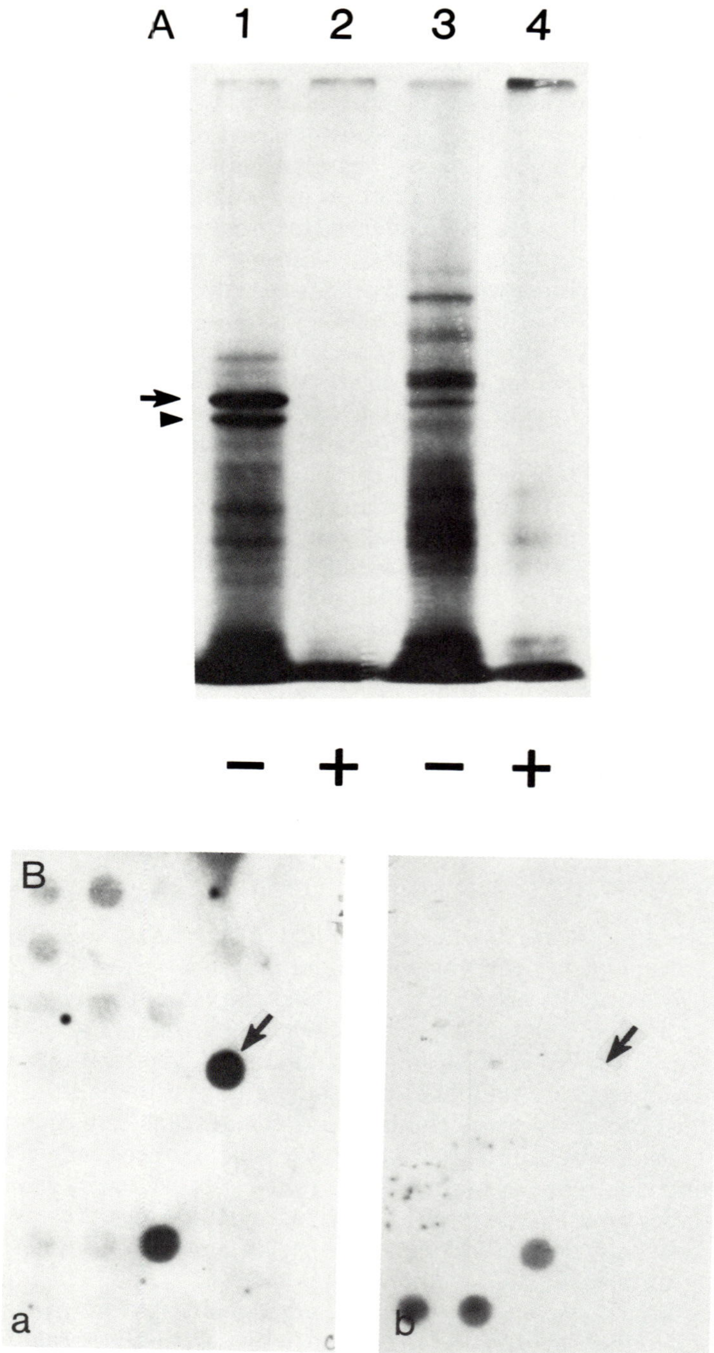
A
1
2
3
4
−
+
−
+
B
a
b

cluded are the use of avian myeloblastosis virus (AMV) (Efstratiadis *et al.*, 1976) or Moloney murine leukemia virus (M-MLV) reverse transcriptase for first strand cDNA synthesis, followed by *Escherichia coli* DNA polymerase I (Efstratiadis *et al.*, 1976; Lehrach *et al.*, 1978), Klenow fragment (Yamamoto *et al.*, 1980), or reverse transcriptase (Myers *et al.*, 1981) for second strand synthesis. The priming of second strand synthesis is accomplished by either the hairpin loop that naturally occurs at the 3′ end of the cDNA after first strand synthesis or by alkali degradation of the RNA template followed by dC-tailing of the first strand and hybridization of an oligo(dG) primer to that tail (Land *et al.*, 1981).

After synthesis of the second strand, S1 nuclease or mung bean nuclease is added to remove the hairpin primer loop as well as to create blunt ends at the termini of the double-stranded cDNA. If an oligo(dG) primer was used for second-strand synthesis, the termini of the cDNA can be made flush either with S1 nuclease or by a reaction with *E. coli* DNA polymerase I or T4 DNA polymerase.

The next step in the cloning process is to create "sticky" termini at the ends of the cDNA so that it may be conveniently cloned into an appropriate vector. For example, homopolymer tails of dC may be added to the cDNA with terminal transferase, allowing it to be cloned into the dG-tailed *Pst*I site of pBR322 (Villa-Komaroff *et al.*, 1978). Ligation of synthetic linkers containing sites for restriction endonucleases, such as *Hin*dIII or *Eco*RI, to the cDNA and cloning into the same site of pBR322 is another alternative (see Maniatis *et al.*, 1982).

In order to increase the percentage of cloned high-molecular-weight cDNAs, size fractionation of the cDNA population can be performed either before or after the creation of sticky ends. One of these methods involves the use of a 5–20% neutral sucrose gradient (Lehrach *et al.*, 1978). Other methods include size fractionation by chromatography on

FIG. 1. (A) Polyacrylamide gel electrophoresis of cell-free translation products synthesized with mRNA isolated from long-term cultured chick tibiotarsal hypertrophic chondrocytes (lanes 1 and 2) and 17-day-old chick embryo sterna (lanes 3 and 4). Arrow indicates the position of type X collagen translation products. The arrowhead indicates a second collagenase-sensitive polypeptide, whose identity is presently unknown, migrating slightly faster than the Type X band. For the cell-free translation analysis a proline-depleted commercial reticulocyte lysate (N150, Amersham) was used with [^{3}H]proline. −, Without bacterial collagenase; +, with bacterial collagenase. (B) Analysis of cDNA clones by dot hybridization using ^{32}P-labeled RNA from long-term cultured chondrocytes (a) and 17-day chick embryonic sterna (b) as probes. The arrows indicate a clone that hybridized strongly to chondrocyte RNA but only weakly to sternal cartilage RNA. Reproduced with permission from Ninomiya *et al.* (1986).

a Bio Gel A-15m column (Wang and Gudas, 1983) or agarose gel electrophoresis.

A variation on the priming of first-strand cDNA synthesis with oligo(dT) is the use of either a synthetic oligonucleotide or a short gene fragment as a primer. This method is feasible only when the amino acid or exon sequence is already known and is most often used to ascertain the signal and 5′-untranslated sequence of the mRNA. In addition, if the primer is end labeled with [γ-^{32}P]ATP and polynucleotide kinase, the products of first-strand synthesis can be electrophoresed in a denaturing polyacrylamide gel, purified by electroelution, and subjected to nucleic acid sequence analysis. In this manner the transcription initiation site of the chicken α2(I) collagen gene (Vogeli *et al.*, 1981a) as well as the nucleotide sequence of the 5′-untranslated region and signal sequence of the chicken α1(I) (Yamada *et al.*, 1983a) and α1(III) (Yamada *et al.*, 1983a) mRNAs were determined. Also, it is possible to clone the primer-extended products as was demonstrated with the chicken α1(I) (Yamada *et al.*, 1983a) and rat α1(II) (Kohno *et al.*, 1984) cDNAs.

In a method developed by Okayama and Berg (1982) for high-efficiency cloning of full-length, double-stranded cDNA, the oligo(dT) primer is part of the plasmid vector (see Fig. 2; Ninomiya *et al.*, 1984a). This primer-vector is annealed to poly(A)$^+$ RNA and the first cDNA strand is synthesized with reverse transcriptase. Terminal transferase is utilized to generate an oligo(dC) tail on the 3′ end of the first strand while it is still hydrogen-bonded to the mRNA. In the next step, the vector is restricted with *Hin*dIII, and a short, double-stranded linker DNA (with a *Hin*dIII sticky site at one end and an oligo(dG) tail at the other) is annealed to the vector–cDNA–mRNA complex so that a circular molecule is formed. In a final step, the mRNA strand is replaced by DNA in a reaction that utilizes RNase H (to remove the mRNA), DNA polymerase I (to repair the second cDNA strand by nick-translation), and DNA ligase (to close the circular molecule). We have utilized the Okayama–Berg method extensively for the cloning of cDNAs of cartilage collagens (Ninomiya *et al.*, 1984b, 1985, 1986; Ninomiya and Olsen, 1984), and the method has worked well in our hands. However, the method is somewhat tedious and involves several steps. Therefore, we have recently used a rapid modification (Gubler and Hoffman, 1983) that includes priming with oligo(dT), first-strand synthesis with reverse transcriptase, second-strand synthesis with RNase H, DNA polymerase I, and DNA ligase as in the Okayama–Berg method, followed by tailing of the cDNA with oligo(dC) and annealing to *Pst*I-cut and oligo(dG)-tailed pBR322 (commercially available).

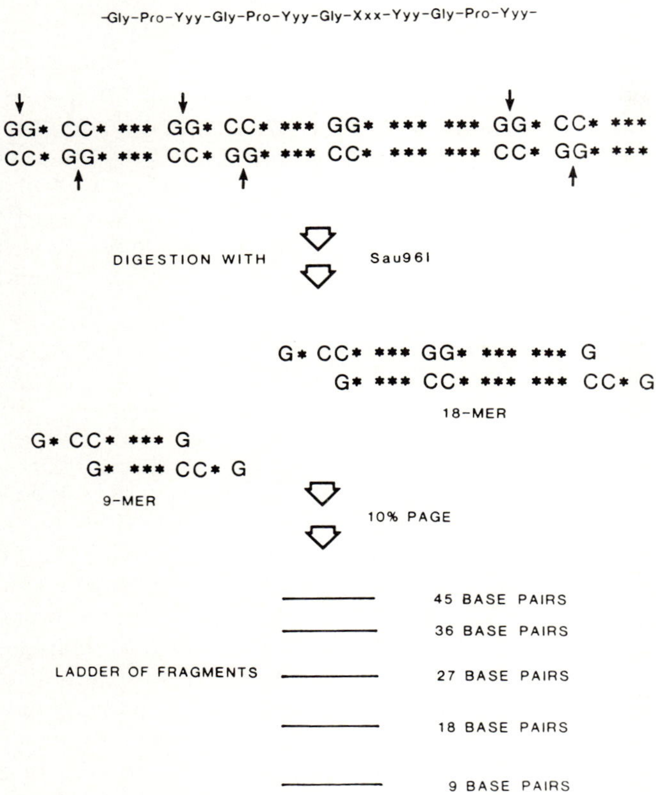

FIG. 2. Diagram illustrating the use of the restriction endonuclease *Sau*96I for screening of collagen cDNAs. The endonuclease cuts within sequences that code for Gly-Pro sequences and produces oligonucleotide fragments that are multiples of 9 bp in length. The fragments produce a characteristic ladder on 10% polyacrylamide gels.

B. Genomic Libraries

1. PURIFICATION OF HIGH-MOLECULAR-WEIGHT DNA

The purification of high-molecular-weight (HMW) DNA of greater than 100 kb is essential for construction of a genomic library. Only in this manner is one insured of having a complete representation of clonable restriction enzyme fragments after partial enzyme digestion. Methods for preparation of HMW genomic DNA from cells or tissue are available (see Perbal, 1984; Kaiser and Murray, 1985).

2. Library Construction

After HMW DNA preparation, an appropriate cloning vector must be selected. Vectors derived from bacteriophage lambda are widely employed, but unfortunately have a cloning capacity of only up to 20 kb. After partial restriction enzyme digestion, genomic DNAs corresponding to approximately 15–20 kb are size fractionated on a 10–40% sucrose gradient (see Kaiser and Murray, 1985). DNAs purified in this manner are then ligated to commercially available phage vector arms, packaged into infectious phage particles using a commercially available kit, and used to infect an appropriate host cell. The phage titers of the pooled packaging reactions should yield a total number of plaque-forming units (PFUs) which roughly correspond to at least three times the number of clones necessary to generate one genome equivalent of DNA. Subsequent amplification of the gene library allows multiple screenings to be performed using a variety of gene probes.

The use of bacteriophage lambda as a cloning vector is based on the fact that, although the linear DNA molecule is 48.6 kb long, only 60% of the lambda genome is needed for propagation of the phage (see Maniatis *et al.*, 1982). Thus, the central region is dispensable and can be replaced by non-lambda DNA. Such replacement vectors are now available in a variety of versions. The original Charon phages, such as Charon 4A (Blattner *et al.*, 1977) and Charon 30 (Rimm *et al.*, 1980), are still being used (we have used them for the construction of chicken genomic libraries) but are increasingly being replaced by vectors that contain "polylinkers" with multiple restriction endonuclease sites. In addition, replacement vectors, such as EMBL3 and EMBL4 contain a functional *gam* gene in the central, dispensable "stuffer" fragment (see Kaiser and Murray, 1985). This makes the parental phage vector unable to form plaques on *E. coli* strains lysogenic for phage P2. Therefore, after cutting out the stuffer fragment with a restriction endonuclease, it is unnecessary with these vectors to physically remove the fragment from vector arms before they are ligated to the exogenous DNA. By growing the library in an *E. coli* (P2) host, only recombinant phages will grow.

The use of cosmid vectors (Collins and Hohn, 1978) in library construction was a significant advance in gene isolation technology because it allowed the isolation of DNA segments of up to 45 kb in size. This provides for a reduction in the number of steps required to walk along a chromosome. In addition, with the larger insert size, cosmids can incorporate particularly long eukaryotic genes (such as collagen genes) within a single insert fragment (Cheah *et al.*, 1985). Cosmids combine the drug selectivity of plasmids, and their ability to replicate autonomously within a bacterial host, with the packageability and infectivity of bacteriophage lambda. Therefore, once the host cell is infected by the phage particle, the cosmid will function entirely as a plasmid, and bacterial colonies will be generated. High-density screening of the clones can be performed as outlined by Hanahan and Meselson (1980). However, colony hybridization for cosmids is less efficient than plaque hybridization for phage. This is, in part, due to the fact that plasmid copy number is reduced when the size is increased. A second disadvantage of cosmids is their tendency to lose insert sequences by deletion. Obviously, since the copy number increases as the insert size is reduced, deleted cosmids outgrow the larger copies.

III. Strategies for Identifying Collagen cDNA and Genomic Clones

A. *Use of Cross-Hybridizing cDNA and Genomic DNA Probes*

Thus, while the synthesis and cloning of cDNA is becoming easier, the screening for a specific clone remains a time-consuming effort whose successful outcome is critically dependent on the screening method used. Therefore, we will discuss in some detail the different approaches one can take for screening of collagen cDNA clones. Some of the methods can also be used for genomic DNA screening, and we will point that out where it is appropriate.

As soon as cDNAs specific for chicken type I collagen chains became available, they were used as cross-hybridizing probes to isolate other collagens from genomic libraries. For example, chicken α1(I) and α2(I) cDNAs were used to isolate collagen genes from *Caenorhabditis elegans* (Kramer *et al.*, 1982) and *Drosophila melanogaster* (Monson *et al.*, 1982) as well as from the human α1(II) (Cheah *et al.*, 1985) and the mouse α1(I) collagen genes (Monson and McCarthy, 1981). Conversely, genomic probes can be used in cross-hybridization experiments to isolate cDNA clones, as was demonstrated by Loidl *et al.* (1984), when a human α1(III) cDNA was identified by cross-hybridization with a chick α1(III) gene probe.

Clearly, the use of cross-hybridizing probes is a preferred method when a probe that shows good cross-hybridization with the sequence of interest is available. Large numbers of cDNA or genomic DNA clones can be easily screened with high-density colony (Hanahan and Meselson, 1980) or plaque hybridization (Benton and Davis, 1977). This approach is limited, however, by the degree of cross-hybridization with the sequence of interest. In addition, due to the large number of homologous collagen genes, unwanted cross-hybridizations between different collagen types and other GC-rich sequences can occur, especially when screening across species boundaries. Finally, the genes of some of the more recently discovered collagens, such as types IX and X, show a codon usage distinctly different from that of fibrillar collagen genes (even in the same species). As a consequence, cDNA and genomic probes for fibrillar collagens did not show hybridization with types IX and X DNA and could not, therefore, be used to screen for types IX and X collagen cDNAs.

Obviously, as the number of DNA probes for specific collagens increases, the likelihood of finding a probe that will detect additional collagen types by cross-hybridization will also increase. Therefore, one can expect that cDNA and genomic screenings for additional collagen types will become easier.

B. *Use of Synthetic Oligonucleotide Probes*

Another method of identification of collagen-containing cDNA and genomic clones involves the use of synthetic oligonucleotide probes (see Szostak *et al.*, 1979). In these cases the amino acid sequence of a peptide has already been determined, while the corresponding DNA sequence is unknown (Suggs *et al.*, 1981; Goldberg *et al.*, 1986). Due to the degeneracy of the genetic code, a mixture of oligonucleotides must be synthesized representing all possible DNA sequences. For example, Kohno *et al.* (1984) used a mixture of 16 tetradecamers, whose sequence was based on the amino acid sequence Phe-Asp-Gln-Lys-Ala from the N-telopeptide of rat proα1(II) collagen, to both synthesize cDNA and screen those cDNAs for a proα1(II) collagen cDNA.

C. *Use of Restriction Endonucleases That Cut at Repetitive Codons*

In situations where cross-hybridizing probes or protein data are not available, we have successfully used a restriction endonuclease method to screen for cDNAs encoding collagenous proteins (Ninomiya and Olsen, 1984; Ninomiya *et al.*, 1984a, 1985). The method is based on the fact that collagens contain Gly-X-Y amino acid repeats, with X frequently being a prolyl residue. Therefore, collagen cDNAs will contain a repeat of the nucleotide sequence GGx-xxx-yyy, in which x and y can be any nucleotide. Whenever an amino acid triplet is Gly-Pro-Y, the corresponding nucleotide sequence is GGx-CCx-yyy. This nucleotide sequence contains the recognition site GGxCC for the restriction endonuclease *Sau*96I (Fig. 2). Within a cDNA that codes for a collagenous sequence, one is therefore likely to find multiple *Sau*96I sites, and the distances between such sites will be multiples of nine nucleotides. By digesting collagen cDNAs with *Sau*96I, end labeling the products using the Klenow fragment of DNA polymerase I to fill in the 5′ overhangs (created by *Sau*96I) with [α-^{32}P]dGTP, a characteristic ladder of bands representing fragments that are multiples of 9 bp in length can be displayed by electrophoresis in 10% polyacrylamide gels (Fig. 3). We have used this approach to screen cDNAs from libraries made with both chondrocyte and fibroblast mRMA. For screening, plasmid DNA is isolated from 1 or 1.5-ml cultures of bacterial clones picked at random from cDNA library plates. The DNA is digested with *Sau*96I, labeled with ^{32}P and electrophoresed on 10% polyacrylamide gels. Within a few weeks, several hundred clones can be screened by this method.

The method has the advantage of only detecting cDNAs that extend far enough from the 3′ end of the mRNA of a collagenous molecule to cover a substantial part of the triple-helical coding region. Also, since the recognition site for *Sau*96I, GGxCC, does not place any restrictions on the nucleotide in the third base positions of the glycyl and prolyl codons, the method allows detection of collagen DNAs that have a codon usage substantially different from those of collagens already available. The usefulness of the method is illustrated by the successful isolation of cDNAs encoding the α1 and α2 chains of type IX collagen (Ninomiya and Olsen, 1984; Ninomiya *et al.*, 1985), and cDNA for a novel type IX-like collagen expressed by chick tendon fibroblasts (Gordon *et al.*, 1987).

The major disadvantage of the restriction endonuclease method is that it is relatively slow and labor intensive as compared with colony and plaque hybridization methods. A second disadvantage is that once a cDNA clone is identified by this method as being collagenous, further characterization (Northern blot analysis, nucleotide sequencing) is necessary to establish the identity of the clones. However, as a method for detection of cDNAs encoding novel collagens, we consider it extremely valuable.

D. Use of Antibodies for Screening of Expression Libraries

While the restriction endonuclease screening method permits the isolation of collagen cDNAs even before isolation and characterization

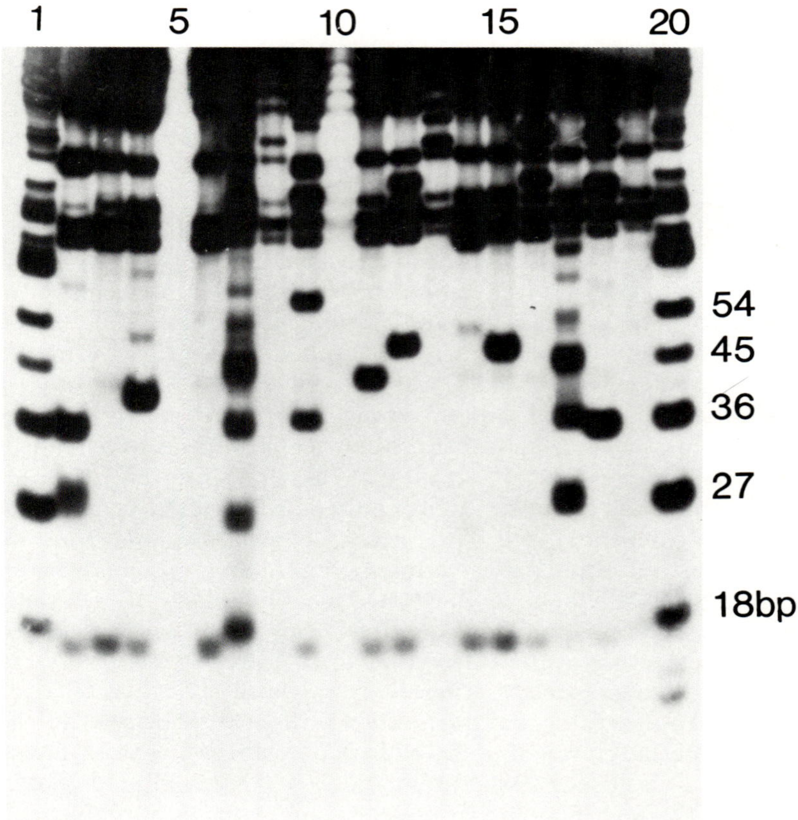

FIG. 3. Polyacrylamide gel electrophoresis of fragments produced by digesting several DNAs with *Sau*96I (courtesy of Dr. Tomoatsu Kimura). DNAs were prepared from plasmid "minipreps" digested with *Sau*96I and end labeled with [α-^{32}P]dGTP and the Klenow fragment of DNA polymerase I, before electrophoresis in a 10% gel. Samples in lanes 1 and 20 are standards (type X collagen) showing characteristic fragments of 18, 27, 36, 45, and 54 base pairs. Samples in lanes 7 and 17 are unknown cDNAs that show a clear collagenous ladder. The conditions for endonuclease digestion and labeling are as follows: (1) To 10 μl total volume of 20 m*M* Tris-HCl, pH 7.4, 5 m*M* $MgCl_2$, 1 m*M* dithiothreitol (DTT), 50 m*M* NaCl (restriction buffer), add plasmid DNA and 1.4 U of *Sau*96I (Bethesda Research Laboratories, Inc.). The amount of DNA per reaction tube is equivalent to DNA prepared from 70 μl of a bacterial culture. (2) Incubate at 37°C for 60 min. (3) Add 1 μl of "Klenow-mix." The Klenow-mix consists of 1.5 μl (15 μCi) of [α-^{32}P]dGTP (NEN, 800 Ci/mmol), 2.0 μl Klenow-fragment (BRL, 3 U/μl), and 20 μl restriction buffer. (4) Incubate at room temperature for 15 min. (5) Add 1.5 μl of "termination buffer" (40% glycerol, 0.2% bromphenol blue, 0.1 *M* Na_4-EDTA). (6) Half of the sample is used for electrophoresis.

of the protein, and therefore usually long before antibodies are available, the use of expression vectors is an alternative in those cases where an antibody is available. The identification and isolation of gene sequences by the use of specific antibody probes have been described by several groups (see Clarke *et al.*, 1979; Ehrlich *et al.*, 1979; Helfman *et al.*, 1985). However, it was not until development of the λgt11 vector system that immunological screening of expressing libraries became a widely popular and successful method for cDNA isolation (Young and Davis, 1983, 1985). The approach has led to the isolation of cDNAs for several noncollagenous extracellular matrix molecules (Schwarzbauer *et al.*, 1983; Doege *et al.*, 1986; Yamada *et al.*, 1985) but has not yet been successful for collagens.

The λgt11 expression vector permits insertion of foreign DNA into the β-galactosidase structural gene under the control of the *lac* operator (Fig. 4). Propagation of the DNA occurs lytically with expression of the foreign DNA repressed by *lac* repressor. Addition of isopropyl-β-D-thiogalactoside (IPTG) induces production of fusion polypeptides containing sequences encoded by the foreign DNA, and these can be detected immunologically. Rapid degradation of foreign polypeptides in the *E. coli* host is prevented by using a host cell carrying the *lon* mutation. For cloning of cDNAs encoding well-characterized proteins against which antibodies are available, the λgt11 system is clearly the method of choice.

E. Hybrid-Select Translation Analysis

Identification of cDNA or genomic DNA clones can, in some cases, be accomplished by hybrid-select translation analysis. For example, Wang and Gudas (1983) used hybrid-select translation to identify cDNA clones for type IV collagen and laminin. In the most widely used form of the selection procedure, DNA is immobilized on nitrocellulose and mRNA is hybridized to the filter (Riccardi *et al.*, 1979). After removal of nonhybridized RNA by washing, bound RNA is eluted and analyzed by cell-free translation. When combined with immunoprecipitation of the cell-free translation products, the approach is a good strategy for identifying cloned DNA. In our experience, however, hybrid-select translation is a difficult procedure, and we do not recommend it as a routine procedure for collagenous proteins.

F. Identification of Exons in Genomic DNA

Ideally, the synthesis and cloning of full-length cDNAs for any collagen mRNA should allow derivation of the primary structure of the

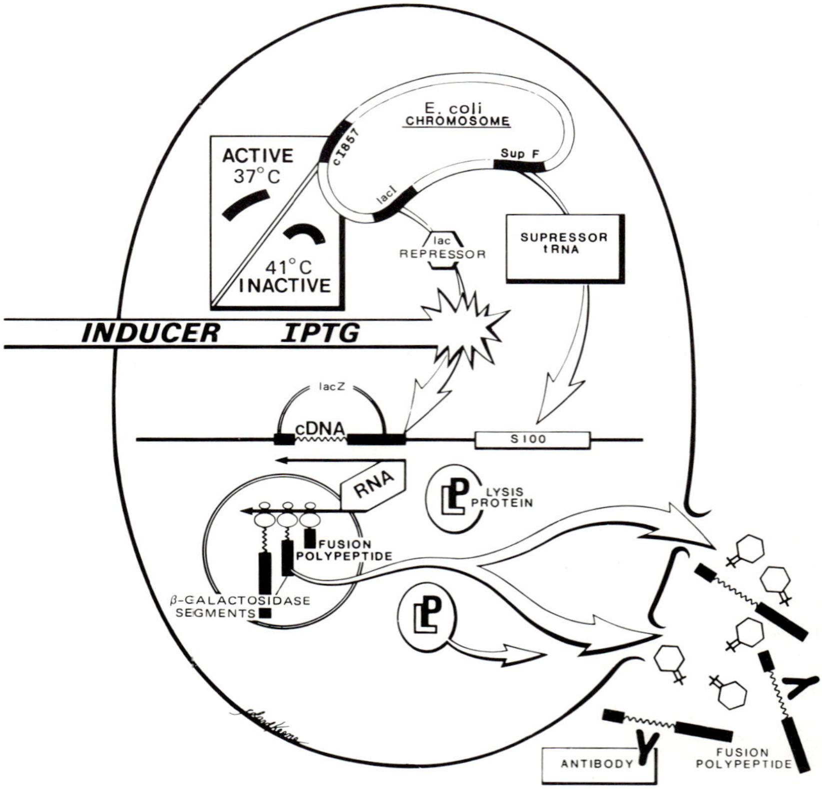

FIG. 4. Diagram illustrating the basis for expression cloning using the λgt11 system. The λgt11 vector allows insertion of cDNA within the β-galactosidase gene. By growing the recombinant phage in a host cell (Y1090) that produces *lac* repressor, synthesis of fusion polypeptides is repressed until induced by overlaying phage plaques with nitrocellulose soaked in IPTG. The host cell produces a suppressor tRNA that suppresses a phage mutation causing defective lysis (S100). To prepare large amounts of a polypeptide specified by a cDNA, the λgt11 recombinant can be expressed as a lysogen in *E. coli* Y1089. This host produces a temperature-sensitive lambda repressor (active at 37°C, inactive at 41°C), it carries a deficiency in the *lon* protease, and contains a mutation which enhances the frequency of phage lysogeny. In addition, Y1089 does not produce suppressor tRNA, so it is not lysed by λgt11. Lysogens (colonies that grow at 32°C, but not at 42°C) are picked, amplified, induced with IPTG, harvested, and lysed by freezing and thawing to produce fusion polypeptides in large amounts.

collagen polypeptides from the nucleotide sequence of the DNA. However, one frequently encounters difficulties in isolation of full-length collagen cDNA sequences. For example, most cDNA clones specific for α1(I) chains isolated in our laboratory (Ninomiya *et al.*, 1984b), as well as those of others (Vuorio *et al.*, 1982; Lukens *et al.*, 1983; Deak *et al.*, 1985; Elima *et al.*, 1985; Sangiorgi *et al.*, 1985), cover only the 3′ untranslated region and the coding sequence for the carboxyl propeptide. The reverse transcriptase sometimes seems to pause or stop at specific sites along the mRNA during synthesis of the first strand of cDNA (Ninomiya *et al.*, 1984b).

One way to circumvent this problem is to isolate genomic clones and deduce the structure of the protein by nucleotide sequence analysis of exons within such clones. Relatively short cDNAs can be used effectively to screen genomic libraries. In fact, the isolation of a series of overlapping clones of the chicken α2(I) collagen gene was accomplished with a cDNA of only 180 bp in one laboratory (Ohkubo *et al.*, 1980).

How can exons within a genomic clone be identified? When cDNAs are available, the problem is, of course, quite simple. Southern blot analysis of restriction endonuclease digests of cloned genomic DNA using the cDNA as probe will identify restriction fragments that contain coding sequences. One is frequently faced, however, with the problem of identifying exons within genomic DNA in regions where cDNAs are not available. Several alternative strategies are available, including R-loop mapping, S1 nuclease protection experiments, Northern blot analysis, and computer analysis of nucleotide sequences. Electron microscopic R-loop mapping has been extremely useful for locating exons in some of the cloned fibrillar collagen genes (Vogeli *et al.*, 1981b; Ohkubo *et al.*, 1980; Wozney *et al.*, 1981; Yamada *et al.*, 1983b; Chu *et al.*, 1984). If highly enriched mRNA is available, this is clearly a fast and simple method of determining gene structures. However, if the RNA of interest only represents a very small fraction of the total mRNA, electron microscopy may not be possible. The alternative is then S1 nuclease protection and/or Northern blot analysis (see Maniatis *et al.*, 1982).

These methods can provide a relatively accurate mapping of exons within genomic clones. However, determination of nucleotide sequences is necessary to precisely define coding sequences and exon/intron boundaries. Once a sequence is determined, several computer methods (see Soll and Roberts, 1984) can be used both to locate likely coding regions within it and to define the position of splice junctions (Staden, 1985). In our own laboratory, a systematic search for intron/exon splice junctions together with a search for sequences encoding collagenous domains have allowed definition of the exon structure of a large portion of the α2(IX) collagen gene (Lozano *et al.*, 1985).

G. Synthesis of Oligopeptides Based on Nucleotide Sequences and Generation of Antipeptide Antibodies

Both monoclonal and polyclonal antibodies can be generated against synthetic oligopeptides coupled to haptens such as hemocyanine (Niman *et al.*, 1983; Tamura *et al.*, 1983). If the peptide sequence has a high hydrophilicity index (Hopp and Woods, 1981), it is likely to be located on the surface of the intact protein, and the anti-peptide antibodies are, therefore, likely to bind to the intact protein as well.

We have used this strategy to generate antibodies against each of the three distinct polypeptide chains of chicken type IX collagen (Konomi *et al.*, 1986), and the antibodies have allowed identification of these polypeptides after their electrophoretic separation on polyacrylamide gels. As cDNA and genomic DNA clones for novel collagens are likely to be isolated before the protein products are identified, it is possible that this approach will become more widely used.

H. Nucleotide Sequence Analysis

Determination of the amino acid sequence of a collagen polypeptide by molecular cloning requires the establishment of nucleotide sequences of cDNAs and genomic DNAs. Nucleotide sequencing can easily be accomplished by either the Maxam–Gilbert chemical method (Maxam and Gilbert, 1980) or by the dideoxy-chain termination method (Sanger *et al.*, 1977). For large sequencing projects where several kilobases need to be sequenced, subcloning into a single-stranded phage vector and sequencing by the dideoxy method is undoubtedly the faster method (Messing, 1983; Barnes *et al.*, 1983) and can be used either with a shotgun or sequential digestion approach. The shotgun strategy relies on fragmenting the DNA with restriction endonucleases (Messing, 1983), DNase (Anderson, 1981), or sonication (Deininger, 1983). The fragments are cloned in M13 and sequenced. The order of the fragments is determined by finding overlaps in the sequences. In the sequential strategy, the DNA is shortened with an exonuclease such as *Bal*31 (Poncz *et al.*, 1982) or exonuclease III followed by S1 nuclease or exonuclease VII (Guo *et al.*, 1983). We have had extensive experience with *Bal*31 and find it extremely useful for producing overlapping fragments, as long as M13 plaques for sequencing are picked on the basis of hybridization to the initial DNA (G. Lozano, D. McCormick and B. R.

Olsen, unpublished results). A drawback with the *Bal*31 method is the need for isolation of double-stranded cloned DNA and the recloning of DNA into M13 after the exonuclease digestion. This obstacle is circumvented in a recently described method (Dale *et al.*, 1985) in which a sequential series of overlapping clones are generated directly from single-stranded M13 clones.

In our experience, however, not all collagen sequences can easily be determined by the dideoxy method, and one must, therefore, rely on the chemical degradation method as well. Also, for limited sequence analysis of smaller DNA fragments (<1000 nucleotides) the Maxam–Gilbert method is probably the method of choice. Several modifications have recently been introduced in the chemical method, so that the sequencing reactions and the subsequent gel electrophoresis of the reaction products can be performed within a day (R. Tisser, personal communication). For both the dideoxy and the chemical methods, wedged sequencing gels that are thicker at the bottom than at the top are most useful. The compressed sequence ladder that forms at the bottom of wedged gels allows more nucleotides to be read than is the case with gels of even thickness. Also, gradient gels (Biggin *et al.*, 1983) can be used with similar results, but they are more difficult to pour than wedged gels.

The use of the dideoxy method and [α-^{35}S]dATP (Biggin *et al.*, 1983), combined with drying of the sequence gels before autoradiography, routinely allows the reading of 300–350 nucleotides from a single gel. With multiple loadings of the gels and extended electrophoresis times, this number can be increased to 500 or more.

Good computer programs are available for analysis and handling of nucleotide sequence data (see Soll and Roberts, 1984). Several programs are available commercially. A sonic digitizer is extremely useful for entering sequence data directly from gels into a microcomputer, thus eliminating the need for writing sequences down and entering them via the keyboard of the computer.

IV. Structural Protein Information That Can Be Easily Obtained from Cloned DNA

A. *Amino Acid Sequences Derived from Nucleotide Sequences*

Obviously, the amino acid sequence of a polypeptide can be determined by conceptually translating the nucleotide sequence of cloned DNA. Figures 5 and 6 illustrate the results obtained when translating the nucleotide sequence of the type X collagen gene. If the available

```
 2
ATT CTT TGA CTC TAA ATG TAT TGA TGT AAA AAA AAT TCT AAA GGC

47
ATC ATA GCT CAC ATT ACA ACT CCC AAC ATA TTT GCA CTT GTT ATA

92
TAC TCA TAT GAG ACT CTT AGG TCA CTT CCA CTA TGA AAG TCA GCT

137                                     167
CAG GTT CTG AGT GAA CAA CAG TGA TGT TAA ACA CTT GAT TTC TGT
                                        Thr Leu Asp Phe Cys
                                         1

182
GAA GAC CAA ATG AAG CTA TAT ATA CTC TTC ACT ATG CAA TCA ACG
Glu Asp Gln Met Lys Leu Tyr Ile Leu Phe Thr Met Gln Ser Thr
 6

227
TGC AAA TCT GGA AGG GCT TTT ACA ACT TAC ATG ATT CTA CAA AAT
Cys Lys Ser Gly Arg Ala Phe Thr Thr Tyr Met Ile Leu Gln Asn
21

272
GTT ATG GCT GAC CTT GTT TCT TCT CAT ACA GGT GTG CAG ATG AGG
Val Met Ala Asp Leu Val Ser Ser His Thr Gly Val Gln Met Arg
 36

317
GGT GAA CAA GGG CCC CCT GGT CCC CCA GGC CCT ATT GGA CCA AGA
Gly Glu Gln Gly Pro Pro Gly Pro Pro Gly Pro Ile Gly Pro Arg
 51

362
GGA CAA CCA GGT CCT GCA GGA AAA CCT GGG TTT GGA AGC CCT GGA
Gly Gln Pro Gly Pro Ala Gly Lys Pro Gly Phe Gly Ser Pro Gly
 66

407
CCC CAA GGT CCT CCT GGT CCC CTA GGA CCA CCT GGA TTC TCC ACT
Pro Gln Gly Pro Pro Gly Pro Leu Gly Pro Pro Gly Phe Ser Thr
 81

452
GTT GGA AAA CTA GGC ATG CCA GGT CTA CCA GGG AAG CCA GGA GAA
Val Gly Lys Leu Gly Met Pro Gly Leu Pro Gly Lys Pro Gly Glu
96

497
AGA GGA TTA AAT GGT GAG AAA GGA GAA GCC GGA CCC GTT GGG CTC
Arg Gly Leu Asn Gly Glu Lys Gly Glu Ala Gly Pro Val Gly Leu
111

542
CCA GGA GCA AGA GGG CCA CAA GGA CCC CCT GGC ATT CCT GGC CCT
Pro Gly Ala Arg Gly Pro Gln Gly Pro Pro Gly Ile Pro Gly Pro
126
```

FIG. 5. See legend on p. 304.

587
GCA GGA CTG TCT GTC CTT GGA AAG CCA GGA CCC CAA GGC CCT CCA
Ala Gly Leu Ser Val Leu Gly Lys Pro Gly Pro Gln Gly Pro Pro
141

632
GGA GCT CAA GGG CCA AGA GGC CCG CCT GGT GAG AAA GGT GAA CCA
Gly Ala Gln Gly Pro Arg Gly Pro Pro Gly Glu Lys Gly Glu Pro
156

677
GGT GTT CCT GGT ATA AAT GGA CAA AAA GGA GAA ATG GGA TTT GGT
Gly Val Pro Gly Ile Asn Gly Gln Lys Gly Glu Met Gly Phe Gly
171

722
GTT CCA GGC CGC CCA GGT AAC AGG GGT CTT CCA GGC CCA CAG GGA
Val Pro Gly Arg Pro Gly Asn Arg Gly Leu Pro Gly Pro Gln Gly
186

767
CCC CAA GGC CTC CCT GGT TCT GCT GGG ATA GGG AAG CCT GGA GAA
Pro Gln Gly Leu Pro Gly Ser Ala Gly Ile Gly Lys Pro Gly Glu
201

812
AAT GGC CTT CCT GGT CAG CCA GGT ATG AAA GGT GAC AGA GGT CTA
Asn Gly Leu Pro Gly Gln Pro Gly Met Lys Gly Asp Arg Gly Leu
216

857
CCA GGA GCA CGT GGA GAA GCT GGT ATC CCA GGC CCC CAG GGT CCA
Pro Gly Ala Arg Gly Glu Ala Gly Ile Pro Gly Pro Gln Gly Pro
231

902
CCA GGA GAA CCT GGA GAA GTT GGC ATC GGC AAG CCT GGG CCA ATG
Pro Gly Glu Pro Gly Glu Val Gly Ile Gly Lys Pro Gly Pro Met
246

947
GGA CCA CCA GGA CCA GCA GGC ATC CCC GGA GCC AAA GGA TTG CCT
Gly Pro Pro Gly Pro Ala Gly Ile Pro Gly Ala Lys Gly Leu Pro
261

992
GGA CCT GCA GGC CTG CCT GGA TCC CCA GGT CTT CCA GGA TTT GGA
Gly Pro Ala Gly Leu Pro Gly Ser Pro Gly Leu Pro Gly Phe Gly
276

1037
AAG CCA GGA TTG CCA GGG ATG AAG GGA CAT AGA GGA CCT GAA GGA
Lys Pro Gly Leu Pro Gly Met Lys Gly His Arg Gly Pro Glu Gly
291

1082
CCT CCT GGC TTT CCA GGA CCT AAA GGA GAC CAA GGC CCA GCT GGT
Pro Pro Gly Phe Pro Gly Pro Lys Gly Asp Gln Gly Pro Ala Gly
306

FIG. 5. See legend on p. 304.

```
1127
 GTA CCA GGA GAA CTA GGT CCT GCT GGG CCA CAA GGA AAT ATG GGC
 Val Pro Gly Glu Leu Gly Pro Ala Gly Pro Gln Gly Asn Met Gly
 321

1172
 CCT CAA GGA CTC AAA GGC TTG CCT GGT GAG AAT GGC TTA CCC GGA
 Pro Gln Gly Leu Lys Gly Leu Pro Gly Glu Asn Gly Leu Pro Gly
 336

1217
 CCC AAA GGT GAC ATG GGC CCT GTA GGC CCT GCA GGA TTC CCA GGA
 Pro Lys Gly Asp Met Gly Pro Val Gly Pro Ala Gly Phe Pro Gly
 351

1262
 GCC AAG GGT GAA AGA GGT CTA CCA GGA TTA GAT GGT AAA CCA GGA
 Ala Lys Gly Glu Arg Gly Leu Pro Gly Leu Asp Gly Lys Pro Gly
 366

1307
 TAT CCA GGT GAG CAG GGT CTT CCT GGT CCT AAG GGA CAT CCA GGT
 Tyr Pro Gly Glu Gln Gly Leu Pro Gly Pro Lys Gly His Pro Gly
 381

1352
 CTC CCT GGT CAA AAA GGA GAC ACT GGC CAT GCT GGC CAT CCT GGC
 Leu Pro Gly Gln Lys Gly Asp Thr Gly His Ala Gly His Pro Gly
 396

1397
 TTG CCT GGT CCA GTT GGA CCA CAA GGA GTT AAG GGA GTT CCA GGT
 Leu Pro Gly Pro Val Gly Pro Gln Gly Val Lys Gly Val Pro Gly
 411

1442
 ATT AAT GGT GAA CCA GGC CCA AGA GGA CCT TCA GGA ATA CCT GGG
 Ile Asn Gly Glu Pro Gly Pro Arg Gly Pro Ser Gly Ile Pro Gly
 426

1487
 GTC AGA GGC CCC ATT GGT CCC CCT GGC ATG CCA GGA GCC CCT GGT
 Val Arg Gly Pro Ile Gly Pro Pro Gly Met Pro Gly Ala Pro Gly
 441

1532
 GCA AAA GGT GAG GCA GGG GCA CCA GGA CTG CCA GGC CCA GCA GGT
 Ala Lys Gly Glu Ala Gly Ala Pro Gly Leu Pro Gly Pro Ala Gly
 456

1577
 ATT GTT ACA AAA GGT TTA AGA GGA CCC ATG GGA CCA CTT GGA CCT
 Ile Val Thr Lys Gly Leu Arg Gly Pro Met Gly Pro Leu Gly Pro
 471

1622
 CCT GGA CCT AAA GGC AAC AGC GGA GAA CCT GGC TTG CCA GGT CCC
 Pro Gly Pro Lys Gly Asn Ser Gly Glu Pro Gly Leu Pro Gly Pro
 486
```

FIG. 5. See legend on p. 304.

```
1667
 CCC GGT CCT CCT GGT CCC CCT GGC CAA TCC ACA ATC CCA GAA GGT
 Pro Gly Pro Pro Gly Pro Pro Gly Gln Ser Thr Ile Pro Glu Gly
501

1712
 TAT GTT AAA GGA GAG TCT AGA GAA CTA TCA GGA ATG TCT TTC ATG
 Tyr Val Lys Gly Glu Ser Arg Glu Leu Ser Gly Met Ser Phe Met
516

1757
 AAA GCA GGA GCA AAT CAA GCT CTC ACA GGA ATG CCA GTG TCT GCC
 Lys Ala Gly Ala Asn Gln Ala Leu Thr Gly Met Pro Val Ser Ala
531

1802
 TTC ACT GTC ATT CTC TCA AAA GCC TAC CCT GGG GCA ACA GTC CCC
 Phe Thr Val Ile Leu Ser Lys Ala Tyr Pro Gly Ala Thr Val Pro
546

1847
 ATC AAA TTT GAC AAA ATC TTG TAC AAC AGA CAG CAA CAC TAT GAC
 Ile Lys Phe Asp Lys Ile Leu Tyr Asn Arg Gln Gln His Tyr Asp
561

                             ↓
1892
 CCC AGG ACA GGA ATC TTT ACC TGC AGG ATC CCT GGT CTA TAC TAT
 Pro Arg Thr Gly Ile Phe Thr Cys Arg Ile Pro Gly Leu Tyr Tyr
576

1937
 TTC TCC TAT CAT GTA CAT GCA AAA GGA ACA AAT GTT TGG GTT GCA
 Phe Ser Tyr His Val His Ala Lys Gly Thr Asn Val Trp Val Ala
591

1982
 CTC TAT AAA AAT GGC TCC CCA GTC ATG TAC ACT TAT GAT GAA TAC
 Leu Tyr Lys Asn Gly Ser Pro Val Met Tyr Thr Tyr Asp Glu Tyr
606

2027
 CAG AAA GGA TAC CTT GAC CAG CCT CAG GCA GTG CTG TCA TTG ATC
 Gln Lys Gly Tyr Leu Asp Gln Pro Gln Ala Val Leu Ser Leu Ile
621

2072
 TCA TGG AGA ACG ATC AAG TGT GGC TCC AGC TGC CAA ATT CAG AAT
 Ser Trp Arg Thr Ile Lys Cys Gly Ser Ser Cys Gln Ile Gln Asn
636

2117
 CCA ATG GTC TCT ATT CCT CTG AAT ATG TTC ATT CTT CTT TCT CAG
 Pro Met Val Ser Ile Pro Leu Asn Met Phe Ile Leu Leu Ser Gln
651

2162
 GTT TCC TAT TTG CTC AAA TCT AAC AAC ATC CCA CTT ACA ATG TCC
 Val Ser Tyr Leu Leu Lys Ser Asn Asn Ile Pro Leu Thr Met Ser
666
```

FIG. 5. See legend on p. 304.

```
2207
  TGA CAT ACT CTC CTT CAG GAC ATT TTA ACT GGG CAC TGA TGT GAT
  ***
 681

2252
  ATG TCC TTA GCA GGT CAT GTA ACT TCA ATT TTG CAA CAG AAA TCG

2297
  AAA AGA GTG AAA AGG TAA AGA AAT TGG TTC AAA ACT AGT AGA AAA

2342
  ATA TTG TAT GTT TTT AGA AAC TTG TCA TGG GAA ATT TGT TTT GAA

2387
  CAA TGG TCA ATT GTC CAA CTG AAT GCA ACA TTA AAA TTT TAT GAA

              ↓
2432
  TTA ATA TAA TAC ACC TGT CCA TCT CTC TTC CTT AAA AAA AGG GGG

2477
  GGG TGA AGG GGG GGG GGG AGG GGG GAG GGC AAA TCA GAT TTT CTA

2522
  AGT ATC ATT TCT CAT TTA ACA GGT AAA GTA GCC ATT CCA AGA CTC

2567
  TAG ACC ACA CTC GGG TAC ACT GAA TGA TAC CTT CAG TTG ATT CA
```

FIG. 5. Nucleotide sequence of the type X collagen gene and the amino acid sequence of the conceptual translation product. Noncollagenous sequences are indicated by boxed areas. Short horizontal one-sided arrows indicate amino acid residues that were confirmed by direct amino acid sequencing of two cyanogen bromide peptides of type X collagen. The two types of imperfections in the collagenous domain are underlined with single lines for the short type (Gly-X-Gly) and double lines for the longer type (Gly-X-Y X-Y-Gly). The two short vertical arrows indicate the 5′ and 3′ ends of the sequence contained within the cDNA pYN3116 [the poly(A) segment excluded]. Within the carboxy-terminal noncollagenous domain, a relatively hydrophobic amino acid sequence of 29 residues between 2 lysyl residues is indicated with a heavy line. Reproduced with permission from Ninomiya *et al.* (1986).

DNA sequence covers the whole translated portion of the mRNA, the complete sequence of the translation product can be deduced quite easily. Residues that frequently are difficult to determine by amino acid sequencing of peptides, such as serine, threonine, asparagine, and glutamine, present no problem at the DNA level. However, since hydroxyproline and hydroxylysine are not encoded by specific codons, these residues can only be determined by protein sequencing. For col-

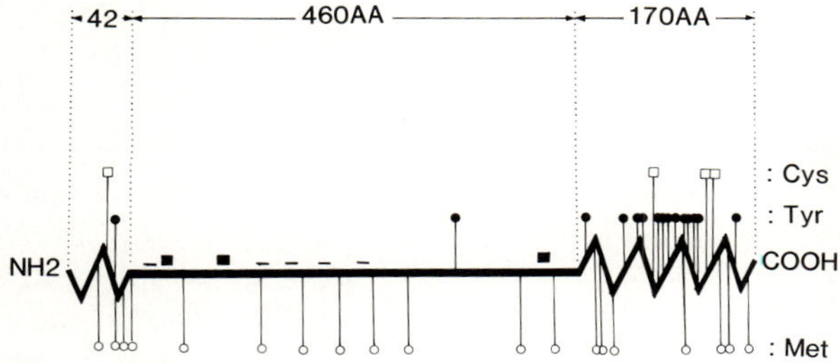

FIG. 6. Diagram summarizing characteristic features of the type X collagen translation product as deduced from the nucleotide sequence of the gene. If we assume that the methionyl codon at nucleotide 191 represents the start of translation, the polypeptide contains 42 amino acid residues of noncollagenous sequence at the amino end and 170 residues of noncollagenous sequence at the carboxyl end. The collagenous domain is 460 amino acid residues long. Short horizontal bars indicate the locations of short imperfections (Gly-X-Gly); solid squares indicate the locations of long imperfections (Gly-X-Y-X-Y-Gly). The polypeptide sequence contains a large number of methionyl residues (open circles) and a concentration of tyrosyl residues (filled circles) within the carboxy-terminal noncollagenous domain. Cysteinyl residues are indicated by open squares. Reproduced with permission from Ninomiya *et al.* (1986).

lagens, therefore, the substantial information introduced into polypeptide chains by posttranslational modifications cannot be determined by nucleotide sequence analysis.

B. *Size of Primary Translation Products*

For proteins undergoing extensive posttranslational modifications, including proteolytic processing, primary structure determination by nucleotide sequencing enables one to determine the size of the primary translation products with relatively little effort. In addition, for collagens that frequently have been characterized as pepsin-resistant proteolytic fragments, it is, of course, extremely valuable to be able to define noncollagenous sequences without having to isolate intact polypeptides from cells or tissues. A good example from our own laboratory is type IX collagen. Only after the isolation of a cDNA for the α1(X) collagen chain (Ninomiya and Olsen, 1984) did the structure of type IX collagen become clear. The sequence of the cDNA immediately suggested that the molecule contained three triple-helical domains separated by noncollagenous domains, and that the collagenous fragments HMW and LMW (see van der Rest and Mayne, this volume) were derived from the same molecule.

V. Further Use of DNA Probes

Although the isolation and characterization of cDNA and genomic DNA specific for various collagen types permits deduction of the primary structure of collagenous polypeptides, the work on the molecular biology of collagens is primarily aimed at answering questions of regulation of gene expression. Also, the analysis of collagen gene structure will permit a genetic classification of collagens and will lead to elucidation of evolutionary relationships between collagen types.

Finally, by modifying cloned genes *in vitro* and inserting them into cells and organisms, it should be possible to directly address the question of what functional roles the different collagen types play in extracellular matrices.

VI. Summary and Conclusions

The use of molecular cloning techniques has, in the past few years, led to rapid yet significant progress in the characterization of genetic types of collagen. Clearly, by their speed and simplicity, these techniques represent a superior method for determination of collagen primary structure. When used in conjunction with protein and peptide chemistry, amino acid sequencing, peptide synthesis, and monoclonal antibody technology, molecular biology promises to provide definitive answers to questions of collagen structure and function both in normal extracellular matrix formation and in disease states.

References

Adams, S. L., Sobel, M. E., Howard, B. H., Olden, K., Yamada, K. M., de Crombrugghe, B., and Pastan, I. (1977). *Proc. Natl. Acad. Sci. U.S.A.* **74,** 3399–3403.

Anderson, S. (1981). *Nucleic Acids Res.* **9,** 3015–3027.

Barnes, W. M., Bevan, M., and Son, P. H. (1983). *Methods Enzymol.* **101,** 98–122.

Benton, W. D., and Davis, R. W. (1977). *Science* **196,** 180–182.

Biggin, M. D., Gibson, T. J., and Hong, G. F. (1983). *Proc. Natl. Acad. Sci. U.S.A.* **80,** 3963–3965.

Blattner, F. R., Williams, B. G., Blechl, A. F., Denniston-Thompson, K., Faber, H. E., Furlong, L. A., Grunwald, D. J., Kiefer, D. O., Moore, D. D., Schumm, J. W., Sheldon, E. L., and Smithies, O. (1977). *Science* **196,** 161–169.

Burnett, W., and Rosenbloom, J. (1979). *Biochem. Biophys. Res. Commum.* **86,** 478–484.

Cheah, K. S. E., Stoker, N. G., Griffin, J. R., Grosveld, F. G., and Solomon, E. (1985). *Proc. Natl. Acad. Sci. U.S.A.* **82,** 2555–2559.

Chirgwin, J. M., Przybyla, A. E., MacDonald, R. J., and Rutter, W. J. (1979). *Biochemistry* **18,** 5294–5299.

Chu, M.-L., de Wet, W., Bernard, M., Ding, J.-F., Morabito, M., Myers, J., Williams, C., and Ramirez, F. (1984). *Nature (London)* **310,** 337–340.

Clarke, L., Hitzeman, R., and Carbon, J. (1979). *Methods Enzymol.* **68,** 436–442.

Collins, J., and Hohn, B. (1978). *Proc. Natl. Acad. Sci. U.S.A.* **75,** 4242–4246.

Dale, R. M. K., McClure, B. A., and Houchins, J. P. (1985). *Plasmid* **13,** 31–40.

Deak, F., Argraves, W. S., Kiss, I., Sparks, K. J., and Goetinck, P. (1985). *Biochem. J.* **229,** 189–196.

Deininger, P. L. (1983). *Anal. Biochem.* **129,** 216–223.

Doege, K., Hassell, J. R., Caterson, B., and Yamada, Y. (1986). *Proc. Natl. Acad. Sci. U.S.A.* **83,** 3761–3765.

Efstratiadis, A., Kafatos, F. C., Maxam, A. M., and Maniatis, T. (1976). *Cell* **7,** 279–288.

Ehrlich, H. A., Cohen, S. N., and McDevitt, H. O. (1979). *Methods Enzymol.* **68,** 443–453.

Elima, K., Makela, J. K., Vuorio, T., Kauppinen, S. Knowles, J., and Vuorio, E. (1985). *Biochem. J.* **229,** 183–188.

Frischauf, A. M., Lehrach, H., Rosner, C., and Boedtker, H. (1978). *Biochemistry* **17,** 3243–3249.

Glover, D. M. (ed.) (1985). "DNA Cloning: A Practical Approach," Vols. I and II. IRL Press, Oxford, England.

Goldberg, G. I., Wilhelm, S. M. Kronberger, A., Bauer, E. A., Grant, G. A., and Eisen, A. Z. (1986). *J. Biol. Chem.* **261,** 6600–6605.

Gordon, M., Gerecke, D., and Olsen, B. R. (1987). Submitted.

Grossman, L. and Moldave, K., (eds.) (1980). *Methods Enzymol.* **65.**

Gubler, U., and Hoffman, B. J. (1983). *Gene* **25,** 263–269.

Guo, L., Yang, R. C. A., and Wu, R. (1983). *Nucleic Acids Res.* **11,** 5521–5540.

Hanahan, D., and Meselson, M. (1980). *Gene* **10,** 63–67.

Helfman, D. M., Feramisco, J. R., Fiddes, J. C., Thomas, G. P., and Hughes, S. H. (1985). *In* "Genetic Engineering" (J. K. Setlow, ed.), Vol. 7, pp. 185–197. Plenum Press, New York.

Hopp, T. P., and Woods, K. R. (1981). *Proc. Natl. Acad. Sci. U.S.A.* **78,** 3824–3828.

Kaiser K., and Murray, N. E. (1985). *In* "DNA Cloning: A Practical Approach" (D. M. Glover, ed.), Vol. I, pp. 1–47. IRL Press, Oxford, England.

Kohno, K., Martin, G., and Yamada, Y. (1984). *J. Biol. Chem.* **259,** 13668–13673.

Konomi, H., Seyer, J. M., Ninomiya, Y., and Olsen, B. R. (1986). *J. Biol. Chem.* **261,** 6742–6746.

Kramer, J. M., Cox, G. N., and Hirsh, D. (1982). *Cell* **30,** 599–606.

Land, H., Grez, M., Hauser, H., Lindenmayer, W., and Schutz, G. (1981). *Nucleic Acids Res.* **9,** 2251–2266.

Lehrach, H., Frischauf, A. M., Hanahan, D., Wozney, J., Fuller, F. Crkvenjako, R., Boedtker, H., and Doty, P. (1978). *Proc. Natl. Acad. Sci. U.S.A.* **75,** 5417–5421.

Loidl, H. R., Brinker, J. M., May, M., Pihlajaniemi, T., Morrow, S., Rosenbloom, J., and Myers, J. C. (1984). *Nucleic Acids Res.* **12,** 9383–9384.

Lozano, G., Ninomiya, Y., Thompson, H., and Olsen, B. R. (1985). *Proc. Natl. Acad. Sci. U.S.A.* **82,** 4050–4054.

Lukens, L. N., Frischauf, A. M., Pawlowski, P. J., Brierley, G. T., and Lehrach, H. (1983). *Nucleic Acids Res.* **11,** 6021–6039.

Maniatis, T., Fritsch, E. F., and Sambrook, J. (1982). "Molecular Cloning: A Laboratory Manual." Cold Spring Harbor Lab. Cold Spring Harbor, New York.

Maxam, A., and Gilbert, W. (1980). *Methods Enzymol.* **65,** 499–560.

Messing, J. (1983). *Methods Enzymol.* **101,** 20–78.

Monson, J. M., and McCarthy, B. J. (1981). *DNA* **1,** 59–69.

Monson, J. M., Natzle, J., Friedman, J., and McCarthy, B. J. (1982). *Proc. Natl. Acad. Sci. U.S.A.* **79,** 1761–1765.

Myers, J. C., Chu, M. L., Faro, S. H., Clark, W. J., Prockop, D. J., and Ramirez, F. (1981). *Proc. Natl. Acad. Sci. U.S.A.* **78,** 3516–3520.

Niman, H. L., Houghton, R. A., Walker, L. E., Reisfeld, R. A., Wilson, I. A., Hogle, J. M., and Lerner, R. A. (1983). *Proc. Natl. Acad. Sci. U.S.A.* **80,** 4949–4953.

Ninomiya, Y., and Olsen, B. R. (1984). *Proc. Natl. Acad. Sci. U.S.A.* **81,** 3014–3018.

Ninomiya, Y., Showalter, A. M., and Olsen, B. R. (1984a). *In* "The Role of Extracellular Matrix in Development" (R. L. Trelstad, ed.), pp. 255–275. Liss, New York.

Ninomiya, Y., Showalter, A. M., van der Rest, M., Seidah, N. G., Chrétien, M., and Olsen, B. R. (1984b). *Biochemistry* **23,** 617–624.

Ninomiya, Y., van der Rest, M., Mayne, R., Lozano, G., and Olsen, B. R. (1985). *Biochemistry* **24,** 4223–4229.

Ninomiya, Y., Gordon, M., van der Rest, M., Schmid, T., Linsenmayer, T., and Olsen, B. R. (1986). *J. Biol. Chem.* **261,** 5041–5050.

Ohkubo, H., Vogeli, G., Mudryj, M., Avvedimento, E., Sullivan, M., Pastan, I., and de Crombrugghe, B. (1980). *Proc. Natl. Acad. Sci. U.S.A.* **77,** 7059–7063.

Okayama, H., and Berg, P. (1982). *Mol. Cell. Biol.* **2,** 161–170.

Payvar, F., and Schimke, R. T. (1979). *Eur. J. Biochem.* **101,** 271–282.

Pelham, H. R. B., and Jackson, R. J. (1976). *Eur. J. Biochem.* **67,** 247–256.

Perbal, B. (1984). "A Practical Guide to Molecular Cloning." Wiley, New York.

Poncz, M., Solowiejczyk, D., Ballentine, M., Schwartz, E., and Surrey, S. (1982). *Proc. Natl. Acad. Sci. U.S.A.* **79,** 4298–4302.

Riccardi, R. P., Miller, J. S., and Roberts, B. E. (1979). *Proc. Natl. Acad. Sci. U.S.A.* **76,** 4927–4931.

Rimm, D. L., Horness, D., Kucera, J., and Blattner, F. R. (1980). *Gene* **12,** 301–309.

Sanger, R., Nicklen, S., and Coulson, A. R. (1977). *Proc. Natl. Acad. Sci. U.S.A.* **74,** 5463–5467.

Sangiorgi, F. O., Benson-Chanda, V., de Wet, W., Sobel, M. E., and Ramirez, F. (1985). *Nucleic Acids Res.* **13,** 2815–2826.

Schwarzbauer, J. E., Tamkun, J. W., Lemischka, I. R., and Hynes, R. O. (1983). *Cell* **35,** 421–431.

Soll, D., and Roberts, R. J. (eds.) (1984). "The Applications of Computers to Research on Nucleic Acids II," Parts 1 and 2. IRL Press, Oxford, England.

Staden, R. (1985). *In* "Genetic Engineering" (J. K. Setlow, ed.), Vol. 7, pp. 67–114. Plenum, New York.

Suggs, S. V., Wallace, B., Hirose, T., Kawashima, E. H., and Itakura, K. (1981). *Proc. Natl. Acad. Sci. U.S.A.* **78,** 6613–6617.

Szostak, J. W., Stiles, J. I., Tye, B.-K., Chiu, P., Sherman, F., and Wu, R. (1979). *Methods Enzymol.* **68,** 419–428.

Tamura, T., Bauer, H., Birr, C., and Pipkorn, R. (1983). *Cell* **34,** 587–596.

Villa-Komaroff, L., Efstratiadis, A., Bromme, S., Lomedico, P., Tizard, R., Naber, S. P., Chick, W. L., and Gilbert, W. (1978). *Proc. Natl. Acad. Sci. U.S.A.* **75,** 3727–3731.

Vogeli, G., Ohkubo, H., Sobel, M. E., Yamada, Y., Pastan, I., and de Crombrugghe, B. (1981a). *Proc. Natl. Acad. Sci. U.S.A.* **78,** 5334–5338.

Vogeli, G., Ohkubo, H., Avvedimento, V. E., Sullivan, M., Yamada, Y., Mudryj, M., Pastan I., and de Crombrugghe, B. (1981b). *Cold Spring Harbor Symp. Quant. Biol.* **45,** 777–783.

Vuorio, E., Sandell, L., Kravis, D., Sheffield, V. C., Vuorio, T., Dorfman, A., and Upholt, W. B. (1982). *Nucleic Acids Res.* **10,** 1175–1192.

Wang, S.-Y., and Gudas, L. (1983). *Proc. Natl. Acad. Sci. U.S.A.* **80,** 5880–5884.

Werner, D., Chemla, Y., and Herzberg, M. (1984). *Anal. Biochem.* **141,** 329–396.

Wozney, J., Hanahan, D., Morimoto, R., Boedtker, H., and Doty, P. (1981). *Proc. Natl. Acad. Sci. U.S.A.* **78,** 712–716.

Wu, R. (ed.) (1979). *Methods Enzymol.* **68.**

Wu, R., Grossman, L., and Moldave, K. (eds.) (1983a). *Methods Enzymol.* **100.**

Wu, R., Grossman, L., and Moldave, K. (eds.) (1983b). *Methods Enzymol.* **101.**

Yamada, Y., Mudryj, M., and de Crombrugghe, B. (1983a). *J. Biol. Chem.* **258,** 14914–14919.

Yamada, Y., Mudryj, M., Sullivan, M., and de Crombrugghe, B. (1983b). *J. Biol. Chem.* **258,** 2758–2761.

Yamada, Y., Sasaki, M., Kohno, K., Kleinman, H. K., Kato, S., and Martin, G. R. (1985). *In* "Basement Membranes" (S. Shibata, ed.), pp. 139–146. Elsevier, Amsterdam.

Yamamoto, T., Sobel, M. E., Adams, S. L., Avvedimento, V. E., DiLauro, R., Pastan, I., de Crombrugghe, B., Showalter, A., Pesciotta, D., Fietzek, P., and Olsen, B. (1980). *J. Biol. Chem.* **255,** 2612–2615.

Young, R. A., and Davis, R. W. (1983). *Proc. Natl. Acad. Sci. U.S.A.* **80,** 1194–1198.

Young, R. A., and Davis, R. W. (1985). *In* "Genetic Engineering" (J. K. Setlow, ed.), Vol. 7, pp. 29–41.

Index

D

E

F

G

R

S

T